Progress in Colloid and Polymer Science · Volume 116 · 2000

W0245928

Springer-Verlag
Berlin Heidelberg GmbH

Progress in Colloid and Polymer Science

Editors: F. Kremer, Leipzig and G. Lagaly, Kiel

Volume 116 · 2000

Surface and Colloid Science

Volume Editors:
V. Razumas, B. Lindman and T. Nylander

 Springer

The series Progress in Colloid and Polymer Science is also available electronically (ISSN 1437-8027)

- Access to tables of contents and abstracts is *free* for everybody.
- Scientists affiliated with departments/institutes subscribing to Progress in Colloid and Polymer Science
  as a whole also have full access to all papers in PDF form. Point your librarian to the LINK access registration form
  at http://link.springer.de/series/pcps/reg-form.htm

ISSN 0340-255X
ISBN 978-3-662-14693-4
ISBN 978-3-540-44941-6 (eBook)
DOI 10.1007/978-3-540-44941-6

The use of general descriptive names, registered names, trademarks, etc. in this publication does not imply, even in the absence of specific statement, that such names are exempt from the relevant protective laws and regulations and therefore free for general use.

© Springer-Verlag
Berlin Heidelberg 2001
Originally published by Springer-Verlag Berlin, Heidelberg, New York in 2001
Softcover reprint of the hardcover 1st edition 2001

Typesetting: SPS, Madras, India

Cover: Estudio Calamar,
F. Steinen-Broo, Pau/Girona, Spain

SPIN: 10719415

Printed on acid-free paper

Progr Colloid Polym Sci (2000) 116: V
© Springer-Verlag 2000

PREFACE

The Scandinavian Symposium on Surface Chemistry has been a regular event since 1963, when the first meeting was organized in Åbo (Turku), Finland by Professor Per Ekvall. The previous meeting, which was the 13th, was held at Umeå University, in Sweden August, 24-26 (1997).

The last years have seen an increasing number of scientific contacts between the Scandinavian and Baltic countries (Estonia, Latvia and Lithuania). Most of the collaborations have so far been based on personal contacts that have shown that many scientists in the Baltic countries have a strong interest in surface and colloid chemistry. The time has come to link them more formally to our Scandinavian surface and colloid chemical community. The 1st Nordic-Baltic Meeting on Surface and Colloid Science was therefore held in Vilnius, Lithuania on August 21-25, 1999, as a continuation of the Scandinavian Symposium on Surface Chemistry. The Lithuanian Academy of Sciences, Vilnius University and the State Research Institutes of Biochemistry and Chemistry organized the meeting.

Some of the leading Nordic-Baltic scientists in surface and colloid science presented recent research on a broad range of topics: Adhesion, adsorption processes, characterization of solid/liquid and solid/polymer interfaces, chemical and particle deposition, colloid stability, emulsification and encapsulation, functionalized surfaces for bio- and chemosensors, interfacial reactions, new surfactants, polymer-surfactant interactions, and self-assembly of lipids and surfactants. In order to promote the contacts between young scientists, one day was reserved for presentation by students and newly graduated scientists. We had the pleasure to see more than 100 participants at the meeting, not only from Denmark, Estonia, Finland, Latvia, Lithuania, Norway and Sweden, but also from Brazil, Italy, France, Germany, Japan, The Netherlands, Poland, Portugal, Russia, and USA.

Vilnius, the capital of Lithuania, founded in 1323 and on United Nations' list of the world heritage, provided the right atmosphere for the meeting. Apart from a fruitful scientific meeting in a very friendly atmosphere, we got a glimpse of Lithuanian history and culture, and of the many the historical buildings in Vilnius. We could also enjoy the famous Lithuanian amber in both solid and liquid phase!

It is the tradition of the Scandinavian Symposium on Surface Chemistry to publish the proceedings in "Progress in Colloid and Polymer Science". It is our pleasure to keep that tradition and we would like to thank the authors for their excellent contributions.

Finally, we would like to thank Ministry of Education and Science of the Republic of Lithuania, Lithuanian State Science and Studies Foundation, Lund University, Sweden, Astra Hässle AB (now AstraZeneca R&D Mölndal), Sweden and Perstorp Flooring AB, Sweden for their generous support. Furthermore, the foreign visitors extend sincere thanks to their Lithuanian colleagues for an excellent organization and a warm hospitality.

Vilnius and Lund, May 2000

Prof. Valdemaras Razumas
Prof. Björn Lindman
Dr. Tommy Nylander

Progr Colloid Polym Sci (2000) 116: VI–VII
© Springer-Verlag 2000

CONTENTS

Biosensor Applications

Contents of Volume 116

Progr Colloid Polym Sci (2000) 116:1–8
© Springer-Verlag 2000

M. C. Sabra
K. Jørgensen
O. G. Mouritsen

Effects of the insecticides malathion and deltamethrin on the phase behaviour of dimyristoylphosphatidylcholine multilamellar lipid bilayers

M. C. Sabra · O. G. Mouritsen (✉)
Department of Chemistry
Technical University of Denmark
2800 Lyngby, Denmark
e-mail: ogm@kemi.dtu.dk
Fax: +45-45-934808

K. Jørgensen
Department of Pharmaceutics
The Royal Danish School of Pharmacy
2100 Copenhagen Ø, Denmark

Abstract The effects of the insecticides malathion and deltamethrin on the phase behaviour of multilamellar dimyristoylphosphatidylcholine (DMPC) bilayers were studied by differential scanning calorimetry. The results show that both insecticides broaden the main transition and suppress the main transition temperature. High concentrations of malathion narrow the main transition compared to lower concentrations of malathion. The C_p curves for lipid bilayers with high concentrations of malathion show several peaks in the main transition region, suggesting that more than one phase is present. For malathion interacting with DMPC multilamellar vesicles a high-temperature feature developed with increasing insecticide concentration. This was not seen for the vesicles with deltamethrin. Malathion affects the pretransition by broadening it and by lowering the transition temperature and the transition enthalpy. Small concentrations of deltamethrin make the pretransition disappear, but a minor peak in the specific heat appears at the same position for higher concentrations of deltamethrin. Although both insecticides are hydrophobic molecules their effects on DMPC multilamellar bilayers are very different. This suggests that details of the molecular structure of hydrophobic molecules are important for their nonspecific interactions with lipid bilayers.

Key words Lipid bilayer · Phase behaviour · Calorimetry · Malathion · Deltamethrin

Introduction

Organic insecticides can be divided into several classes [1]: the chlorinated hydrocarbons such as 1,1,1-trichloro-2,2-bis(p-chlorophenyl)ethane, better known as DDT, and lindane, the organophophorous derivatives, including compounds such as parathion and malathion, and the pyrethroids, which are similar in structure to the naturally occuring pyrethrins, which are also insecticides. Examples of the pyrethroids include allethrin and deltamethrin. Finally, substances of miscellaneous chemical structure have been used as insecticides, for example, naphtalene and rotenone [1].

Owing to the strong hydrophobicity of all these different compounds it is widely accepted that their site of action is somewhere at the biological membrane. For the pyrethroids, specifically deltamethrin, a direct stereospecific interaction with the γ-aminobutyric acid (GABA) receptor–ionophore complex has been found [2]; however, effects of deltamethrin on other membrane-bound proteins such as sodium channels [3] and voltage-dependent chloride channels [4] have also been reported. For the organophophorous compounds it seems likely that the insecticidal effect is due to the inhibition of acetylcholinesterase, although physiological effects not related to this enzyme have also been observed [5]. Also, the chlorinated hydrocarbons have been reported to affect several different membrane-bound enzymes. For example, the insecticidal effect of lindane is believed to be due to action on the GABA$_A$ receptor chloride channel complex [6], but effects of this insecticide have also been reported on, for example, Ca^{2+}–ATPase and 5′-nucleotidase [7].

The fact that effects of some insecticides have been found on several different membrane-bound proteins might suggest that direct specific interactions with each of these different proteins are not the only important interaction in the membrane–insecticide system, but that more nonspecific interactions with the lipid bilayer component of the membrane also play a major role.

The biological membrane is an extremely complex many-particle system. In order to identify the effects of insecticides on the lipid bilayer component of the membrane, it is therefore necessary to use model systems such as artificial lipid bilayers. In this article we report a comparative study of the effects of two insecticides on the thermotropic phase behaviour of dimyristoylphosphatidylcholine (DMPC) lipid membranes. The two insecticides are malathion, which is of the organophosphorous type, and deltamethrin, which is a pyrethroid. Malathion has previously been studied in relation to lipid bilayer effects, and it was observed that the partition coefficient has a maximum at the main transition [8] as for the organophosphorous insecticide parathion [9]. Furthermore, parathion and methylparathion have been reported to order the fluid phase of lipid bilayers [10, 11]. Effects of several pyrethroids on lipid bilayers have also been studied [12–16], and some of these seem to have the effect of ordering the fluid phase whilst the gel phase becomes disordered [15, 16].

The chemical structure of the two insecticides considered here are shown in Fig. 1. Since both insecticide molecules are strongly hydrophobic they are interesting, not only from an insecticide point of view but also from the point of view of the interaction of hydrophobic molecules with lipid bilayer membranes [17].

Materials and methods

Chemicals

DMPC was obtained from Avanti Polar Lipids (Alabaster, Ala., USA). Malathion, at least 99% pure, was a gift from Cheminova Agro (Lemvig, Denmark). Deltamethrin, at least 99% pure, was a gift from Hoechst Schering Agrevo (Hattersheim, Germany). All chemicals were used without further purification.

Preparation of vesicles

Multilamellar vesicles of DMPC with various amounts of insecticide were prepared by cosolubilising lipid and insecticide in chloroform and blowing off the chloroform under dry nitrogen. The resulting lipid–insecticide mixture was then left under vacuum at room temperature for at least 12 h. The dry mixture was then hydrated in a solution of 50 mM KCl and 1 mM NaN$_3$ at 40 °C for at least 1 h. The final lipid concentrations of the samples were determined using an enzymatic colourimetric test from Boehringer

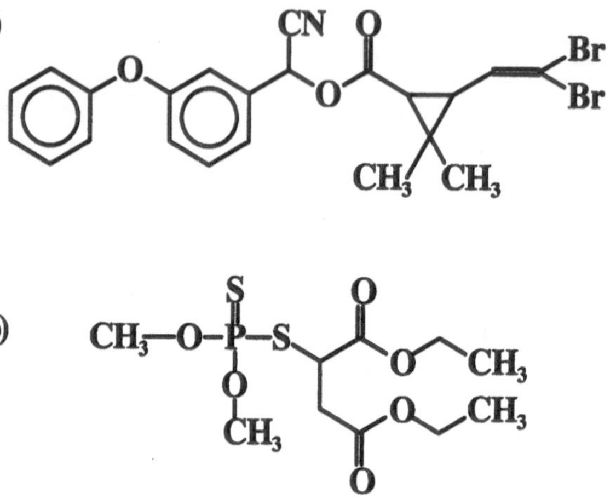

Fig. 1a, b Chemical structure of the two insecticides used in this study. a The pyrethroid deltamethrin and b the organophosphorous derivative malathion

Mannheim (Mannheim, Germany). The final insecticide concentrations in the membranes were not controlled because partitioning between the water and lipid phases is expected to occur. The insecticide concentrations cited are the overall insecticide/lipid ratios in the samples.

Differential scanning calorimetry

Differential scanning calorimetry (DSC) was performed on an MC-2 ultrasensitive calorimeter from Microcal (Northampton, Mass., USA). The instrument is of the power-compensating type with cell volumes of 1.2 ml. Three successive upscans were performed on each sample using a scan rate of 20 °C/h. The reference cell contained the KCl solution. Before each upscan the sample was allowed to equilibrate for 50 min at the starting temperature (5 or 10 °C). The samples were subject to a hydrostatic pressure of 2.5 atm (absolute) during scanning.

Data analysis

Appropriate baselines were subtracted from the specific heat curves before they were normalised and integrated to obtain the area under the curve, ΔH, which is a measure of the transition enthalpy. ΔH is subject to errors due to the sensitivity to the precise choice of baseline. We shall return to this issue in the Discussion. The temperature at which the specific heat has its global maximum, T_m, and the width of the transition at half maximum height, $T_{1/2}$, were also determined.

Results

DSC was performed on DMPC multilamellar lipid bilayers containing various amounts of either malathion or deltamethrin. It would be desirable to be able to determine the complete phase diagram for the lipid/insecticide mixtures, but this is not possible using only DSC; however, DSC does provide some indirect insight into the phase behaviour of these mixtures and into how small hydrophobic molecules affect lipid bilayers.

Malathion

The results for DMPC multilamellar lipid bilayers containing malathion are shown in Fig. 2. Compared to the pure sample, which is also shown, the addition of malathion in small amounts (about 10 mol% or less) causes the peak of the main transition to broaden and to be shifted towards lower temperatures. At higher concentrations (15 mol%) the transition extends over several degrees and two maxima are clearly visible at 19 and at 21.5 °C. Furthermore, a closer examination of the graph for 15 mol% malathion reveals a small third peak at 19.7 °C. When the concentration of malathion is increased to 20 mol%, the maximum at 21.5 °C disappears and the transition is narrowed. The intensities of the two other peaks are concomitantly increased, especially the peak at 19.7 °C, and so the transition finally consists of one dominating peak at 19.7 °C and a smaller secondary peak at about 19 °C. Malathion concentrations higher than 20 mol% are found to have no further effect on the shape and intensity of the C_p curves.

Addition of malathion also has an effect on the high-temperature side of the transition as seen in Fig. 3, which shows details of a minor part of Fig. 2 (the curve for 3 mol% malathion is omitted for clarity in Fig. 3). At this magnification, the C_p curve for the sample without malathion reveals a very small peak on the high-temperature side of the main peak. At 5 mol% malathion, this peak disappears and instead there is a shoulder on the main peak, which becomes more pronounced at 10 and 15 mol% malathion. At 20 mol% malathion, the shoulder has developed a maximum. It is noted that, although the main peak is shifted several degrees towards

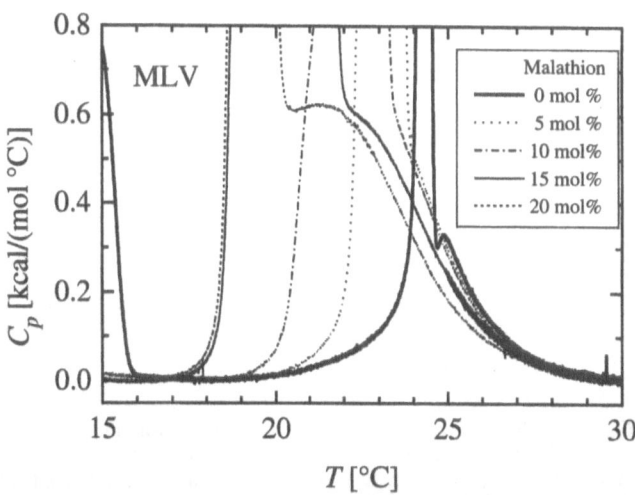

Fig. 3 Detail of the excess heat capacities of multilamellar vesicles of DMPC with malathion. This figure is a magnification of Fig. 2 except that the curve for 3 mol % malathion is omitted for clarity. The peak near 15 °C signals the pretransition in pure DMPC. The curves were obtained by differential scanning calorimetry as upscans at a scan rate of 20 °C/h. The lipid concentrations were 5 mM

lower temperatures by malathion, the shoulder (or broad peak) on the high-temperature side always seems to reach the same level at approximately the same temperature. This makes the shoulder very wide, approximately 5 °C for the highest malathion concentrations, as also seen in Fig. 2, where the shoulder appears over an extended temperature range in which C_p is nonzero and approximately constant.

The effects of malathion are summarised in Fig. 4 in terms of the transition temperature, T_m, the peak width, $T_{1/2}$ and the transition enthalpy, ΔH. In Fig. 4a it is seen that T_m, which is defined as the position of the global maximum in the C_p curve, decreases linearly with the malathion concentration up to about 10 mol%. At higher concentrations of malathion, the transition temperature decreases faster until 20 mol% malathion, where further increase in insecticide concentration has no effect. Figure 4b shows that $T_{1/2}$ increases linearly by addition of up to 10 mol% malathion. Owing to the appearance of multiple peaks, the transition becomes very wide at 15 mol%. At 20 mol% malathion, the width has decreased again and higher concentrations have no further effect. It should be noted that the width of the peak is defined as the width at half maximum and that the shoulder on the high-temperature side of the peak is therefore not included. The main transition enthalpy, which is taken to be the area under the C_p curves in the temperature interval 17–27 °C, is shown in Fig. 4c as a function of the malathion concentration. Even for the highest concentrations of malathion there is no significant change in ΔH upon addition of malathion. The values of around 7 kcal/mol compare well with the value

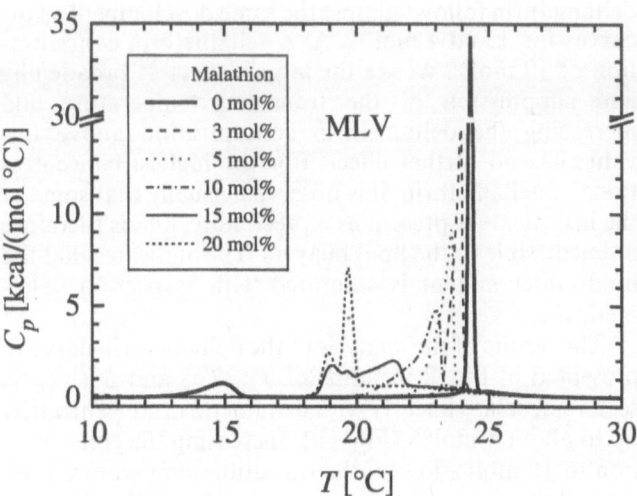

Fig. 2 Excess heat capacities of multilamellar vesicles of dimyristoylphosphatidylcholine (DMPC) with malathion. The curves were obtained by differential scanning calorimetry as upscans at a scan rate of 20 °C/h. The lipid concentrations were 5 mM

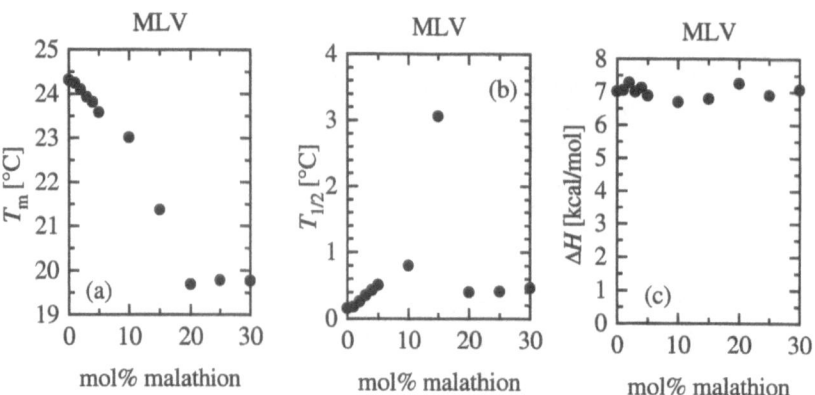

Fig. 4a–c Data extracted from the C_p curves for multilamellar DMPC vesicles with malathion. **a** Dependence of the main transition temperature, T_m, on the malathion concentration, as determined by the position of the peak. **b** Variation of the width of the main transition, $T_{1/2}$, at half peak height, with the malathion concentration. **c** The main transition enthalpy, ΔH, as a function of the malathion concentration. ΔH was determined as the area under the peak. See text for details

of 6.5 kcal/mol found in the literature for pure DMPC bilayers [18].

For the low malathion concentrations, the C_p curves in region of the pretransition are shown in Fig. 5. It is seen that there is a broadening and a depression of the pretransition with increasing malathion concentrations. At 5 mol% malathion, the pretransition is hardly visible anymore. The data extracted from the C_p curves in the range of the pretransition are shown in Fig. 6. In Fig. 6a, the pretransition temperature, T_p, is seen to be shifted towards lower temperatures as a linear function of the malathion concentration. The shift back to a higher

temperature at 5 mol% malathion is probably not significant since it is based on a rather broad peak and a very weak signal. The width of the pretransition increases for 1 and 2 mol% malathion but decreases again slightly for higher concentrations (Fig. 6b). As for T_p, the width obtained for 5 mol% is probably not very accurate. The enthalpy associated with the pretransition is shown in Fig. 6c. It is seen that malathion decreases the pretransition enthalpy linearly with the concentration.

Deltamethrin

The C_p curves obtained for DMPC multilamellar lipid bilayers containing deltamethrin are shown in Fig. 7. As for malathion, deltamethrin causes the transition to broaden and to be shifted towards lower temperatures. At 5 mol% deltamethrin, the transition becomes very broad, in particular because a shoulder appears on the high-temperature side of the transition. On the low-temperature side of the peak, the curve for 5 mol% deltamethrin follows almost the same development as the curves for 3 and 4 mol%. At a deltamethrin concentration of 10 mol% we see the largest effect of broadening and suppression of the transition temperature and increasing the deltamethrin concentration above this value has no further effect. For the highest concentrations of deltamethrin, it is observed visually that some of the insecticide is present as a precipitate, and is therefore not accessible to the lipid bilayers. This indicates that the lipid–water system is saturated with respect to deltamethrin.

The results of the analysis of the deltamethrin data are presented in Fig. 8 in terms of T_m, $T_{1/2}$ and ΔH. T_m is seen to decrease linearly with deltamethrin concentration up to about 4 mol% (Fig. 8a). Increasing the concentration to 10 mol% lowers the transition temperature even more, but at a much lower rate. At 15 mol%, T_m is the same as at 10 mol%, but at 20 mol% T_m increases to a higher value and remains constant for higher concentrations. This increase might be related to the precipitation of deltamethrin as discussed later. The influence of

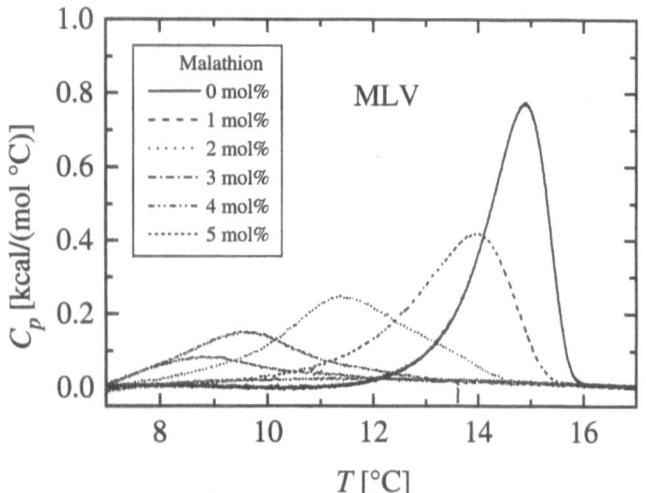

Fig. 5 Excess heat capacities in the pretransition region of multilamellar vesicles of DMPC with malathion. The curves were obtained by differential scanning calorimetry as upscans at a scan rate of 20 °C/h. The lipid concentrations were 5 mM

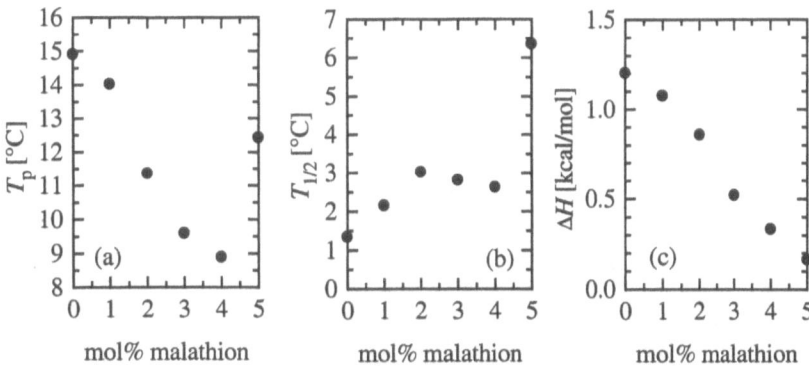

Fig. 6a–c Data extracted from the C_p curves for multilamellar DMPC vesicles with malathion. **a** Dependence of the pretransition temperature, T_p, on the malathion concentration, as determined by the position of the peak. **b** Variation of the width of the pretransition, $T_{1/2}$, at half peak height, with the malathion concentration. **c** The pretransition enthalpy, ΔH, as a function of the malathion concentration. ΔH was determined as the area under the peak. See text for details

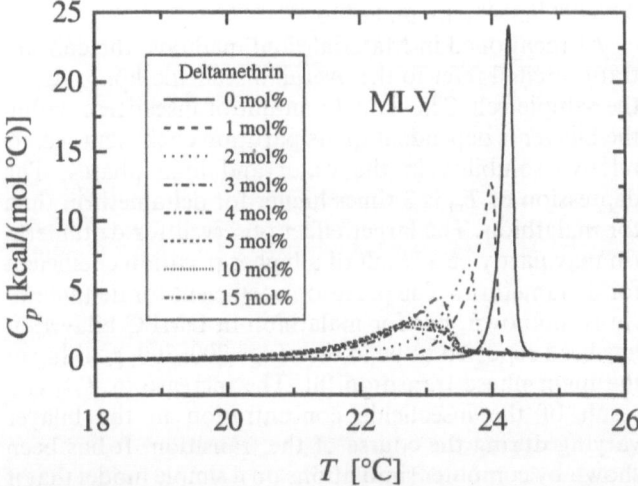

Fig. 7 Excess heat capacities of multilamellar vesicles of DMPC with deltamethrin. The curves were obtained by differential scanning calorimetry as upscans at a scan rate of 20 °C/h. The lipid concentrations were 5 mM

deltamethrin on $T_{1/2}$ is shown in Fig. 8b. Between 0 and 5 mol% deltamethrin, $T_{1/2}$ increases linearly with the insecticide concentration and for higher concentrations it remains approximately constant. The data for the transition enthalpy are shown in Fig. 8c. At first glance, deltamethrin does not seem to affect ΔH; however, in the light of the precipitation of the deltamethrin and of the sudden jump in T_m at 20 mol%, one might choose to discard the data for concentrations above 15 mol%. In that case, it seems as if the transition enthalpy decreases slightly with deltamethrin concentration.

The extended shoulder seen on the high-temperature side of the peak in the case of malathion is absent in the

DMPC–deltamethrin sample. However, the small peak on the high-temperature side of the transition in the case of the pure sample is present (not shown), but it disappears completely when even the smallest amounts of deltamethrin are added.

The effect of deltamethrin on the pretransition is shown in Fig. 9. At the lowest concentrations of deltamethrin the pretransition disappears, but it reappears at 5 mol% as a very small peak which grows only insignificantly when the deltamthrin concentration is increased further. The pretransition enthalpy at the high concentrations is only about 10% of the enthalpy for the pure sample.

Discussion

We have presented the results from a calorimetric study of the effects of the insecticides malathion and deltamethrin on DMPC multilamellar lipid bilayers. For low concentrations of both insecticides some common effects on the main phase transition are revealed, which we are going to discuss first. Secondly, we will turn to a discussion of the effects of high insecticide concentrations, which for some features differ significantly for the two insecticides. Finally, the effects of both insecticides on the pretransition will be considered.

Low insecticide concentrations

For low concentrations of both insecticides, the main phase transition is broadened and shifted towards lower temperatures in a concentration-dependent way. Such a broadening and suppression of the transition temperature has been observed for a number of other compounds interacting with lipid membranes. These include, for example, general and local anaesthetics such as halothane [19] and dibucaine [20] as well as other insecticides such as lindane [21, 22]. The depression of the transition temperature can be understood thermodynamically as a freezing-point depression, while the broadening is a result of the ability of the insecticides to stabilise the

6

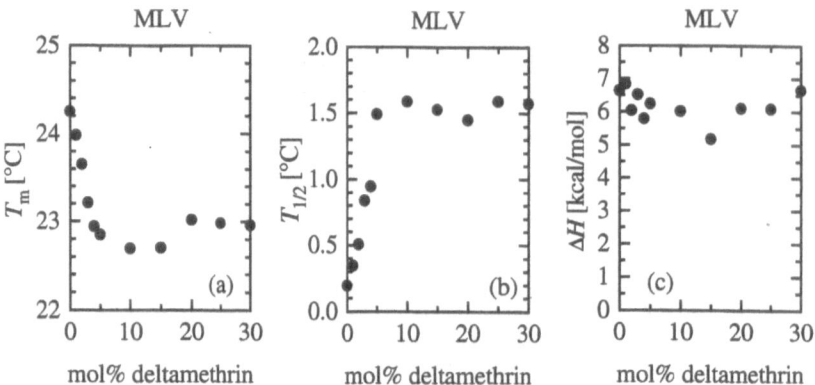

Fig. 8a–c Data extracted from the C_p curves for multilamellar DMPC vesicles with deltamethrin. **a** Dependence of the main transition temperature, T_m, on the malathion concentration, as determined by the position of the peak. **b** Variation of the width of the main transition, $T_{1/2}$, at half peak height, with the malathion concentration. **c** The main transition enthalpy, ΔH, as a function of the deltamethrin concentration. ΔH was determined as the area under the peak. See text for details

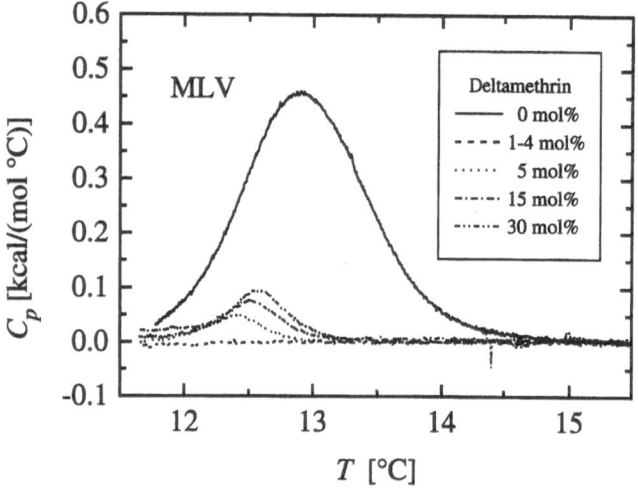

Fig. 9 Excess heat capacities in the pretransition region of multilamellar vesicles of DMPC with deltamethrin. The curves were obtained by differential scanning calorimetry as upscans at a scan rate of 20 °C/h. The lipid concentrations were 5 mM

interface between gel and fluid domains. This makes fluid domains more favourable in the gel phase and gel domains more favourable in the fluid phase, The broadening therefore indicates that near the main transition the bilayer becomes less ordered in the gel phase and more ordered in the fluid phase. This has been measured directly by fluorescence anisotropy techniques in the case of deltamethrin [16].

The effect of broadening the transition and depressing the transition temperature is clearly larger for the interaction with deltamethrin than for the interaction with malathion. This is seen when comparing Figs. 4 and

8. In the linear range (0–4 mol% for deltamethrin and 0–10 mol% for malathion), where the transition consists of only one dominating peak without any major secondary peaks or shoulders, deltamethrin shifts the transition temperature by 0.34 °C/mol%, whereas malathion only shifts it by 0.14 °C/mol%. In the same insecticide concentration range, the width of the transition is increased by 0.20 °C/mol% by deltamethrin and by only 0.067 °C/mol% by malathion.

As mentioned in Materials and methods, the concentrations cited refer to the overall insecticide/lipid ratio in the sample cell. The actual amount of insecticide within the bilayer is dependent on its partition coefficient, i.e. its relative solubility in the water and lipid phases. The depression of T_m is 3 times higher for deltamethrin than for malathion. The larger effect observed for deltamethrin may partly be a result of a higher partition coefficient for deltamethrin. The partition coefficient for deltamethrin is unknown, but for malathion in DMPC bilayers it has been reported to be in the range 200–300, peaking at the main phase transition [8]. The increase in $T_{1/2}$ is a result of the insecticide concentration in the bilayer varying during the course of the transition. It has been shown by computer simulations on a simple model that if the insecticide concentration in the bilayer is constant there is a narrowing of the transition rather than a broadening [23]. The model is valid in the phase-transition region and assumes that the insecticide molecules occupy interstitial sites in the bilayer, thereby precluding them from taking up any excluded volume. This implies that the transition enthalpy is independent of the presence of insecticides in the bilayer.

The effect of malathion and deltamethrin on the transition enthalpy is shown in Figs. 4c and 8c. In both cases ΔH seems to be independent of the insecticide content, although it cannot be excluded that low concentrations of deltamethrin cause the transition enthalpy to decrease slightly. If there is such a decrease in ΔH, it is likely that it is a result of a combination of the size of the insecticide molecule and of the partition coefficient, and it is reasonable that the larger deltamethrin molecule has a larger effect on ΔH than the smaller malathion molecule. If ΔH is not affected by the presence of foreign molecules, these molecules do not take up any excluded volume in the membrane.

High insecticide concentrations

The effects of high concentrations of the two insecticides are very different. For deltamethrin it appears that very high concentrations cannot be obtained. When more than 15 mol% of this insecticide is added, precipitation is observed, but saturation probably occurs at even lower deltamethrin concentrations. It is seen from Fig. 7 that when the deltamethrin concentration is 5 mol% further increases have almost no effect. There is also a limit to the amount of malathion which can be incorporated into the lipid bilayer; however, this limit seems to be much higher than for deltamethrin. Increasing the concentrations of malathion has an effect on the bilayer up to about 20 mol%. This might be the reason why the samples containing malathion show some effects which are weaker than, or not present in, the samples with deltamethrin.

One of these effects is the narrowing of the transition for higher concentrations of malathion compared to lower concentrations. This is not observed for deltamethrin. Another such effect is the splitting of the main peak into at least two clearly distinguishable peaks induced by malathion. It can be argued that a similar splitting is seen for deltamethrin. Although two separate peaks are not clearly visible, the transition is very broad and is characterised by a very pronounced shoulder. The appearance of multiple peaks could be a sign of multiple phases, for example, gel and fluid phases coexisting in the temperature range between the peaks. The splitting of the main peak could also be related to the three-dimensional structure of multilamellar lipid bilayers, since such a splitting is not observed for unilamellar bilayers.

The most remarkable effect of malathion not observed for deltamethrin is probably the appearance of a plateau or a peak on the high-temperature side of the transition. This peak could possibly originate from a very small peak in the pure sample.

Effects on the pretransition

Both malathion and deltamethrin affect the pretransition of DMPC multilamellar vesicles. It should be noted that the kinetics of the pretransition is very slow [24] and that the data, therefore, cannot be considered to be equilibrium data; however, since the pretransition is affected out of equilibrium, it may very well also be affected in equilibrium.

Malathion causes the pretransition to be broadened and to be shifted towards lower temperatures. The shifting of the pretransition temperature is much larger than the shifting of the main transition by the same amount of malathion. The reason for this may partly be that the transition enthalpy for the pretransition is much lower than for the main transition. Furthermore, malathion causes the pretransition enthalpy to decrease linearly with concentration. The effect of deltamethrin on the pretransition is very different from that of malathion. At very low concentrations of deltamethrin, the pretransition disappears, but it reappears again at higher concentrations as a very small peak. At present, we have no explanation as to why. It seems, however, not surprising that there is a larger difference between the effects of the two insecticides at the pretransition than at the main transition, since packing properties are more likely to be important in the gel phase than in the fluid phase.

Concluding remarks

In conclusion, we have studied the effects of the insecticides malathion and deltamethrin on the thermotropic phase behaviour of DMPC lipid bilayers. In small amounts, both insecticides broaden the phase-transition region and suppress the transition temperature of multilamellar lipid bilayers of DMPC. Small amounts of both insecticides do not change the transition enthalpy significantly. This implies that these insecticides do not take up any excluded volume in the membrane. Higher concentrations of malathion narrow the transition compared to lower concentrations of malathion. Furthermore, several peaks are observed in the C_p curves in the transition region, suggesting the presence of a coexistence region where more than one phase is present. For malathion, a high-temperature peak developed with increasing insecticide concentration. This peak was not seen in the vesicles with deltamethrin. Although both insecticides are hydrophobic molecules their effects on the phase transition of DMPC lipid bilayers are very different at high concentrations. This suggests that details of the molecular structure of the insecticides are important for their nonspecific interactions with lipid bilayers.

Acknowledgements This work was supported by the Danish Natural Science Research Council and the Danish Technical Research Council. Cheminova Agro (Lemvig, Denmark) and Hoechst Schering Agrevo (Hattersheim, Germany) are thanked for their gifts of malathion and deltamethrin, respectively. The authors are affiliated with the Danish Centre for Drug Design and Transport, which is supported by the Danish Medical Research Council.

8

References

1. Crossland J (1980) Lewis's pharmacology, 5th edn. Churchill Livingstone, Edinburgh, pp 901–906
2. Lawrence LJ, Casida JE (1983) Science 221:1399
3. Eells JT, Rasmussen JL, Bandettini PA, Propp JM (1993) Toxicol Appl Phamacol 123:107
4. Forshaw PJ, Lister T, Ray DE (1993) Neuropharmacology 32:105
5. Ohkawa H (1982) In: Coats JR (ed) Insecticide mode of action. Academic, New York, pp 169–171
6. Narahashi T (1996) Pharmacol Toxicol 79:1
7. Srivastava SC, Kumar R, Prasad AK, Srivastava SP (1995) Toxicol Lett 75:153
8. Antunes-Madeira MC, Madeira VMC (1987) Biochim Biophys Acta 901:61
9. Antunes-Madeira MC, Madeira VMC (1984) Biochim Biophys Acta 778:49
10. Blasiak J (1993) Pestic Biochem Physiol 45:72
11. Antunes-Madeira MC, Videira RA, Madeira VMC (1994) Biochim Biophys Acta 1190:149
12. Stelzer KJ, Gordon MA (1985) Biochim Biophys Acta 812:361
13. Sarkar SN, Balasubramanian SV, Sikdar SK (1993) Biochim Biophys Acta 1147:137
14. Moya-Quiles MR, Muños-Delgado E, Vidal CJ (1994) Arch Biochem Biophys 312:95
15. Moya-Quiles MR, Muños-Delgado E, Vidal CJ (1996) Chem Phys Lipids 79:21
16. Moya-Quiles MR, Muños-Delgado E, Vidal CJ (1996) Chem Phys Lipids 83:61
17. Sikkema J, de Bont JAM, Poolman B (1995) Microbiol Rev 59:201
18. Lasic D (1991) Liposomes: from physics to applications Elsevier, Amsterdam
19. Mountcastle DB, Biltonen RL, Halsey MJ (1978) Proc Natl Acad Sci USA 75:4906
20. van Osdol WW, Ye Q, Johnson ML, Biltonen RL (1992) Biophys J 63:1011
21. Sabra MC, Jørgensen K, Mouritsen OG (1995) Biochim Biophys Acta 1233:89
22. Sabra MC, Jørgensen K, Mouritsen OG (1996) Biochim Biophys Acta 1282:85
23. Jørgensen K, Ipsen JH, Mouritsen OG, Bennett D, Zuckermann MJ (1991) Biochim Biophys Acta 1062:227
24. Biltonen RL, Lichtenberg D (1993) Chem Phys Lipids 64:129

Progr Colloid Polym Sci (2000) 116:9–15
© Springer-Verlag 2000

J. Engblom
Y. Miezis
T. Nylander
V. Razumas
K. Larsson

On the swelling of monoolein liquid-crystalline aqueous phases in the presence of distearoylphosphatidylglycerol

J. Engblom (✉)
Bioglan AB, P.O. Box 50310
202 13 Malmö, Sweden
e-mail: johan.engblom@bioglan.se
Tel.: +46-40-287580; Fax +46-40-291955

Y. Miezis
Food Technology, Lund University
P.O. Box 124, 221 00 Lund, Sweden

T. Nylander
Physical Chemistry 1, Lund University
P.O. Box 124, 221 00 Lund, Sweden

V. Razumas
Institute of Biochemistry, Mokslininkų 12
2600 Vilnius, Lithuania

K. Larsson
Camurus Lipid Research, Ideon, Gamma 1
Solvegatan 41, 223 70 Lund, Sweden

Abstract The phase behaviour of the liquid-crystalline phases in the monoolein (MO)–distearoylphosphatidylglycerol (DSPG)–water system was investigated with the addition of up to 10 wt% DSPG using small-angle X-ray diffraction. Generally, the introduction of this anionic phospholipid leads to an increased hydration of the liquid-crystalline phases compared to the binary MO–water system. In the binary MO–water system, the cubic phase could be swollen to about 40 wt% water in accordance with previous studies. At high water content the diamond (C_D) phase (cubic space group $Pn3m$) persists, while at lower water content the gyroid (C_G) type (cubic space group $Ia3d$) exists. It was found that at about 1 wt% DSPG and above 45 wt% water the primitive (C_P) phase (cubic space group $Im3m$) replaces the C_D phase. The C_P phase could be swollen up to at least 70 wt% water. Another effect of the addition of DSPG is that the C_G phase is favoured and expands on the expense of the C_D phase. Furthermore, at a high DSPG content (about 5 wt%) the C_G phase can take about 60% water. This cubic phase is transformed into a highly swollen lamellar phase, which exists from just below 70 wt% water and up to about 80 wt% water.

Key words Monoolein · Liquid-crystalline phase · Cubic phase · Distearoylphosphatidylglycerol · Swelling of liquid-crystalline phases

Introduction

A common feature of polar lipids is their tendency to form liquid-crystalline phases, and these phases play a very significant role in nature. The lamellar phase, consisting of planar lipid bilayers separated by water, mimics biomembranes and has therefore been used by biologists and biophysicists as a membrane model system as such or in excess water as vesicles. There are now numerous examples indicating that liquid-crystalline phases with cubic symmetry also occur frequently in living organisms [1–5]. The bicontinuous cubic phase in these systems is considered to be comprised of curved, nonintersecting bilayers organised to form two disjointed continuous water channels [3–5]. The mid-surface of

the lipid bilayer of such a cubic phase can be described by an infinite periodic minimal surface (IPMS) [4, 5]. The curvature of any surface is given by the two principal radii of curvature, R_1 and R_2, and the average curvature $1/2(1/R_1 + 1/R_2)$ at any point of a minimal surface is zero. Three types of IPMS, describing different cubic space groups, are important in lipid systems [4, 5]: the diamond (C_D) type [primitive lattice ($Pn3m$)], the gyroid (C_G) type [body-centered lattice ($Ia3d$)] and the so-called primitive (C_P) type [body-centered lattice ($Im3m$)].

It has been suggested that acylglycerols assume a cubic structure in the intermediate state of fat digestion [2, 6, 7]. Recent studies demonstrated that cubic phases can accommodate a wide range of globular proteins

[2, 8–12]. A possible application is to use lipid–water based bicontinuous cubic phases with entrapped enzymes to produce electrochemical biosensors [10, 12]. The bicontinuous cubic phases can accommodate lipophilic as well as hydrophilic substances and can therefore be used to study the redox activity of membrane bound compounds, like vitamin K_1 and ubiquinone 10 [13, 14]. In addition cubic phases can be used in pharmaceutical applications for controlled drug release [1].

One natural lipid that forms cubic phases under physiological conditions on addition of water is monoolein (MO). The phase behaviour of the MO aqueous binary system was determined by Hyde et al. [15] and was later confirmed [16, 17]. At room temperature, MO forms a reverse micellar phase (L_2) at very low water content (4–5 wt% water) and a lamellar phase (L_α) upon further addition of water (8–22 wt%). In the region 25–40 wt% water, two different bicontinuous cubic structures, belonging to the body-centred space group $Ia3d$ (C_G phase) and to the primitive space group $Pn3m$ (C_D phase), appear at low and high water content, respectively. At a water content higher than 40 wt% the C_D phase coexists with excess water. In the region 20–30 wt% water a reverse hexagonal (H_{II}) phase forms at $T > 80$ °C. The addition of a third component, which promotes a reverse hexagonal structure, can have the same effect as an increase in the temperature [13, 18, 19]. It has also been shown that the formation of free oleic acid occurs in the binary MO–water system, owing to hydrolysis of the MO, which can induce a phase transition from the cubic to the H_{II} phase [19].

In applications, where the dimensions of the water channels in the cubic phases are important, it is desirable to be able to vary the maximal swelling. This can be achieved by introducing amphiphilic nonionic polymers [16] or ionic surfactants such as cetyltrimethylammonium bromide [20] or sodium cholate [21]. In this study we used an anionic lipid, distearoyl phosphatidyl glycerol (DSPG). The advantage is that this lipid, with its very low water solubility, will not be released in the excess water as can occur for more soluble surfactants. The introduction of DSPG into MO aqueous cubic phases has previously been shown to modify the release of ionic drugs entrapped in this matrix [22]. We have previously shown that the introduction of DSPG in the MO aqueous system can induce a phase transition from the diamond cubic phase (C_D), with primitive space group $Pn3m$, to the so-called primitive cubic phase (C_P), with body-centred space group $Im3m$ [22, 23]. The results from Raman scattering studies also indicated that incorporation of DSPG into the lipid bilayer decreased the mobility of the acyl chains and increased the number of hydrogen-bonded C=O groups of MO [23]. Here we present an outline of the phase diagram for the ternary MO–DSPG–water system.

Experimental

MO ($M_r = 356.6$ g/mol) from batch TS-ED 173, containing 98.1% monoglycerides, was kindly provided by Danisco Ingredients, Brabrand, Denmark. The (mono)sodium salt of DSPG ($M_r = 778.1$ g/mol) from lot B12142/2 was kindly provided by Genzyme Pharmaceuticals and Fine Chemicals, UK. The water was distilled and passed through a Milli-Q water purification system (Millipore).

The samples were prepared in glass ampoules, which after sealing were centrifuged 6 times at 1000g for 5 min. The samples were allowed to equilibrate at room temperature until no change could be detected when viewed between crossed polarisers. Normally equilibrium was obtained after 2–3 weeks. A one-phase region of the cubic phase is easier to identify than other mesophases, since this phase is completely transparent, very viscous and optically isotropic. In addition microscopic examination of some of the samples and their phase transitions was performed using an Olympus Vanox polarising microscope (Olympus, Japan). A temperature gradient, ranging from 25 to 115 °C was achieved using a Mettler FP52 heating table (Mettler Instrumente, Switzerland), with a heating rate of 10 °C/min. The low-angle X-ray diffractograms for all the samples were obtained using a Guinier camera modified after Luzzati et al. [24]. The specimen-to-film distance was approximately 20 cm, and Cu $K\alpha$ nickel-filtered radiation ($\lambda = 1.542$ Å) was used. The samples were held in a sample holder constructed according to Hernqvist [25]. An exposure time of 24 h was used and measurements were performed at 22 °C.

Results and discussion

A partial phase diagram of the ternary MO–DSPG–water system is shown in Fig. 1. This diagram was obtained by visual inspection of the samples between crossed polarisers, polarising microscopy and small-angle X-ray diffraction (SAXD). The SAXD data are presented in Table 1. The lattice parameters obtained were compared with the dimensions expected from the appropriate swelling laws as discussed later. In this way we could check whether a sample contained only one phase or if the particular phase was in excess water. The main difference in relation to the binary MO–water system is the expansion of the monophasic regions towards the water-rich corner at a relatively low DSPG content (5 wt% or less). This leads to the formation of an additional bicontinuous cubic phase, which corresponds to Schwartz primitive IPMS (C_P). It has a body-centred space group, $Im3m$, and appears at about 1 wt% DSPG and above 40 wt% water, where it replaces the C_D phase. Among the bicontinuous cubic phases, the C_P phase has the largest geometrical constraint, with regard to water encapsulation, which means that it only appears at high water content [4]. As the DSPG content increases to about 5 wt%, the C_G phase is stabilised at the expense of the C_D phase. It can take substantially more water than in the binary MO–water system. In fact in the presence of DSPG the C_G phase could take up about 60 wt% water. Here we note that DSPG by itself in water forms a lamellar liquid-crystalline phase above the chain melting

Fig. 1 The partial phase diagram of the monoolein– distearoylphosphatidylglycerol (*DSPG*)–water system. The experimental data points are inserted, where the *open symbols* denote samples containing more than one phase. The lamellar (L$_\alpha$, ○) cubic gyroid (C$_G$,⌘), cubic diamond (C$_D$, □) and cubic primitive (C$_P$, ↖) phases are indicated. The tie lines are given as *dotted lines*

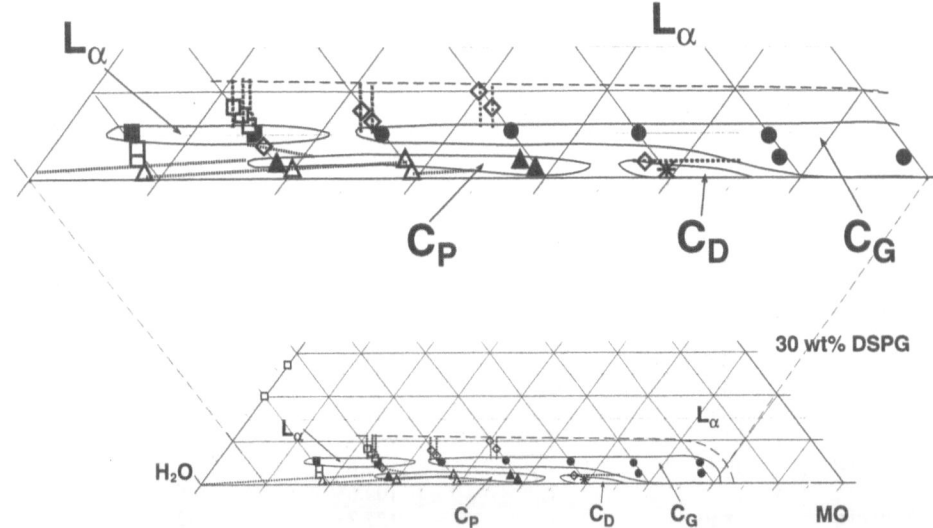

transition at about 54 °C [26]. At room temperature, DSPG exists as a gel phase with a bilayer structure similar to the lamellar liquid crystal, but with frozen chains (see later). Consequently, the presence of DSPG will favour a lamellar phase or structures that border a lamellar region, like the C$_G$ phase in the binary MO–water system.

The formation of a particular phase can be rationalised on the basis of the geometrical packing properties of the lipid in the particular environment [27]. These properties can be easily expressed by the packing parameter, V/AL, where V is the volume of the hydrophobic chain, A the head group area and L the chain length [27, 28]. For direct curved structures (curved towards the apolar region), like micellar (L$_1$) and hexagonal (H$_I$) phases, the packing parameter assumes values $V/AL < 1$, while V/AL assumes values around unity for lamellar phases and above 1 for phases of reversed morphology, for example, bicontinuous cubic (V$_2$) and reversed hexagonal (H$_{II}$) phases. The value of V/AL for the different cubic (V$_2$) phases increases in the order C$_G$ < C$_D$ < C$_P$ [4]. The C–H stretching modes in the Raman spectra of the MO–DSPG–water system, reported in an earlier study [23], clearly show a decreased average mobility of the acyl chains compared to the binary MO–water system. This means that the V/L ratio decreases in the ternary system. However, the Raman study gave no indications of any difference in the intrachain ordering between the binary and ternary cubic liquid-crystalline phases [23]. In addition the Raman spectra revealed an increase in the relative number of hydrogen-bonded C=O groups in the ternary system. This implies an increase in the effective head group area, A, which in turn leads to a decrease in the packing parameter.

The stabilisation of the C$_G$ phase at the expense of the C$_D$ phase has also been observed upon adding other lipids that form lamellar phases, for example, soy bean phosphatidylcholine [12]. At higher DSPG content a large region of a lamellar phase is indicated. The extent of this region was not investigated further; however, at high enough concentration of DSPG compared to MO the lamellar phase is expected to transform into a gel phase. Pure DSPG forms a gel phase that is able to swell to an interlayer spacing of 62 Å in the presence of water (Table 1). Assuming a lipid bilayer thickness of 42 Å gives a water layer thickness of 20 Å. The solid state of DSPG also shows a long-spacing of 42 Å (Table 1). It is interesting to note that in this system we observe two lamellar phases. This has also been observed in related system with MO, for example, the ternary MO–dioleoylphosphatidylcholine–water system [29].

The SAXD data are summarised in Table 1. We first calculated the expected amount of water on the basis of the experimentally determined lattice parameters. If the lamellar phases shows the ideal one-dimensional swelling, the repeat distance, d, from the SAXD measurements is proportional to the inverse of the volume fraction of the lipid, Φ_l, according to

$$d = \frac{2l}{\Phi_l} \ , \tag{1}$$

where $2l$ is the average bilayer thickness. Alternatively we can calculate the volume fraction of water, Φ_w, expected from a certain interlayer spacing, d, as

$$\Phi_W = \frac{d - 2l}{d} \ . \tag{2}$$

The water volume fraction can be transferred to the weight fraction of water, φ_w, by

$$\varphi_w = \frac{1}{1 + \rho_l \frac{1 - \Phi_w}{\Phi_w}} \ , \tag{3}$$

where the density of water is assumed to be 1 and the density of the lipid is ρ_l. In the present study we assumed

Table 1 Lattice parameters for the monoolein (*MO*)–distearoylphosphatidylglycerol (*DSPG*)–water system from small-angle X-ray diffraction (*SAXD*) together with related hydration levels, as well as the maximum lattice parameter calculated from the total amount of water available

Composition MO/DSPG/H₂O (wt%)	MO/DSPG (molar ratio)	Phase	a (Å)[a]	Φ_w (vol%)[b]	φ_w (wt%)[c]	a_{max} (Å)[d]
18.5/0.9/80.6	42.9	C_P	196.8	58.7	60.2	406.8
17.5/2.0/80.5	19.2	L_α	130.6	73.8	75.0	166.4
		L_c	42.5			
16.8/3.3/79.9	11.1	L_α	127.3	73.3	74.5	161.3
		L_c	42.4			
15.5/5.1/79.4	6.5	L_α	114.0[e]	70.2	71.5	157.2
0/20.0/80.0	0	L_c	61.7	44.9/31.9[f]	46.4/33.3[f]	162.1/200.3[f]
29.8/1.2/69.0	56.5	C_P	191.9	57.7	59.2	254.8
28.2/1.9/69.9	31.6	C_P	268.4	69.3	70.6	261.7
25.0/5.0/70.0	10.9	L_α	75.6	55.0	56.6	108.7
		L_α	100.6[g]	66.2	67.5	
24.1/5.9/70.0	8.9	L_α	127.2	73.3	74.5	108.6
		L_α	104.0	67.3	68.6	
23.0/7.0/70.0	7.2	C_D	153.0	58.5	59.9	205.6
		L_α	127.0	73.2	74.4	108.7
21.9/8.1/70.0	5.9	C_D	146.0	56.6	58.1	205.6
		L_α	124.7	72.7	73.9	108.7
0/27.4/72.6	0	L_α	57.5	40.9/27.0[f]	42.4/28.2[f]	118.9
29.1/6.0/64.9	10.5	L_α	126.0[e]	73.0	74.2	93.3
39.1/1.0/59.9	85.2	C_P	154.6	48.3	49.8	195.5
38.3/2.1/59.6	41.0	C_P	165.2	51.4	52.9	194.3
34.9/5.0/60.1	15.3	C_G	263.8	61.9	63.3	242.0
33.3/6.6/60.1	10.9	C_G	266.0	62.2	63.6	242.1
		L_α	108.8	68.8	70.1	82.2
31.9/7.7/60.4	9.2	C_G	274.7	63.3	64.7	244.3
		L_α	114.2	70.2	71.5	83.0
48.7/1.3/50.0	83.5	C_P	154.3	48.2	49.7	155.3
47.1/2.1/50.8	48.6	C_P	161.4	50.3	51.8	157.8
44.9/5.2/49.9	18.7	C_G	208.1	52.4	53.9	190.6
41.8/8.1/50.1	11.3	C_G	216.4	54.1	55.6	191.5
		L_a	87.4	61.1	62.6	66.2
40.0/9.9/50.1	8.7	C_G	217.5	54.3	55.8	191.5
		L_a	91.4	62.8	64.2	66.2
59.0/0.9/40.1	136.9	C_D	103.7	40.8	42.2	99.6
56.9/2.0/41.1	63.2	C_G	135.7	30.7	32.0	160.1
		C_D	106.0	41.9	43.4	101.6
55.0/4.9/40.1	24.4	C_G	167.9	42.2	43.6	157.0
67.5/2.0/30.5	73.7	C_G	139.6	32.3	33.3	132.3
65.2/4.7/30.1	30.1	C_G	139.2	32.1	33.4	131.4
77.0/2.1/20.9	80.2	C_G	117.2	22.3	23.3	112.6
75.5/4.7/19.8	34.9	C_G	117.6	22.5	23.5	110.4

[a] Lattic parameter from SAXD measurements
[b] Volume fraction of water expected from SAXD calculated from Eqs. (2) and (6) for cubic and lamellar phases, respectively
[c] The corresponding weight fraction of water calculated from Eq. (3)
[d] The maximum lattice parameter calculated from the total amount of water available from Eqs. (1) and (4) for cubic and lamellar phases, respectively
[e] Weak SAXD reflection pattern, calculated interlayer spacing uncertain
[f] Calculated for a bilayer thickness corresponding to the one observed for DSPG, 42 Å
[g] Calculated on second reflection

that $\rho_1 = 0.942$, which is the value for MO, independent of the content of DSPG. The swelling of bicontinuous MO aqueous cubic phases is more complex [30, 31], and the linear swelling is just a first approximation. The following expression for the unit cell dimension, a, can be derived for reversed bicontinuous cubic phase according to Engblom and Hyde [31]:

$$a = l \left(\frac{-2\pi\chi}{H} \right)^{1/3} \left[\sqrt{3}\sin\left(\frac{\Delta}{3}\right) - \cos\left(\frac{\Delta}{3}\right) \right]^{-1}, \quad (4)$$

where

$$\Delta = \tan^{-1}\left(\frac{\sqrt{1-\Phi_l^2}}{-\Phi_l}\right) . \tag{5}$$

Here χ is a topology index of the surface known as the Euler–Poincaré characteristics and H is a dimensionless characteristic, the "homogeneity index", which combines the surface-to-volume ratio of a hyperbolic surface with its topology. The parameters χ and H assume the values -8, -2 and -4, and 0.7665, 0.7498 and 0.7163 for the gyroid, diamond and primitive surface, respectively. As for the lamellar phase we can invert Eq. (5) and calculate the expected volume fraction of water, $\Phi_w = 1 - \Phi_l$, as

$$\Phi_w = 1 - \sqrt{\frac{1}{\tan^2\Delta + 1}} , \tag{6}$$

with

$$\Delta = 3\sin^{-1}\left[\frac{l}{2a}\left(\frac{-2\pi\chi}{H}\right)^{\frac{1}{3}}\right] - \frac{\pi}{2} . \tag{7}$$

By applying Eq. (3) we can then obtain the corresponding weight fraction of water. We assume a lipid bilayer thickness, $2l$, of 34 Å, a value obtained for swelling of a MO lamellar phase [21]. We neglect the effect of DSPG on the bilayer thickness when the concentration of DSPG compared to MO is low. A typical bilayer thickness of DSPG is expected to be about 42 Å [32], and this value was used for some of the calculations (Fig. 2).

The previous equations were used to calculate the values in Table 1. First the value of the lattice parameter, a, for the lamellar ($a=d$) and cubic phases, respectively,

Fig. 2a, b The ratio between the lattice parameter and the monolayer thickness, a/l, plotted versus the lipid weight fraction, $\varphi_l = 1 - \varphi_w$, on the basis of Eqs. (4) and (5) for the cubic phases and Eq. (1) for the lamellar phase. The experimental data points assuming $l = 17$ Å, using same symbols as in Fig. 1, are also inserted. The *bar* corresponds to a monolayer thickness equal to that of DSPG in the crystalline state ($l = 21$ Å). *Open symbols* indicate two-phase samples. **a** The C_G and L_α samples. **b** The C_D and C_P samples

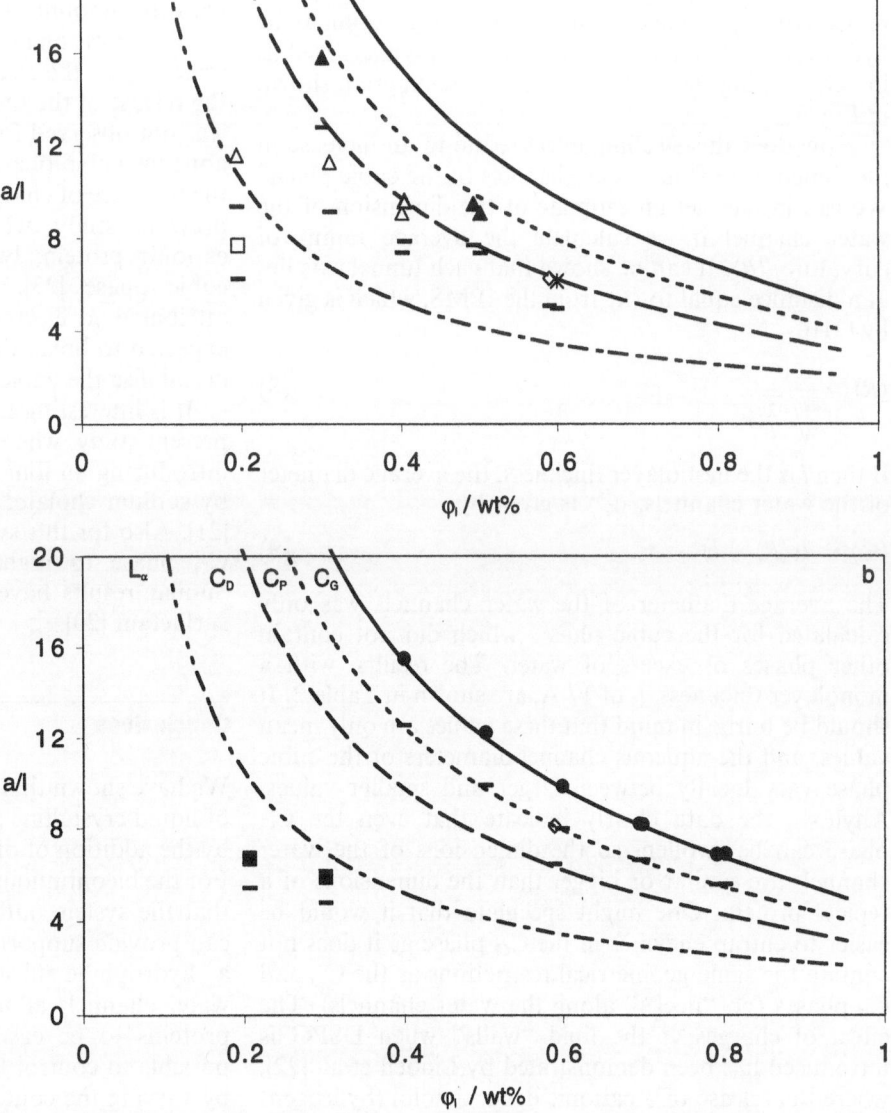

calculated from a given sample composition was compared with the experimentally obtained composition. A significant deviation indicates that the sample is not a one-phase sample, but rather contains one phase in excess water. This is illustrated in Fig. 2, where a/l is plotted versus the lipid weight fraction, φ_l, on the basis of Eqs. (4) and (5). The experimental data points are also inserted, where the bar corresponds to a monolayer thickness equal to that of DSPG in the crystalline state. Open symbols indicate two-phase samples. The C_G and L_α samples are shown in Fig. 2a, while C_D and C_P samples are shown in Fig. 2b. The experimental data could be adequately described by the theory, using a bilayer thickness of 34 Å (Fig. 2). This enabled us to check whether the assigned space groups of the homogenous cubic phases were correct. Furthermore, we could use Fig. 2 to determine if a sample contained more than one phase (i.e., excess water). For the two-phase samples the lattice parameter of the fully swollen phase from theory was compared with the experimentally determined lattice parameter, which allowed us to estimate the hydration level in the individual phase. This also aided us in constructing the tie lines in the phase diagram shown in Fig. 1.

How does this swelling effect relate to an increase in the dimension of the water channels of the cubic phase? We can in fact get an estimate of the dimension of the water channel if we calculate the average radius of curvature, $\langle R \rangle$. It can be shown that each tunnel axis lies at a distance equal to $\langle R \rangle$ from the IPMS, which is given by [31])

$$\langle R \rangle = \frac{a}{\sqrt[3]{\frac{-2\pi\chi}{H}}} \ . \tag{8}$$

If then l is the monolayer thickness, the average diameter of the water channels, $\langle \varnothing \rangle$ is given by

$$\langle \varnothing \rangle = 2(\langle R \rangle - l) \ . \tag{9}$$

The average diameter of the water channels was only calculated for the cubic phases which did not contain other phases or excess of water. The results, with a monolayer thickness, l, of 17 Å, are shown in Table 2. It should be borne in mind that these values are only mean values, and the aqueous channel diameters of the cubic phase vary locally between larger and smaller values. Anyhow, the data clearly indicate that even the C_G phase can be swollen, so the dimensions of the water channels are similar or bigger than the dimensions of a typical protein. One might speculate that it would be easier to entrap enzymes in the C_G phase as it does not contain the same geometrical restrictions as the C_D and C_P phases (no "necks" along the water channels). The effect of charges at the lipid "walls" when DSPG is introduced has been demonstrated by Lindell et al. [22], where the release of a cationic drug, timolol (hydrogen)

Table 2 The average diameter of the water channel for the different cubic phases as a function of the degree of swelling

Composition (wt%) MO/DSPG/water	IPMS	a (Å)[a]	$\langle R \rangle$ (Å)[b]	$\langle \varnothing \rangle$ (Å)[c]
47.1/2.1/50.8	C_P	161.4	49	65
28.2/1.9/69.9	C_P	268.4	82	130
59.0/0.9/40.1	C_D	103.7	41	47
75.5/4.7/19.8	C_G	117.6	29	24
65.2/4.7/30.1	C_G	139.2	35	35
55.0/4.9/40.1	C_G	167.9	42	49
44.9/5.2/49.9	C_G	208.1	52	69
34.9/5.0/60.1	C_G	263.8	65	97

[a] Obtained from SAXD data, see Table 1
[b] Calculated from Eq. (8)
[c] Calculated from Eq. (9), taking the monolayer thickness to be 17 Å

maleate from a MO aqueous cubic with varying amounts of DSPG was investigated. It was found that the amount released as well as the release rate decreased with the amount of DSPG included in the cubic phase. Furthermore, an increase in the ionic strength, that is the screening of the electrostatic attractive forces, increased the release of the timolol (hydrogen) maleate. This effect was not observed for the release of the drug from a MO aqueous cubic phase without DSPG. Further effects of the presence of charges at the bilayer were presented in a previous study, where we tried to entrap a strongly cationic protein, lysozyme, in a MO–DSPG aqueous cubic phase [23]. The failure of this attempt was attributed to electrostatic interaction which should be expected to break the integrity of the mixed bilayer and destabilise the cubic phase.

It is interesting to note that the phase diagram in the present study, where increased swelling was achieved by introducing an ionic lipid, resembles the effect obtained by sodium cholate, a water-soluble anionic amphiphile [21]. Also for this system extensions of the lamellar and C_G phase to higher water contents were observed. Similar results have also been observed for a cationic surfactant [20].

Conclusions

We have shown that it is possible to increase the swelling of liquid crystalline phases of MO to a significant degree by the addition of only a small amount of an ionic lipid. For the bicontinuous cubic phases in particular, it means that the system, although highly hydrated, is rigid and can provide support for entrapped hydrophobic as well as hydrophilic substances. The large dimensions of the water channels allows even such large biomolecules as proteins to be easily encapsulated. In this way it is possible to control the dimension of the water channels by varying the content of the anionic lipid.

Acknowledgements It is our pleasure to acknowledge Katarina Lindell for valuable discussions and suggestions. The support from Sven Engström and Stephen Hyde is also gratefully acknowledged. Financial support was obtained from the Swedish Research Council for Engineering Sciences the and Royal Swedish Academy of Sciences.

References

1. Larsson K (1994) Lipids – molecular organization, physical functions and technical applications. The Oily Press, Dundee
2. Mariani P, Luzzati V, Delacroix H (1998) J Mol Biol 204:165–189
3. Lindblom G, Rilfors L (1989) Biochim Biophys Acta 988:221–256
4. Larsson K (1989) J Phys Chem 93:7304–2314
5. Hyde S, Andersson S, Larsson K, Blum Z, Landh T, Lidin S, Ninham BW (1997) The language of shape. The role of curvature in condensed matter: physics, chemistry and biology. Elsevier, Amsterdam
6. Patton JS, Carey MC (1979) Science 204:145–148
7. Patton JS, Vetter RD, Hamosh B, Borgstrøm B, Lindstrøm M, Carey MC (1985) Food Microstructure 4:29–41
8. Ericsson B, Larsson K, Fontell K (1983) Biochim Biophys Acta 729:23
9. Portmann M, Landau EM, Luisi PL (1991) J Phys Chem 95:8437–8440
10. Razumas V, Kanapieniené J, Nylander T, Engström S, Larsson K (1994) Anal Chim Acta 289:155–162
11. Razumas V, Larsson K, Miezes Y, Nylander T (1996) J Phys Chem 100:11766–11774
12. Nylander T, Mattisson C, Razumas V, Miezes Y, Håkansson B (1996) Colloids Surf A 114:311–320
13. Caboi F, Nylander T, Razumas V, Talaikyté Z, Larsson K (1997) Langmuir 13:5476–5483
14. Razumas V, Talaikyté Z, Barauskas J, Nylander T, Miezis Y (1998) Prog Colloid Polymer Sci 108:76–82
15. Hyde ST, Andersson S, Ericsson B, Larsson K (1984) Z Kristallogr 168:213–219
16. Landh T (1994) J Phys Chem 98:8453–8467
17. Briggs J, Chung H, Caffrey M (1996) J Phys II 6:723–751
18. Barauskas J, Razumas V, Nylander T (1999) Chem Phys Lipids 97:167–179
19. Caboi F, Amico GS, Pitzalis P, Monduzzi M, Nylander T, Larsson K (2000) Chem Phys Lipids (in press)
20. Gustafsson J, Orädd G, Nyden M, Hansson P, Almgren M (1998) Langmuir 14: 4987–4996
21. Gustafsson J, Nylander T, Almgren M, Ljusberg-Wahren H (1998) J Colloid Interface Sci 211:326–335
22. Lindell K, Engblom J, Jonströmer M, Carlsson A, Engström S (1998) Prog Colloid Polymer Sci 108:111–118
23. Razumas V, Talaikyte Z, Barauskas J, Larsson K, Miezis Y, Nylander T (1996) Chem Phys Lipids 84:123–138
24. Luzzati V, Mustacchi H, Skoulios A, Husson F (1960) Acta Crystallogr 13:660–667
25. Hernqvist L (1984) PhD thesis. Lund University, Sweden
26. Kristensen A, Nylander T, Paulsson M, Carlsson A (1997) Int Dairy J 7:87–92
27. Mitchell DJ, Ninham BW (1981) J Chem Soc Faraday Trans 2 77:601–629
28. Israelachvili JN, Mitchell DJ, Ninham BW (1976) J Chem Soc Faraday Trans 2 72:1525–1567
29. Gutman H, Arvidsson G, Fontell K, Lindblom G (1984) In: Mittal KS, Lindman B (eds) Surfactants in solution, vol 1. Plenum, New York, pp 143–152
30. Chung H, Caffrey M (1994) Biophys J 66:377–381
31. Engblom J, Hyde ST (1995) J Phys II 5:171–190
32. Cevc G (ed) (1993) Phospholipids handbook. Dekker, New York

Progr Colloid Polym Sci (2000) 116 : 16–20
© Springer-Verlag 2000

J. Barauskas
V. Razumas
T. Nylander

Entrapment of glucose oxidase into the cubic Q^{230} and Q^{224} phases of aqueous monoolein

J. Barauskas (✉) · V. Razumas
Institute of Biochemistry, Mokslininkų 12
2600 Vilnius, Lithuania
e-mail: justas.barauskas@bchi.lt
Tel.: +370-2-729186; Fax: +370-2-729196

T. Nylander
Department of Physical Chemistry 1,
Center for Chemistry
and Chemical Engineering
University of Lund, P.O. Box 124
221 00 Lund, Sweden

Abstract Using X-ray diffraction measurements, the entrapment of glucose oxidase (GOD) from *Aspergillus niger* was investigated in the reversed bicontinuous cubic phases (Q_{II}) of aqueous monoolein (MO, 86 wt% monooleoylglycerol). At 25 °C, a partial phase diagram of the MO/GOD/H_2O system indicated that the Q^{230} phase can accommodate only up to 0.5 wt% of the commercial GOD preparation. At higher GOD contents, this cubic phase transforms into a reversed hexagonal phase (H_{II}). On the other hand, the Q^{224} phase of MO can accommodate at least up to 6 wt% GOD. The enzyme preparation promotes the thermotropic $Q^{230} \rightarrow H_{II}$ phase transition, whereas it has no effect on the thermal stability of the Q^{224} phase. On the basis of an examination of the calculated geometric parameters of the pseudobinary and pseudoternary cubic phases (thickness of lipid monolayer, diameter of water channels), it is concluded that the native enzyme molecules cannot be located exclusively in the water channels of the phases. The results are also discussed in terms of their possible application in the construction of biosensors.

Key words Cubic phases · Aqueous monoolein · Glucose oxidase · X-ray diffraction · Phase transitions

Introduction

Among other liquid-crystalline phases of lipids, the reversed bicontinuous cubic phases (Q_{II}) are the most complex. It is now widely accepted that a three-dimensional structure of the Q_{II} phases is formed by the curved continuous lipid bilayer which separates two continuous channels of water [1–3]. Since the diameter of the water channels can range up to about 50–100 Å, a number of studies on the protein-containing cubic phases of lipids have been conducted [4–9]. These investigations have demonstrated that various proteins and enzymes retain their native structure and activity in the cubic mesophases. Therefore, there is good reason to believe that the Q_{II} phases of lipids can serve as flexible and biocompatible matrixes for the immobilization of practically important enzymes.

Indeed, some of us have shown that, when constructing bioelectrodes and electrochemical biosensors, the outer bioactive layer of the working electrode can be formed by the enzyme-containing Q_{II} phase [7–9]. In this respect, the glucose oxidase (GOD) from *Aspergillus niger* was among the first enzymes entrapped in the reversed bicontinuous cubic phase Q^{224} (space group type $Pn3m$) of aqueous monoolein (MO). This GOD-containing phase of MO has been successfully used in the construction of the amperometric biosensor for the determination of glucose [7]. In the bioactive layer of the electrode, GOD catalyzes the oxidation of β-D-glucose by molecular oxygen to D-glucono-1,5-lactone and hydrogen peroxide. The latter compound is monitored by the current of its oxidation on the Pt electrode surface.

However, GOD from *A. niger* is a rather complex enzyme. It is a homodimeric glycoprotein of molecular weight 160 kDa (about 20 wt% carbohydrate), having a hydrodynamic radius of about 40 Å and two tightly bound flavin adenine dinucleotide FAD/FADH$_2$ redox centers. Therefore, to optimize the performance of our glucose biosensor, it is important to know the influence of GOD content on the lyotropic and thermotropic properties of the MO-based immobilization matrix.

Considering the foregoing, in the present work, using small-angle X-ray diffraction (SAXD) measurements, a partial phase diagram of the pseudoternary MO/GOD/H$_2$O system at 25 °C is constructed and discussed. Besides, the effect of GOD content on the thermal stability of the MO cubic phases is also investigated. Here, it should be pointed out that the MO/GOD/H$_2$O system is called a "pseudoternary" system since from the practical point of view the typical commercial preparations of MO and GOD are used without further purification.

Materials and methods

A mixture of mono- and diglycerides (about 25:1 by weight), denoted as MO, was produced and kindly provided by Danissco Ingredients (Brabrand, Denmark) with the following fatty acid composition (batch TS-ED 173): 90 wt% oleic acid, 5 wt% linoleic acid, 2.7 wt% stearic acid, 1 wt% palmitic acid, 0.3 wt% linolenic acid, and 1 wt% of other fatty acids. GOD (EC 1.1.3.4, from *A. niger*, type X-S, lot 48H3778, containing about 75 wt% of protein) was purchased from Sigma Chemical Co. The preparations of MO and GOD were used as received. The water was ion-exchanged, distilled, and passed through a Milli-Q water purification system (Millipore, Bedford, Mass., USA).

The samples were prepared by weighing appropriate amounts of MO into 6 mm (internal diameter) glass ampoules, melting MO at 40 °C, adding a GOD solution in water, and then briefly mixing the content using a glass stick. The ampoules were immediately flame-sealed, then centrifuged for 1 h at 3000 *g* and 25 °C, and allowed to equilibrate at 25 °C for at least 2 weeks before measurements.

SAXD measurements were performed at 25 °C on a Kratky compact small-angle system equipped with an OED 50 M (MBraun, Graz, Austria) position-sensitive detector containing 1024 channels of width 53.1 μm. Cu K$_α$ nickel-filtered radiation of wavelength 1.542 Å was provided by a Seifert ID 3000 X-ray generator (Rich. Seifert, Ahrensburg, Germany) operating at 50 kV and 40 mA. The exposure times were 10–30 min at the sample-to-detector distance of 277 mm. Temperature control within 0.1 °C was achieved by using a Peltier element. A stepwise increase of temperature (typically 2–4 °C in every step with an equilibration time at a given temperature of 10 min) was used to induce phase transitions.

Results and discussion

According to the earlier investigations of the pure 1-monooleoylglycerol/H$_2$O system [10, 11], at 25 °C, two reversed bicontinuous cubic phases can be obtained over the H$_2$O content range of 19–42 wt%: the Q^{230} phase of space group *Ia3d* (19–38 wt% H$_2$O) and the

Q^{224} phase of space group *Pn3m* (37–42 wt% H$_2$O; at higher water content, this phase coexists with bulk water). Besides, it has also been shown that both cubic phases transform to the reversed hexagonal phase (H$_{II}$) above 90 °C [10, 11].

Considering the previously mentioned results and aims of the present study, the effects of GOD on the lyotropic and thermotropic phase behavior of aqueous MO were investigated over the water content range 15–42 wt%. On the other hand, the content of the GOD preparation was limited to about 6 wt%.

Influence of GOD content on the lyotropic properties of the MO-based cubic phases

The composition of the samples used to construct a partial phase diagram of the pseudoternary MO/GOD/H$_2$O system at 25 °C is shown in Fig. 1. On the basis of SAXD measurements and visual inspection, the individual phases and their mixtures were identified, and they are labeled by different symbols. From these data the approximate phase boundaries were hand-drawn.

Firstly, it is seen from Fig. 1 that, when compared to the pure 1-monooleoylglycerol/H$_2$O system [10, 11], our pseudobinary MO/H$_2$O mixture displays a similar sequence of cubic phases, but their boundaries are shifted towards lower water content: Q^{230} (up to 30 wt% water) → Q^{224} (30–37 wt% water; at higher water content, the phase coexists in equilibrium with excess H$_2$O). Such shift of the boundaries is not surprising if it is remembered that the MO preparation contains admixed diglycerides and polyunsaturated monoglycerides that promote formation of the reversed mesophases [12].

As evident from Fig. 1, even small amounts of the GOD preparation are capable of inducing drastic changes in the lyotropic phase behavior of MO. The phase diagram also clearly shows that the cubic phases of MO exhibit different sensitivity to the addition of GOD. Thus, the Q^{230} phase can accommodate only up to 0.5 wt% GOD. With increasing enzyme content, this cubic phase transforms to the H$_{II}$ phase. It is interesting to note that the GOD-containing Q^{230} and H$_{II}$ phases are stable up to about 30 wt% water, whereas they transform to the Q^{224} phase at higher water content. The latter cubic phase is much less sensitive to the entrapment of GOD. At a water content higher than 34 wt%, the Q^{224} phase can accommodate at least up to 6 wt% of the GOD preparation. Besides, the excess water boundary for the enzyme-containing Q^{224} phase is almost the same as for the pseudobinary MO/H$_2$O system.

The unit cell dimensions (*a*) of the cubic phases, calculated from the SAXD data, were used to establish

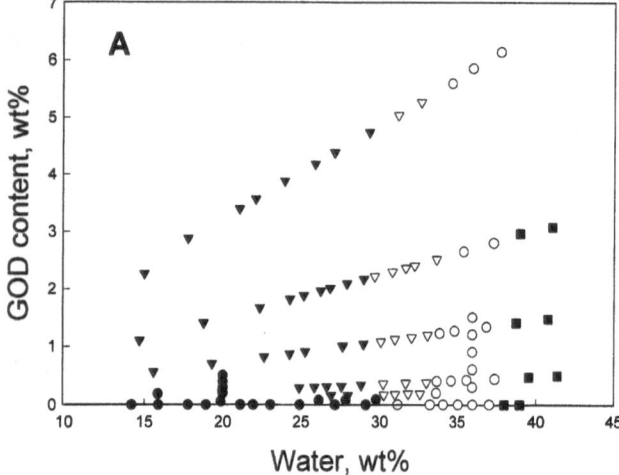

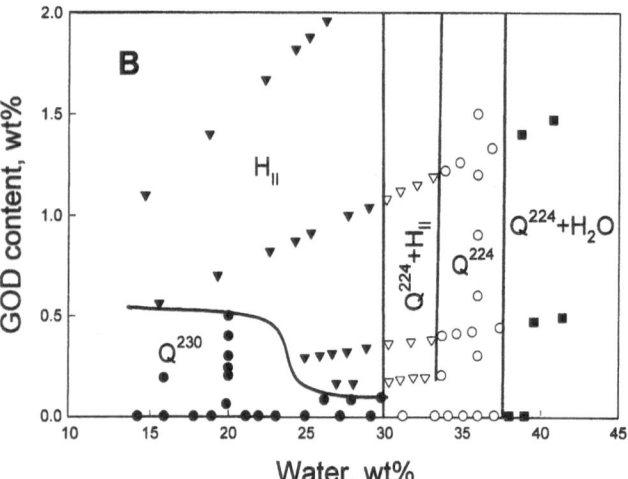

Fig. 1 A Schematic phase diagram of the pseudoternary monoolein (*MO*)/glucose oxidase (*GOD*)/H₂O system at 25 °C. The phases are labeled as follows: *filled circles*, reversed bicontinuous cubic phase of space group *Ia3d* (Q²³⁰); *open circles*, reversed bicontinuous cubic phase of space group *Pn3m* (Q²²⁴); *filled, inverted triangles*, reversed hexagonal phase (H$_{II}$); *open, inverted triangles*, Q²²⁴ + H$_{II}$; *filled squares*, Q²²⁴ + H₂O. **B** An enlarged image of the phase diagram at low GOD content

the effect of the GOD preparation on the structural characteristics of the phases.

It is believed that the reversed bicontinuous cubic phases consist of a curved lipid bilayer draped onto an infinite periodic minimal surface (IPMS) which is located in the middle of the bilayer. The geometric characteristics of such a bilayer can be modeled assuming that all other molecularly defined surfaces lie parallel to the IPMS. According to this model, once the monolayer thickness is constant throughout the structure, the lipid length (l) is determined by solving the following equation [13]:

$$\phi = 2A_0 \left(\frac{l}{a}\right) + \frac{4\pi\chi}{3}\left(\frac{l}{a}\right)^3 , \qquad (1)$$

where ϕ is the volume fraction of lipid (equal to the weight fraction when the specific gravity equals unity), A_0 is the ratio of the IPMS area to (unit cell volume)$^{2/3}$ (3.09 for the Q²³⁰ IPMS, and 1.92 for the Q²²⁴ IPMS), and χ is the Euler characteristic for a given cubic phase (−8 and −2 for the Q²³⁰ and Q²²⁴ IPMSs, respectively).

The cross-sectional area of a lipid molecule at a distance ξ from the IPMS, integrated over one of the two monolayers within the unit cell, $A(\xi)$, can be calculated from Eq. (2) [14]:

$$A(\xi) = A_0 a^2 + 2\pi\chi\xi^2 . \qquad (2)$$

Equation (2), in turn, can be applied to estimate the water channel radius in the cubic phases. Thus, at the center of the water channel, where ξ is the sum of the water channel radius (r_w) and the lipid length, $A(\xi)$ reduces to zero. Therefore, by making appropriate substitutions, Eq. (2) transforms to [11]

$$r_w = -\left(\frac{A_0}{2\pi\chi}\right)^{\frac{1}{2}} a - l . \qquad (3)$$

In our study, assuming that the lipid length is independent of the cubic phase composition over the water content range studied, l was calculated using Eq. (1) for the Q²³⁰ and Q²²⁴ phases containing different amounts of GOD. The values of l are presented in Table 1.

The results in Table 1 indicate that, when compared to the reference MO/H₂O system, the GOD preparation does not influence the lipid monolayer thickness in the cubic phases. Moreover, both cubic phases are characterized by the same l value of about 16 Å. Here it should be pointed out that this monolayer thickness agrees closely with the parameter determined for the cubic phases of aqueous 1-monooleoylglycerol [11, 15].

The l values calculated were then used to determine the water channel radius in the cubic phases by means of Eq. (3). The calculated values of r_w for the pseudobinary and pseudoternary cubic phases are shown in Fig. 2 against the water content. As is evident form Fig. 2, the radius of the water channels in the cubic phases increases with increasing water content. Thus, r_w ranges from

Table 1 The calculated lipid monolayer thickness (l) for the cubic phases of the pseudobinary monoolein (*MO*)/H₂O and pseudoternary MO/glucose oxidase (*GOD*)/H₂O systems

Cubic phase	GOD content (wt%)	l (Å)
Q²³⁰	0.0	16.0 ± 0.2
Q²³⁰	0.07 ± 0.01	16.2 ± 0.2
Q²³⁰	0.25 ± 0.03	16.4 ± 0.1
Q²²⁴	0.0	15.6 ± 0.1
Q²²⁴	0.09 ± 0.02	16.1 ± 0.3
Q²²⁴	0.44 ± 0.04	15.7 ± 0.2
Q²²⁴	1.31 ± 0.15	16.2 ± 0.4
Q²²⁴	2.80 ± 0.32	16.5 ± 0.2
Q²²⁴	5.70 ± 0.55	15.9 ± 0.1

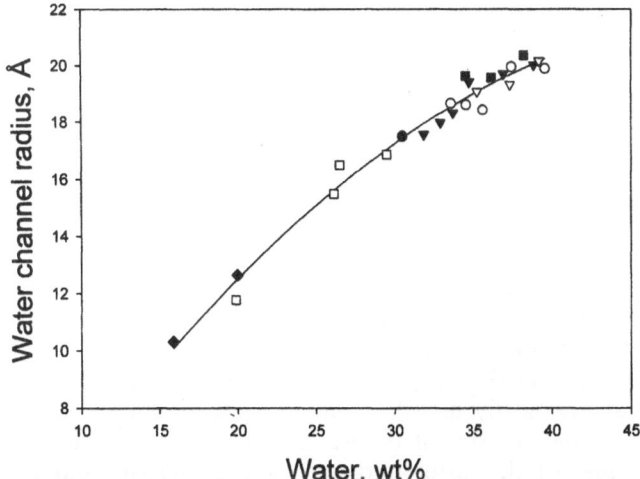

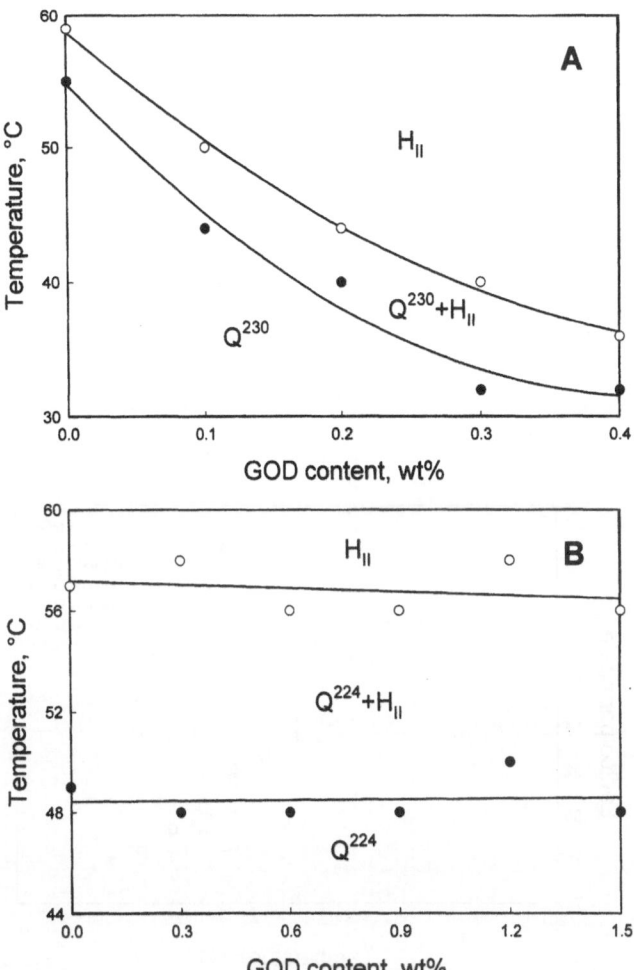

Fig. 2 Dependence of the water channel radius on the water content as a function of GOD content in the MO Q_{II} phases. For the GOD-containing Q^{230} phase, *open squares* and *filled diamonds* correspond to 0.07 and 0.25 wt% GOD. For the GOD-containing Q^{224} phase, *filled circles, open circles, filled, inverted triangles, open, inverted triangles* and *filled squares* correspond to 0.09, 0.44, 1.31, 2.80 and 5.70 wt% GOD, respectively. The *line* is drawn to guide the eye

Fig. 3 **A** Dependence of the $Q^{230} \to H_{II}$ transition temperature on the GOD content in the MO/H$_2$O sample containing 20 wt% water. **B** Dependence of the $Q^{224} \to H_{II}$ transition temperature on the GOD content in the MO/H$_2$O sample containing 36 wt% water. The *lines* are drawn to guide the eye

about 10 to 20 Å at 16 and 37 wt% water, respectively. Interestingly, despite the fact that one cubic phase transforms to another, the r_w values change continuously over the transition region. On the other hand, in both cubic phases, r_w is independent of the GOD preparation content.

Thermal stability of the MO-based cubic phases with entrapped GOD

The thermotropic behavior of the MO-based cubic phases as a function of the GOD content was also investigated by the SAXD measurements. The dependencies of the $Q^{230} \to H_{II}$ and $Q^{224} \to H_{II}$ transition temperatures on the GOD content are shown in Fig. 3.

The results in Fig. 3A indicate the onset of the H_{II} phase formation in the Q^{230} phase of the pseudobinary MO/H$_2$O system at about 55 °C; however, this transition temperature decreases with the addition of the GOD preparation. As may be seen, even a GOD content of 0.4 wt% reduces the $Q^{230} \to H_{II}$ transition temperature by almost 25 °C.

On the other hand, Fig. 3B shows that, for the pseudobinary MO/H$_2$O system, the $Q^{224} \to H_{II}$ transition starts at about 48 °C. In contrast to the Q^{230} phase, the entrapment of up to 1.5 wt% of the GOD preparation into the Q^{224} phase leaves the transition temperature practically unaltered.

In addition to the results in Fig. 3, Fig. 4 provides the dependence of the experimental unit cell dimension (a) of

the pseudobinary and pseudoternary cubic phases on temperature. In all cases, with increasing temperature, the lattice parameter decreases. The results in Fig. 4 also indicate that, regardless of the cubic phase type, the entrapped preparation of GOD has no effect on the value of a everywhere over the temperature range being studied.

Final remarks

The results reported here indicate that the Q^{224} phase of aqueous MO can accommodate high amounts (at least up to 6 wt%) of the commercial preparation of *A. niger* GOD. This GOD-containing pseudoternary cubic phase of space group $Pn3m$ can coexist in equilibrium with bulk water and also exhibits high thermal stability with respect

20

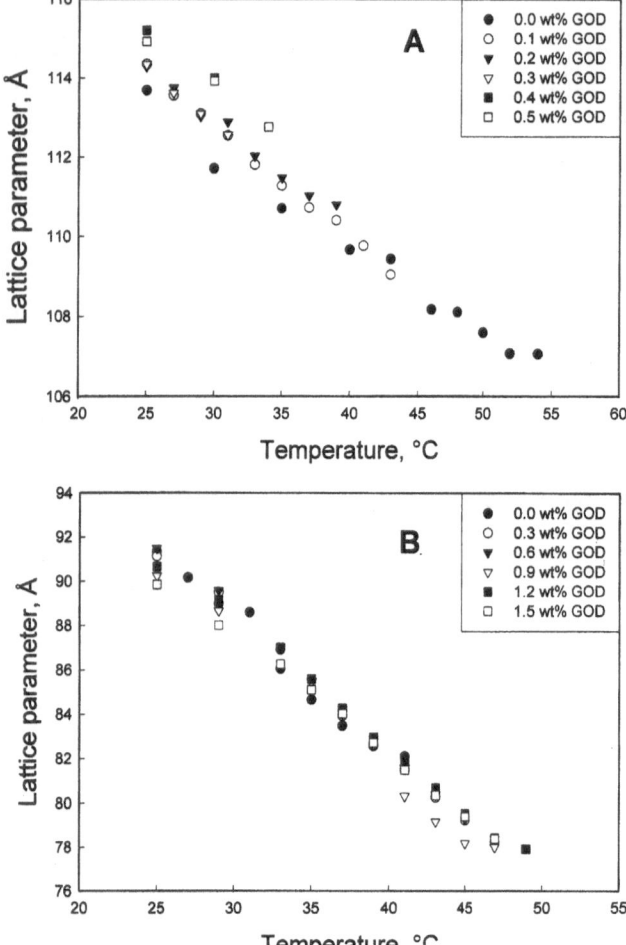

Fig. 4 A Dependence of the lattice parameter of the Q^{230} phase on temperature as a function of GOD content in the MO/H$_2$O sample containing 20 wt % water. **B** Dependence of the lattice parameter of the Q^{224} phase on temperature as a function of GOD content in the MO/H$_2$O sample containing 36 wt% water

to the Q^{224} → H$_{II}$ phase transition. Therefore, as far as the construction of the GOD-based biosensors is concerned, these results are very important.

On the other hand, our data show that, at 25 °C, the Q^{230} phase of MO transforms to the H$_{II}$ phase even at a GOD preparation content lower than or equal to 0.5 wt%. Moreover, in contrast to the Q^{224} phase, the entrapment of the enzyme preparation into the MO cubic phase of space group $Ia3d$ significantly decreases the Q^{230} → H$_{II}$ transition temperature.

Although the experimental results clearly indicate a sharp distinction between the capacities of the MO Q^{230} and Q^{224} phases to accommodate the GOD preparation, at the moment, it is not possible to give a satisfactory explanation of this feature. A mathematical analysis of the SAXD data suggests that the enzyme preparation has practically no effect on the lipid monolayer thickness and the size of the water channels. Moreover, the values of the latter parameter (10–20 Å) rule out the possibility that the native enzyme dimers of hydrodynamic diameter of about 80 Å are located exclusively in the water channels of the cubic phases. Since Li et al. [16] have shown that GOD is capable of penetrating into the neutral glycolipid monolayers, we might hypothesize that a similar molecular process takes place in the GOD-containing cubic phases of MO. However, it is evident that a great deal needs to be done before the genuine molecular structures of the MO/GOD/H$_2$O mesophases become fully understood. We can only conclude by listing some remaining problems of importance:

1. Structural characteristics and catalytic properties of the entrapped GOD molecules should be established.
2. Possible effects of the GOD preparation purity on the phase behavior of aqueous MO must be examined.
3. Physical factors which determine the Q$_{II}$ → H$_{II}$ transitions in the MO/GOD/H$_2$O system need to be understood.

Studies along these lines are in progress.

2

Acknowledgements The authors gratefully acknowledge the financial support given by the Royal Swedish Academy of Sciences.

References

1. Larsson K (1989) J Phys Chem 93:7304–7314
2. Lindblom G, Rilfors L (1989) Biochim Biophys Acta 988:221–256
3. Luzzati V (1997) Curr Opin Struct Biol 7:661–668
4. Ericsson B, Larsson K, Fontell K (1983) Biochim Biophys Acta 729:23–27
5. Mariani P, Luzzati V, Delacroix H (1988) J Mol Biol 204:165–189
6. Wallin R, Engström S, Mandenius CF (1993) Biocatalysis 8:73–80
7. Razumas V, Kanapieniené J, Nylander T, Engström S, Larsson K (1994) Anal Chim Acta 289:155–162
8. Nylander T, Mattisson C, Razumas V, Miezis Y, Håkansson B (1996) Colloids Surf A 114:311–320
9. Razumas V, Larsson K, Miezis Y, Nylander T (1996) J Phys Chem 100:11766–11774
10. Hyde ST, Andersson S, Ericsson B, Larsson K (1984) Z Crystallogr 168:213–219
11. Briggs J, Chung H, Caffrey M (1996) J Phys II 6:723–751
12. Seddon JM (1990) Biochim Biophys Acta 1031:1–69
13. Turner DC, Wang Z-G, Gruner SM, Mannock DA, McElhaney RN (1992) J Phys II 2:2039–2063
14. Anderson DM, Gruner SM, Leibler S (1988) Proc Natl Acad Sci USA 85:5364–5368
15. Chung H, Caffrey M (1994) Biophys J 66:377–381
16. Li J-R, Du Y-K, Boullanger P, Jiang L (1999) Thin Solid Films 352:213–217

Progr Colloid Polym Sci (2000) 116: 21–25
© Springer-Verlag 2000

A. A. Zinchenko
V. G. Sergeyev
O. A. Pyshkina

DNA–surfactant complexes in a water/alcohol mixture

A. A. Zinchenko · V. G. Sergeyev
O. A. Pyshkina (✉)
Department of Polymer Science
Faculty of Chemistry
Moscow State University
Vorob'evy Gory, Moscow
119899 Russia
e-mail: stazi@mail.ru
Tel.: +7-095-9393124
Fax: +7-095-9390174

Abstract The interaction between DNA (and poly[di(carboxylatophenoxy)phosphazene]) and cationic surfactants in a water/2-propanol mixture was studied. By means of UV and IR spectroscopy and flame photometry the stoichiometric polyelectrolyte–surfactant complexes were characterized. The data of high-rate sedimentation and UV spectroscopy show that nonstoichiometric soluble complexes can be formed in the presence of 40% (v/v) water/2-propanol.

Key words DNA · (poly[di(carboxylatophenoxy)phosphazene])–surfactant complexes · Water/2-propanol mixture · UV and IR spectroscopy · Sedimentation · Soluble nonstoichiometric complexes

Introduction

Interaction between polymers and surfactants has received considerable interest during the last decades [1–3]. Investigation of these reactions is motivated by the opportunity to synthesize a new type of polymer–colloid complexes with unique properties on the mesoscopic scale [4, 5]. Among various polymer–surfactant systems the interaction between polyelectrolytes (PE) and oppositely charged surfactants (S) holds special interest because of its significance in life science and biomedical applications. This interaction proceeds cooperatively in water solution with the formation of PE–S complexes. The complexes are formed owing to electrostatic interaction of the PE chain units with S counterions and hydrophobic interactions of the hydrocarbon "tails" stabilize complex molecules [6].

In the present article we describe a systematic investigation of complexation of S molecules with DNA and ordinary flexible polyanion poly[di(carboxylatophenoxy)phosphazene] (PCPP) in a water/alcohol medium.

To clarify the composition and physicochemical characteristics of DNA (and PCPP)–S complexes in a water/alcohol medium we examined the effect of S on DNA (PCPP) properties.

Experimental

DNA from salmon sperm, $M_r = (1.9–3.2) \times 10^5$ (b.p. 300–500) was purchased from Soyuzkhimreaktiv, Russia, and was purified using standard techniques [7]. Salmon sperm native DNA was used for measurements requiring large amounts of the sample.

Cetyltrimethylammonium bromide (CTAB) and distearyldimethylammonium chloride (DSDAC) were obtained from Tokyo Kasei Kogyo Co. and were recrystallized twice from acetone and dried under vacuum. PCPP, $M_w = 10^6$ (see structural formula), was synthesized by A.K. Andrianov (Virus Research Institute, USA) and was kindly provided for the investigation. The concentration of the polymers in solution was about 10^{-4} M.

Deionized water doubly distilled in glass and 2-propanol (spectral grade) were used throughout. Water/alcohol mixtures (solution) were prepared on a volume basis and if not specified contained 40% (v/v) 2-propanol.

Absorption spectra were obtained with a Specord M-40 spectrophotometer. The concentration of DNA in solution was calculated assuming that the molar extinction coefficient for native DNA in water is 6500 $M^{-1}cm^{-1}$. The circular dichroism (CD) spectra were recorded with a Jasco 500c spectropolarimeter. Cells with an optical path of 1 cm were used for the UV and CD measurements. All measurements were carried out at 20 °C. The IR spectra were recorded with a Specord M-80 spectrophotometer using dried samples in KBr pellets at 20 °C. Ultracentrifugation was carried out with Beckman E analytical ultracentrifuge at 60,000 rpm, 20 °C, in scanning mode. Scanning was performed at 260 or 235 nm, corresponding to a maximum in the DNA or the PCPP absorption spectrum.

Bacteriofage T4 DNA (166 kilobase pairs) was purchased from Nippon Gene and was mainly used for single-molecule observation by fluorescence microscopy. The fluorescent dye 4',6-diamidino-2-phenybdole (DAPI) and the antioxidant 2-mercaptoethanol (ME) were obtained from Wako Pure Chemical Industries. DAPI was used as a fluorescent label. ME was used as a free-radical scavenger to reduce fluorescence fading and light-induced damage of DNA molecules. The sample solutions, microscope slides, and cover slips were carefully prepared as in previous studies [8]. Microscopic fluorescence measurements were performed as follows. 2-Propanol was added to the DNA solution in half-diluted standard buffer made from tris(hydroxymethyl)aminoethane base, boric acid and ethylenediaminetetraacetic acid containing DAPI and ME to obtain a concentration of 40% (v/v) 2-propanol. The final concentration of DNA in the nucleotides was 0.6 μM, the DAPI concentration was 0.6 μM, and the ME concentration was 4% (v/v). The samples were illuminated with 365-nm UV light, and fluorescence images of DNA molecules were observed using a Zeiss Axiovert 135-TV microscope equipped with a 100× oil-immersed lens and recorded on S-VHS videotape through a high-sensitivity Hamamatsu SIT-TV camera. The observations were carried out at 25 °C. The apparent length of the long axis, L, which was defined as the longest distance in the outline of the DNA image, was calibrated with an Argus 10 image processor (Hamamatsu Photonics). The blurring effect was estimated to be of the order of 0.3 μm, and the data for L are given in the text without correction. The stock S solution in 40% (v/v) 2-propanol solution was added into the DNA 40% (v/v) 2-propanol solution to obtain the desired final concentration of S. The mixture was shaken in a gentle manner and kept for 15 min before observation.

Results

On addition of the S solution to the polyanion solution containing 40% (v/v) 2-propanol the reaction mixture remained homogeneous in a S-to-PE concentration ratio (Z) of less than 6 in the case of CTAB and ofless than 0.8 in the case of DSDAC, but progressively turned turbid if Z exceeded these values owing to the formation of insoluble complexes. The insoluble complexes were separated by centrifugation and the concentration of the PE remaining in the supernatant was measured spectrophotometrically. The corresponding precipitation curves are shown in Fig. 1. It is seen that a noticeable decrease in the PE concentration in the supernatant is actually observed at a CTAB-to-PE concentration ratio around 7 (Fig. 1A). Further addition of CTAB leads to a drastic decrease of PE absorbance. Finally, at a CTAB-to-PCPP concentration ratio close to 16 and at a CTAB-to-DNA concentration ratio close to 12 the aqueous solution does not contain any determinable amount of PE owing to the formation of insoluble PE–CTAB complexes. In the case of more hydrophobic S (DSDAC), the noticeable decrease in DNA or PCPP concentration in the supernatant starts at a DSDAC-to-PE concentration ratio of about 0.9, and at a DSDAC-to-PE concentration ratio close to 1 the solution does not contain any determinable amount of PE (Fig. 1B).

The IR spectrum of a DNA–DSDAC nonsoluble complex obtained from a water/alcohol mixture is

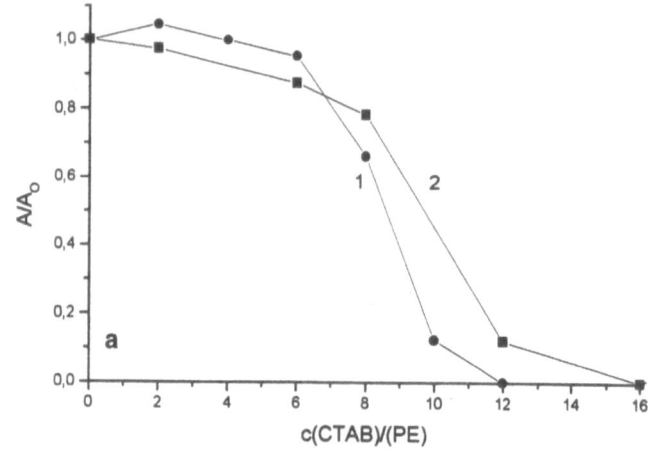

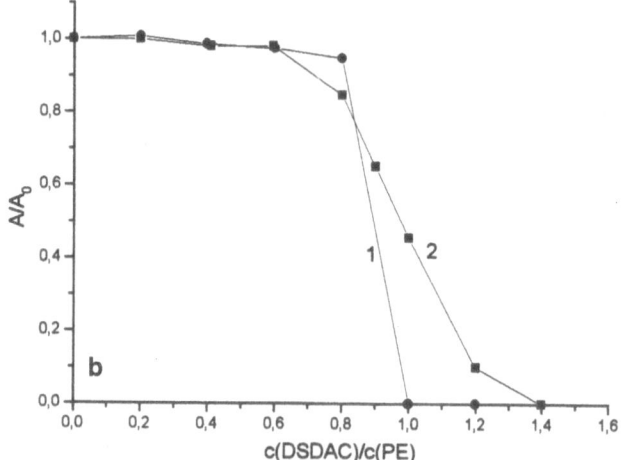

Fig. 1 A The dependences of DNA (*curve 1*) and poly[di(carboxylatophenoxy)phosphazene] (*PCPP*) (*curve 2*) absorbance in the supernatant on the [cetyltrimethylammonium bromide (*CTAB*)]/ [polyelectrolyte (*PE*)] ratio in a 40% 2-propanol/water mixture. The *concentration* of PE is about 10^{-4} M. **B** The dependences of DNA (*curve 1*) and PCPP (*curve 2*) absorbance in the supernatant on the [distearyldimethylammonium chloride (*DSDAC*)]/[PE] ratio in a 40% 2-propanol–water mixture. The concentration of PE is about 10^{-4} M

compared with that of the initial native DNA in Fig. 2. Both IR spectra display a native DNA structure, which is characterized by absorbance bands around 1250 and 1100 cm^{-1} (1000–1250 cm^{-1}), corresponding to valence oscillations of phosphate groups, and absorption bands around 1650 and 1700 cm^{-1} (1550–1720 cm^{-1}), corresponding to valence oscillations of nitrous bases [9]. The IR spectrum of the DNA–DSDAC complex (Fig. 2, curve 1) has a distinct doublet absorption band in the 2830–2910 cm^{-1} region, corresponding to valence oscillations of CH groups in hydrophobic fragments of S molecules included in the complex. Thus, we can conclude that the double-helical structure of the DNA molecules is conserved in the DNA–DSDAC complex obtained from water/2-propanol solution.

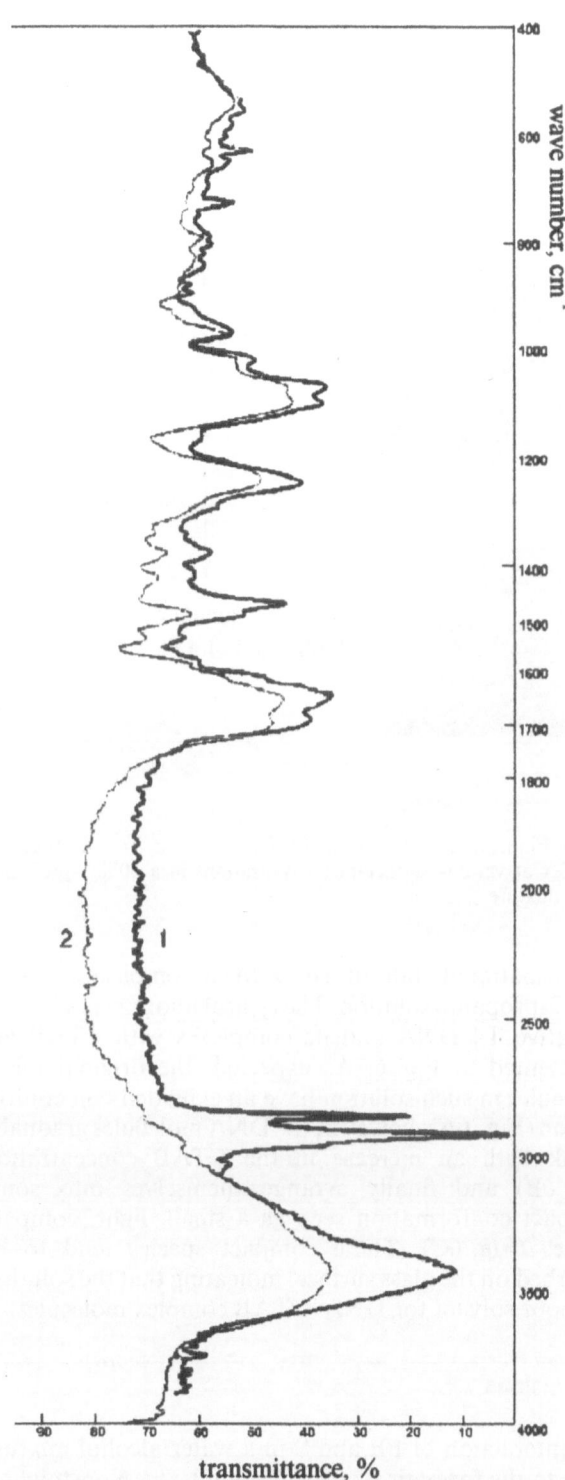

Fig. 2 IR spectra of a DNA–DSDAC complex (*curve 1*) obtained from a 2-propanol/water mixture and the initial DNA (*curve 2*)

In order to estimate the composition of the PE–S complexes obtained from water/alcohol solution the insoluble complexes were separated by centrifugation and the amount of Na⁺ in the supernatant was measured

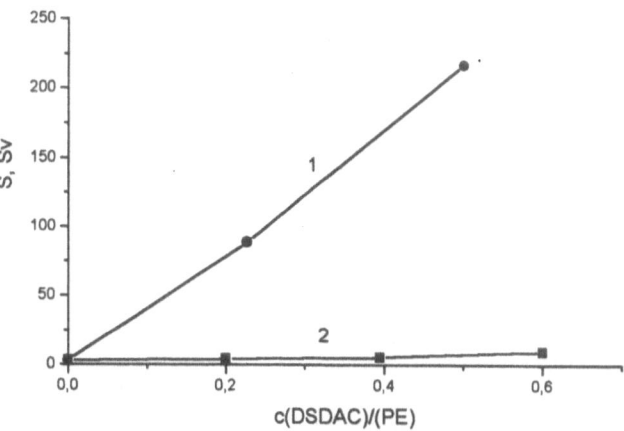

Fig. 3 Dependences of sedimentation coefficients of DNA–DSDAC (*curve 1*) and PCPP–DSDAC (*curve 2*) complexes versus [DSDAC]/[PE] ratio. The concentration of PE is about 10^{-4} M

using flame photometry. For all the complexes the supernatant after centrifugation contained the number of Na⁺ ions corresponding to the nearly stoichiometric ratio of phosphate groups and S cations in the complex. In other words, PE–S complexes of nearly equimolar composition apparently formed in water/alcohol media.

However, as seen in Fig. 1, in the less than critical range of the ratio of the concentration of S to PE the reaction mixture remained homogeneous, most likely owing to the formation of soluble PE–S complexes.

The typical sedimentation profiles of the DNA–CTAB or DNA–DSDAC and the PCPP–CTAB or PCPP–DSDAC complexes in 40% 2-propanol/water mixtures have the only sharp step indicating the (constancy) monodispersity of the composition of the complex. Moreover, the value of sedimentation coefficient, S, increased with increasing S-to-PE concentration ratio and did not depend on the nature of S (Fig. 3). These results indicate that the interaction between PE and S in a water/alcohol mixture leads to the formation of soluble complexes with variable composition and can be represented by the following reaction scheme.

The shift in the adsorption maximum is sensitive to changes in the environment of the fluorophobic groups [10]. Thus the examination of the adsorption maximum $\Delta\lambda = \lambda - \lambda_0$, where λ is wavelength of the absorbance maximum of the PCPP–S complex in the water/2-propanol mixture and λ_0 is the wavelength of initial PE in the same mixture, suggests the formation of a soluble complex (Fig. 4, curve 1). However, the interaction between DNA and DSDAC does not lead to the appearance of $\Delta\lambda$ (Fig. 4, curve 2) because fluorophoric groups of DNA molecules are shielded by the carbohydrate–phosphate skeleton of the DNA molecule and cannot sense the changes in the environment.

The formation of soluble DNA–S complexes may cause the local conformational changes within the DNA

24

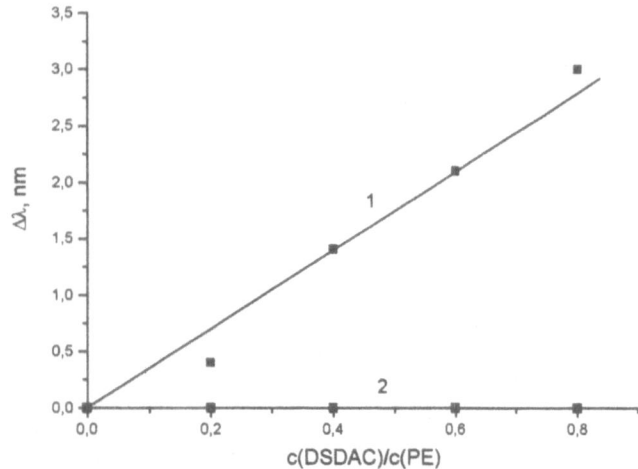

Fig. 4 The dependences of the wavelength shift of the absorbance maximum of PCPP–DSDAC (*curve 1*) and DNA–DSDAC (*curve 2*) complexes on the [DSDAC]/[PE] ratio in a 40% 2-propanol/water mixture. The concentration of PE is about 10^{-4} M

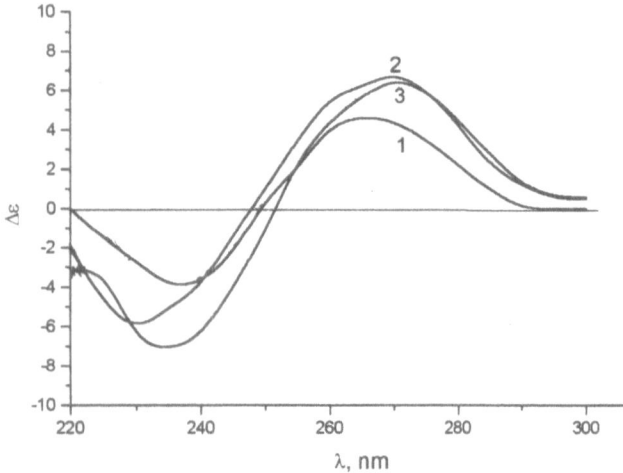

Fig. 5 Circular dichroism spectra of DNA (*curve 1*), DNA–DSDAC complex (composition 1/0.2) (*curve 2*) and DNA–DSDAC complexes (composition 1/0.4) (*curve 3*) in a 40% (v/v) 2-propanol/water mixture. The concentration of PE is about 10^{-4} M

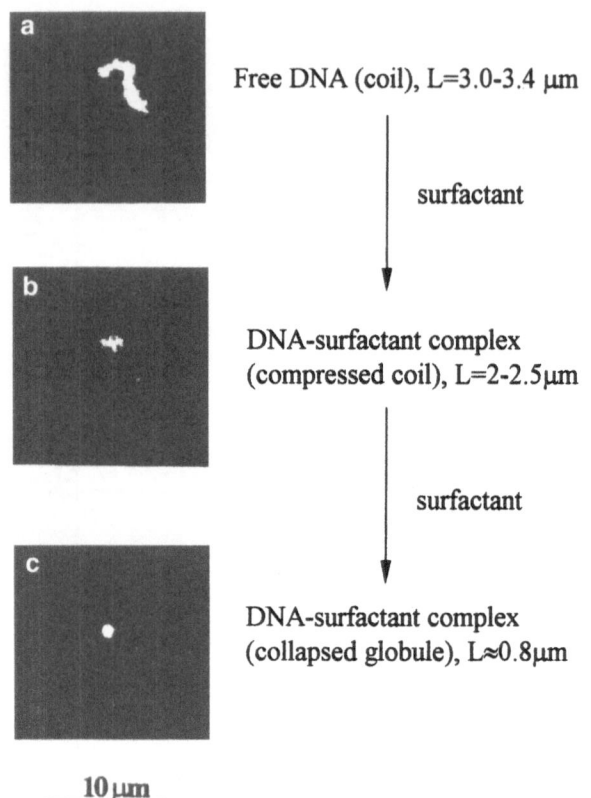

Free DNA (coil), L=3.0-3.4 μm

surfactant

DNA-surfactant complex (compressed coil), L=2-2.5μm

surfactant

DNA-surfactant complex (collapsed globule), L≈0.8μm

10 μm

Fig. 6 Fluorescence images of T4 DNA and a T4 DNA–CTAB complex at various surfactant concentrations in a 40% 2-propanol/water mixture

secondary structure. To gain insight into the change in the secondary structure of DNA in the soluble complex, we performed CD spectroscopic measurements of DNA–DSDAC complexes of different composition (Fig. 5). The CD spectra of DNA and DNA–DSDAC complexes in a 40% 2-propanol/water mixture correspond to the B form that resembles the CD spectra of DNA in buffer solution where DNA molecules are also known to be in the B form.

Earlier it was demonstrated that fluorescence microscopy was a useful tool to monitor the shape of individual ultra-high-molecular-mass DNA molecules and the change induced by various factors in dilute solutions [11]. Thus, we used fluorescence microscopy to clarify the

conformational state of T4 DNA–S complexes in 40% (v/v) 2-propanol solution. The typical fluorescence images of native T4 DNA and its complexes with CTAB are represented in Fig. 6. As expected, the original DNA molecules in such solution have an extended-coil conformation (Fig. 6A); however, the DNA molecules gradually shrink with an increase in the CTAB concentration (Fig. 6B) and finally arrange themselves into some compact conformation seen as a small, light, compact species (Fig. 6C). These compact species tend to be absorbed on the glass surface, indicating that the solution is a poor solvent for DNA–CTAB complex molecules.

Conclusions

The interaction of PE and S in a water/alcohol mixture leads to the formation of PE–S complexes. Nonstoichiometric soluble and nearly stoichiometric complexes were prepared. Moreover, we have demonstrated that DNA included in polymer–S complexes formed in water/alcohol solution preserves its native double-helix structure.

Acknowledgement This work was supported by grant no. 99-03-33337a from the Russian Foundation for Basic Research.

References

1. Hayakawa K, Santerre JP, Kwak JCT (1983) Biophys Chem 17:175
2. Goddard ED (1986) Colloids Surf 19:301–329
3. Ananthapadmanabhan KP (1993) In: Goddard ED, Ananthapadmanabhan KP (ed) Interaction of surfactants with polymers and proteins. CRS, Boca Raton, pp 5–58
4. Mel'nikov SM, Sergeyev VG, Yoshikawa K (1995) J Am Chem Soc 117:2401–2408
5. Mel'nikov SM, Sergeyev VG, Yoshikawa K (1995) J Am Chem Soc 117:9951–9956
6. Bakeev KN, Shu YM, Zezin AB, Kabanov VA, Lezov AV, Mel'nikov AB, Kolomiets IP, Rjumtsev EI, MacKnight WJ (1996) Macromolecules 29:1320–1325
7. Sambrook J, Fritsch TF, Maniatis T (1989) Molecular cloning. Cold Spring Harbor Laboratory Press, New York
8. Sergeyev VG, Mikhailenko SV, Pyshkina OA, Yaminsky IV, Yoshikawa K (1999) J Am Chem Soc 121:1780–1785
9. Sukhorukov GB (1996) Biosens Bioelectron 9:913–922
10. Reeves RL, Harkawey ShA (1970) In: Mittel KK (ed) Micelle formation, solubilization and microemulsions. Mir, Moscow, pp 499–514
11. Yoshikawa K, Takahashi M, Vasilevskaya V, Khokhlov AR (1996) Phys Rev Lett 76:3029

Progr Colloid Polym Sci (2000) 116:26–32
© Springer-Verlag 2000

I. Johansson
C. Strandberg
B. Karlsson
G. Karlsson
K. Hammarstrand

Environmentally benign nonionic surfactant systems for use in highly alkaline media

I. Johansson (✉) · C. Strandberg
B. Karlsson · G. Karlsson
K. Hammarstrand
Akzo Nobel Surface Chemistry AB
44485 Stenungsund, Sweden
e-mail: ingegard.johansson@akzonobel.com
Tel.: +46-303-85108
Fax: +46-303-84371

Abstract The increasing use of surfactants has put more and more emphasis on environmental requirements for the chemicals used. It is also important to follow the whole lifetime of the products and to offer possibilities for efficient wastewater treatment. Good cleaning of oily soil from hard surfaces and efficient separation of the oily soil from the wastewater within the same surfactant system is a challenge. In this article a system is described which, by using the specific cloud-point profile that can be achieved when mixing alkyl ethoxylates with alkyl glucosides, offers a possibility to meet that challenge. Alkyl ethoxylates are known to be good wetting agents, emulsifiers and cleaning agents. They are, however, usually not soluble in concentrated electrolytes and/or alkali, which may be needed in the practical use with different water qualities, etc. The more recently developed nonionic surfactant named alkyl glucoside, which has a polyhydroxyl functionality as its hydrophilic part arising from natural sources such as carbohydrates, offers better solubility as well as strong surface activity. When combining these two categories of surfactants good solubility in, for instance, alkali, is achieved and a special concentration-dependent cloud-point curve results. This phenomenon can be used both to fine-tune the cleaning and wetting efficiency and to stimulate the separation of the emulsified oil from the wastewater.

Introduction

Nonionic surfactants used to be more or less synonymous with alkyl alkoxylates. Today a whole range of polyhydroxyl surfactants of nonionic type have been developed from natural raw materials because of environmental considerations and they are currently being used in increasing amounts. Application areas are as wide as for the conventional nonionics, including both their use in general cleaning and specific uses in agrochemical formulations as wetting and penetrating agents. The most widely investigated polyhydroxyl-based surfactants are the alkyl polyglucosides. The name covers a mixture of oligomers. In this article they are referred to as glucosides.

The hydroxyl functions that provide hydrophilicity to the glucosides, substituting the ether linkages of the alkoxylates, give potentially different properties to these products. The differences are currently being looked at in many groups within universities and industry [1–9].

This work deals with combinations of alkyl alkoxylates and alkyl glucosides in both water and highly concentrated electrolyte solutions as well as highly alkaline systems for use in cleaning of hard surfaces.

The most obvious differences between alkoxylates and glucosides are found in the temperature dependence of the phase behaviour, with alkoxylates having cloud points and glucosides sometimes being considered not to have any. Combinations will influence the temperature-dependent solubility in varying ways.

Further the electrolyte effect differs. Alkoxylates typically are not very soluble in alkali, whereas glucosides sometimes can be more soluble in alkali than in pure water.

Experimental

Methods

The hydrotrope efficiency was determined on the basis of the minimum amount that is needed to get a clear solution of 5% [C10(EO)4] in different amounts of NaOH at room temperature.

The cloud points were measured with a thermometer by following the clouding behaviour of the solution in a test tube by eye while the temperature was varied until the same temperature for clouding was achieved. All surfactant-concentration-dependent clouding curves for the situation without oil were measured with the same concentration of electrolytes throughout. In the separation test, when oil was present, however, the solutions were diluted with water of medium hardness, 3.8° dH, to mimic the actual wastewater situation.

Materials

The C4 glucoside used was Simulsol SL4 (DP = 1) from SEPPIC, the C8 straight and C8 branched glucosides were AG6201 and AG6202, respectively, from Akzo Nobel Surface Chemistry. The C5, C6 and C7 glucosides were laboratory products made with the Fischer process technique [10] at Akzo Nobel Surface Chemistry from glucose and isoamyl alcohol, hexanol (Sigma) and Exxal-7 alcohol (Exxon). The ethoxylate, C10(EO)4, narrow range, is a commercial product from Akzo Nobel Surface Chemistry. Two conventional hydrotropes, sodium cumene sulfonate (Hüls) and octyl imino dipropionate (Ampholak YJH-40 from Akzo Nobel Surface Chemistry), were used as references. NaOH (Merck) and the surfactant mixtures were dissolved in water of medium hardness (3.8° dH).

In all the investigations of the cloud-point curves deionized water was used, except in the case described previously. Nitrilotriacetic acetate sodium salt (NTA, Monsanto), tetrapotassium pyrophosphate (TKPP, BASF) and sodium metasilicate (Fluka), were used as electrolytes. Halpasol 190/240 (Haltemann) was the model oil in the separation tests. A commercial cationic hydrotrope, Berol 556, was used as a reference in the oil-separation investigation.

Use of alkyl glucosides as hydrotropes

As early as 1979 alkyl glucosides were recommended for use as hydrotropes, for example, to solubilize defoamers or low-foaming alkoxylates of blockpolymer type into highly alkaline solutions intended for use in metal cleaning [11].

The hydrotrope concept

Friberg and coworkers [12, 13] have discussed the influence of a hydrotrope in water solution, as a solubilizer and/or as a destabilizer of the liquid-crystal-line phases in the solution. Friberg has shown that the action of a hydrotrope as a disturbance in the ordered phases may well be the way it enlarges the solubilization of hydrophobic elements such as oils or non-water-soluble surfactants; however, there may be other mechanisms in specific cases.

Typical hydrotropes are rather water soluble molecules with bulky hydrophobes but more surface-active types, such as ethoxylated quaternary alkyl ammonium compounds, are also being used as cosurfactants. The ammonium compounds mentioned are very efficient but are not totally benign to the environment. Thus, there is a need for greener products such as alkyl glucosides.

Alkyl glucosides as hydrotropes

We chose to study one specific model formulation with 5% of a hydrophobic alkyl ethoxylate of narrow-range type, C10(EO)4, in different amounts of NaOH and glucoside as solubilizer with the aim of producing highly concentrated alkaline cleaners. The minimum amounts of hydrotrope to get a clear solution at room temperature are shown in Fig. 1.

The products chosen for comparison were glucosides with different chain lengths, ranging from C4 to C8, with straight as well as branched hydrophobe structures and most of them with the same degree of glucosidation, DP = 1.4–1.6. Two commercial hydrotropes were used as references.

The results show an unexpectedly high efficiency for the hexyl glucoside, which is the only one which functions in 30 and 40% NaOH at reasonable amounts. Also within the glucoside family the hexyl glucoside is found to be more efficient than glucosides with shorter as well as longer hydrocarbon chains.

Results of similar investigations by Matero et al. [14] suggest that C4 and C8 glucosides are also efficient though with different mechanisms. In short, the medium-chain (C8 and C10) glucosides are seen to be most effective as solubilizers of nonionics (raising the cloud point), while the short-chain one (C4) destabilizes the crystalline phases most efficiently. Thus, the judgement of hydrotrope efficiency is very much dependent on the test method used. We did not look at the destabilization mechanism or the phase diagram in detail but regard the hexyl glucoside efficiency as the result of its intermediate size, maybe acting both as a destabilizer and as a solubilizer.

However, since the results given here and by Matero et al. [14] refer to one specific formulation or concentration we decided to look at concentration effects as well as at the influence of varying the ratios of ethoxylate and glucoside. Thus, an extensive investigation of the influence of concentration and electrolyte on the cloud points was made.

28

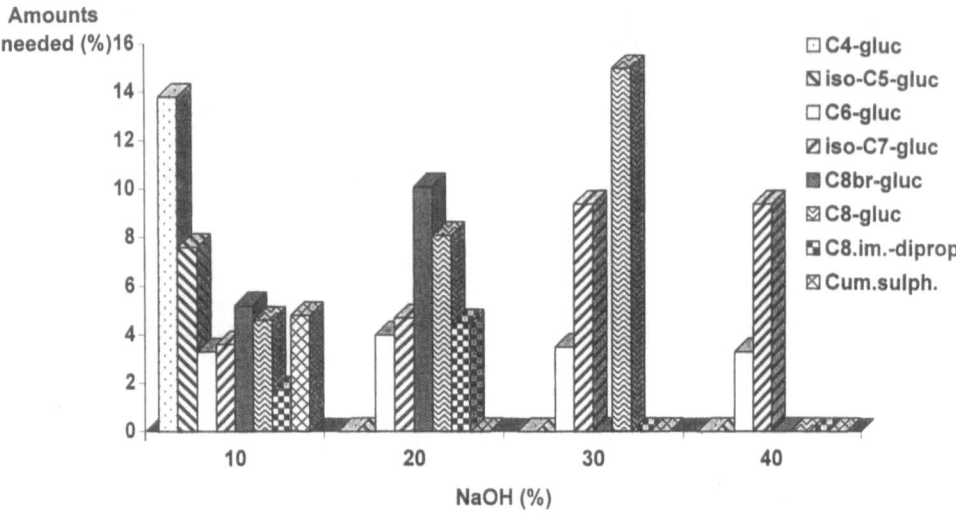

Fig. 1 Hydrotrope efficiency. The formulations contain 5% [C10(EO)4] different NaOH concentrations and the minimum amount of hydrotrope needed to get a transparent solution at room temperature. This amount is shown as *columns*. Zero means not possible to formulate clear solutions

Cloud points

Temperature-dependent phase behaviour for glucosides and ethoxylates

Owing to the very steep temperature dependence of the border lines in the phase diagrams [1–8] the glucosides have often been referred to as not having cloud points. However, as has been shown by Rohm and Haas [11] and Balzer [15] there is a clouding phenomenon that appears for very specifically balanced glucosides. A change in aglycon chain length of one carbon can decrease the cloud point from above 100 °C to below 0 °C [8]. A typical phase diagram for the lower concentration range is shown in Fig. 2 [15] for a glucoside, [C12-14(G)1.8]

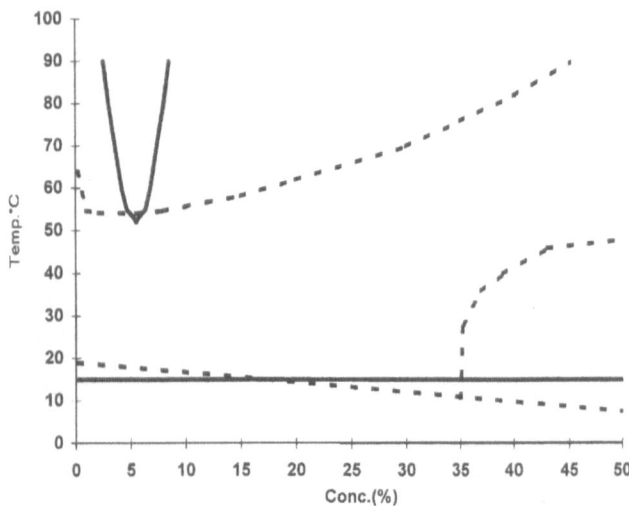

Fig. 2 Phase diagrams (low concentrations) of ■ (*full line*) and ■ (*dotted line*) from ref. 15

together with that of an alkoxylate. Here the typical very strong concentration dependence of the cloud point for the glucoside is in contrast to the shallow curve of the ethoxylate.

In industry nonionics, i.e. alkoxylates, are often characterized by their cloud point. This value is measured according to a procedure given in the norm ISO1065, in which five ways of measuring are described using different media for nonionics which are extremely water soluble, water soluble, sparingly water soluble and not at all water soluble. In each case only one concentration is recommended (even if it is important to state which concentration has been used), a procedure which relies on the fact that the clouding of a nonionic is rather insensitive to the concentration as can be seen in Fig. 2. Since the clouding of the glucosides is extremely concentration dependent, the minimum of the cloud-point curve can be given as a characterization of the glucoside.

What will happen when two products with widely different characteristics such as these are mixed?

Since they are often used together in cleaning of hard surfaces it is important to know what to expect. We measured the cloud point at different total concentrations of the surfactant mixture (glucoside and ethoxylate) keeping the ratio constant. The results for three different short-chain glucosides mixed with a C10(EO)4 narrow-range ethoxylate are given in Figs. 1–3. The glucosides are all fully water soluble and have a cloud point above 100 °C. The alkoxylate is not water soluble at low concentrations and has a cloud point below 0 °C. Three different glucoside/alkoxylate weight ratios were tested for each of the glucosides. Mixing the two types gives a clouding behaviour that for high amounts of glucoside is similar to what is typical for a balanced glucoside mixture [Fig. 2, C12-14(G)1.8] and for a higher nonionic content becomes more like the ethoxylate clouding. What is seen in all three cases is the dip at low concentrations. If the formulation being balanced in the practical application only is tested and characterized at one concentration, for instance, at 1%, the answer can be very misleading.

Electrolyte effects

Glucosides or alkoxylates in highly concentrated electrolytes

In practical applications the formulations very often contain electrolytes as thickening aids (e.g. NaCl in personal care or household cleaning), as complexing agents (detergents and industrial cleaning agents), to regulate pH (highly alkaline cleaning formulations, textile treatment additives) or to achieve specific effects (agrochemical additives).

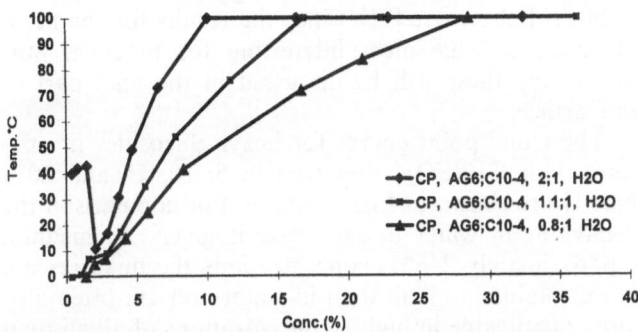

Fig. 3 Cloud point of C6 glucoside and C10(EO)4 in water

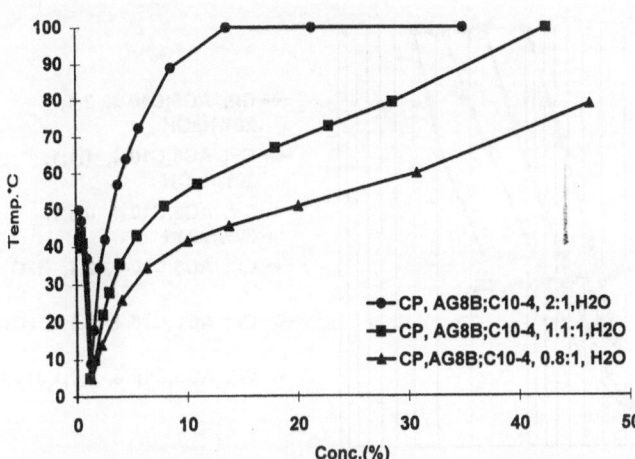

Fig. 4 Cloud point of C8 branched glucoside and C10(EO)4 in water

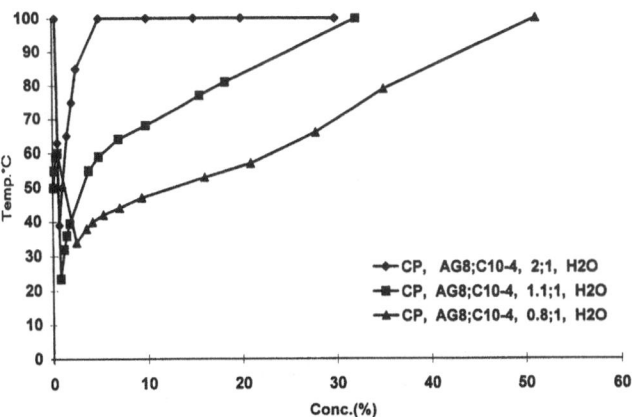

Fig. 5 Cloud point of C8 glucoside and C10(EO)4 in water

The salting-out and salting-in effects depending on the type of salt for the alkoxylates are well known [16, 17]. Similar effects are seen for the glucosides, although they are much less pronounced. Overall the alkyl glucosides are rather insensitive to electrolytes, except for the first small amounts that may screen the minor charges that seem to be present in the technical glucosides [15, 18, 19]. One exception is the sensitivity to alkali. Addition of, for example, NaOH, results in an increase in the cloud-point temperature for the glucoside and a decrease for the ethoxylate. The specific effect on the glucosides may be due to the deprotonation of weakly acidic OH groups on the glucose unit (pK_a = 12.35 for glucose and 12.51 for sucrose) [20].

What is the effect of electrolytes on mixtures of glucosides and alkoxylates?

The trend towards highly concentrated formulations for environmental reasons is strong and well known, for instance, in the detergent industry; therefore, we tested the behaviour of mixtures of C6 or C8 glucosides or C8 branched glucosides and a narrow-range C10(EO)4 alkoxylate in extremely high concentrations of different complexing agents (builders) as well as alkali over a concentration range from 30 to 40% total surfactant down to around 0.5%. Three different glucoside/ethoxylate weight ratios were studied for each surfactant combination, 2:1, 1.1:1 and 0.8:1.

Nitrilotriacetic acid

The influence on the cloud-point curve of NTA (a complexing agent) in rather high amounts is given in Figs. 6 and 7. Figure 6 shows the behaviour of the two C8 glucoside/ethoxylate mixtures, which both seem to be salted-out together with the ethoxylate. The C6 glucoside

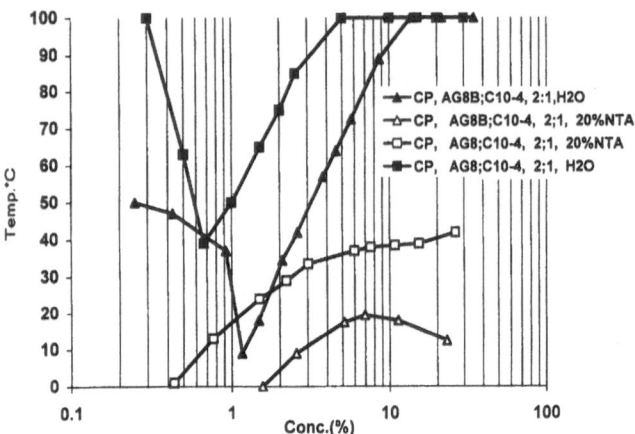

Fig. 6 Cloud point of C8 glucoside and branched and straight C10(EO)4 in water or 20% nitrilotriacetic acid (*NTA*)

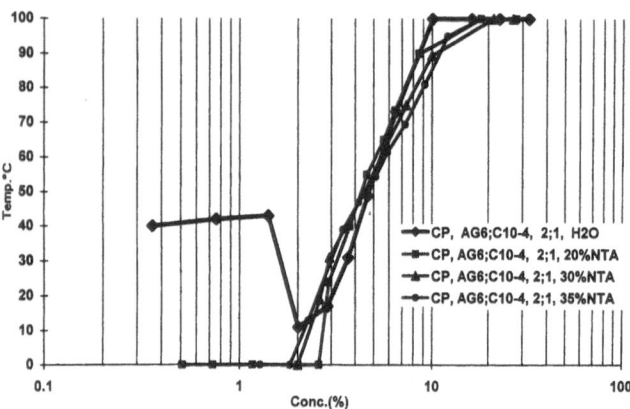

Fig. 7 Cloud point of C6 glucoside and C10(EO)4 in water or different NTA concentrations

mixture (Fig. 7), however, can stand even higher amounts of NTA, up to 30–35%. In this case the preferred choice should be the short C6 glucoside.

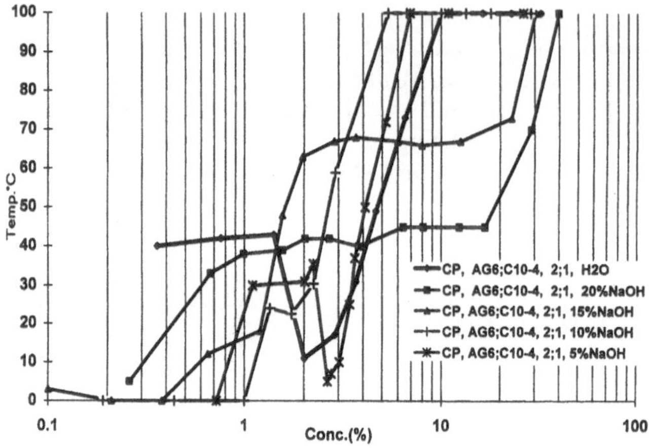

Fig. 8 Cloud point of C6 glucoside and C10(EO)4 in water or different NaOH concentrations

Sodium hydroxide

A very different picture is seen when glucoside/alkoxylate mixtures are dissolved in highly alkaline solutions. The results for the C8 glucosides have been published elsewhere [21]. Since the results for the hexyl glucoside are the most interesting for practical purposes only these will be discussed in the final part of this article.

The cloud-point curves for hexyl glucosides in mixtures with C10(EO)4 dissolved in 5, 10, 15 and 20% NaOH are shown in Figs. 8 and 9. For comparison the behaviour in water in each case is given. At medium (approximately 2–6%) concentrations the mixtures are more soluble in alkali than in water and are often also more dissolvable in higher concentrations of alkali than in lower ones. Typically the dip in the cloud point for the surfactant in water is cut off and the solution is stable at

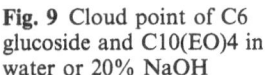

Fig. 9 Cloud point of C6 glucoside and C10(EO)4 in water or 20% NaOH

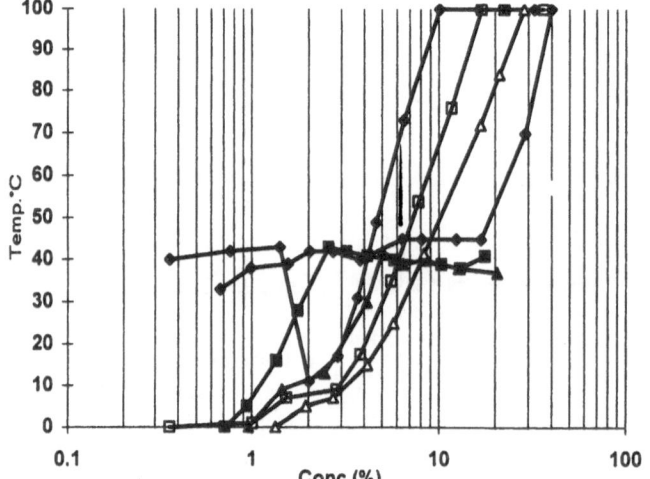

temperatures up to 30–40 °C. At higher surfactant concentrations the cloud point is often depressed by higher alkali quantities.

Of special interest is the behaviour seen in Fig. 9, where for each ratio the mixture at low concentrations is more soluble in 20% NaOH than in water.

TKPP and sodium metasilicate

A formulation with a commonly used concentration of conventional builders giving an elevated pH around 13 as compared to 11 for NTA and approximately 14 for the NaOH solutions was tested as a medium for the same mixtures as before. The results are shown in Fig. 10.

Comparing the results in all three electrolytes gives the impression that two effects are operating: one due to the high pH and one to a more common salt effect (Fig. 11). The octyl glucosides seem to be more sensitive to the "salting-out" phenomenon than the hexyl glucoside. This can be understood in terms of the longer hydrocarbon chains. The effect at very high pH can be interpreted as a change in character of the glucosides towards more and more anionic behaviour. All the OH groups of the glucose moiety are said to be weakly acidic [20].

Use in applications

The balance between surfactants and other additives in a cleaning formulation is important for its function. The complex process of cleaning, be it of fabrics [22] or of hard surfaces [23], has been studied and analysed in terms of the packing of the surface-active agents in the cleaning solution or when adsorbing to oily soil the change in its contact angle. Different mechanisms are related to different aspects of surface activity, but one dominating feature is the proximity to zero spontaneous

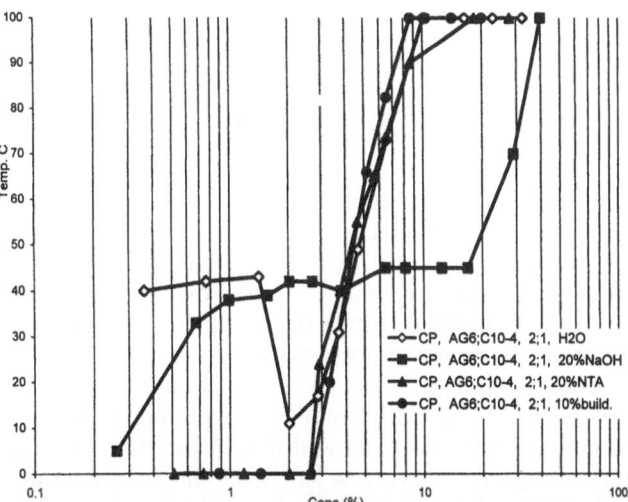

Fig. 11 Cloud point of C6 glucoside and C10(EO)4 with water, NTA, NaOH or TKPP and metasilicate

curvature (packing parameter close to 1), which also gives a minimum in interfacial tensions to the oil phase involved, which is beneficial for spontaneous emulsification and for the creation of a microemulsion state. The proximity to a cloud-point border line usually indicates growing aggregates in the solution with less surface curvature.

The mixtures described perform excellently in cleaning and wetting situations when using the knowledge presented here in the formulation [21, 24].

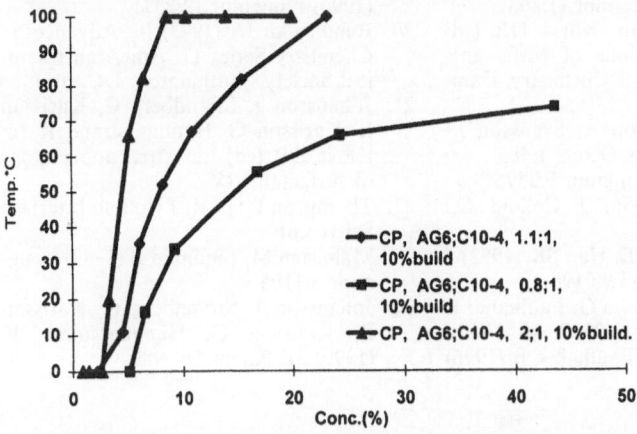

Fig. 10 Cloud point of C6 glucoside and C10(EO)4 with 6% tetrapotassium pyrophosphate (TKPP) and 4% metasilicate

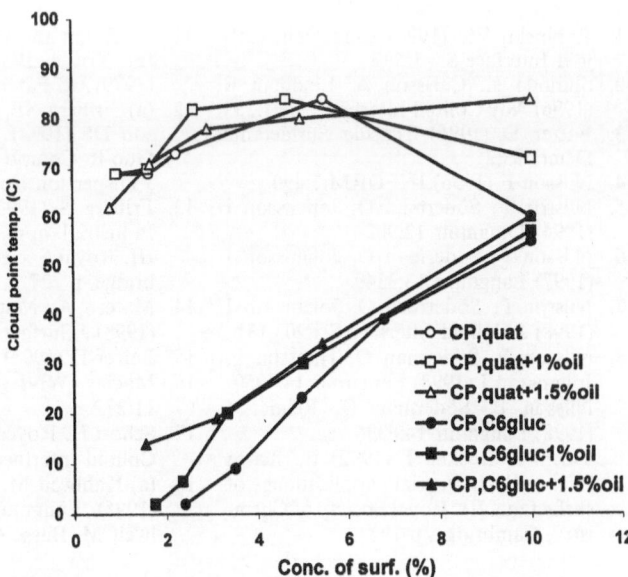

Fig. 12 Cloud point of C10(EO)4/C6 glucoside (1/1) and C10(EO)4/quaternary (commercial mixture) with TKPP and metasilicate, with and without model oil (Halpasol)

Wastewater treatment

When the formulations are used for cleaning oily soil from hard surfaces they contain oil that has to be separated. As little oil as possible should be left in the water phase after dilution with water and separation of the oil phase. This was tested according to the norms of the Swedish environmental authorities (Svenska Natur-vårdsverket) SNV 1975:10. For comparison a very efficient commercial cationic hydrotrope was used with the same C10(EO)4 as with the hexyl glucoside mixture. In both cases the ratio between hydrotrope and ethoxy-late was optimized to get the best cleaning result. They are used as such as commercial products. The oil separation test results showed that in the glucoside case only 3 mg oil/l water was left, while in the quaternary case 24 mg oil/l remained.

Could this be due to differences in the cloud point when the two oil-containing solutions were diluted? To shed some light onto this problem the following test was made. A 10% solution of the two surfactant blends, respectively, in water and 6% TKPP and 4% metasilicate was mixed with 1% oil (Halpasol) and 1.5% oil. This gives clear solutions at room temperature. A concentration of 2% oil is above saturation at room temperature in both cases. The solutions with and without oil were then diluted in steps with tap water (3.8 dH) and the cloud points were measured. The resulting cloud-point curves are shown in Fig. 12.

The overall impression is that the cloud point decreases with dilution for the glucoside-based formulation, while for the quaternary -based one it does not. The explanation for this at a molecular level needs a more sophisticated analysis, which is not the aim of this article. However, since the cloud point indicates the onset of a two-phase region, one would expect better separation in the glucoside blend since the diluted wastewater obviously reaches the two-phase situation. This is the case which has also been seen in full-scale use.

Conclusions

- Alkyl glucosides are good hydrotropes for ethoxylates in highly alkaline and electrolyte-containing media. Hexyl glucoside seems to be the overall most efficient type.
- The cloud-point curve for mixtures of an alkyl ethoxylate and an alkyl glucoside can be varied from very strongly concentration dependent ("steep") to shallow as for an ethoxylate, depending on the mixing ratio.
- Good cleaning and wetting in highly alkaline media can be achieved with these mixtures.
- Oil separation from wastewater is efficient when the system is diluted owing to the steep cloud-point curve.
- Nonionic cleaning systems of this type thus offer an environmentally suitable alternative.

References

1. Rybinskij Wv (1996) Curr Opin Colloid Interface Sci 1:587
2. Shinoda K, Carlsson A, Lindman B (1996) Adv Colloid Interface Sci 64:253
3. Balzer D (1996) Tenside Surfactants Deters 33:2
4. Nilsson F (1996) INFORM 7:490
5. Nilsson F, Söderman O, Johansson I (1996) Langmuir 12:902
6. Nilsson F, Söderman O, Johansson I (1997) Langmuir 13:3349
7. Nilsson F, Söderman O, Johansson I (1998) J Colloid Interface Sci 203:131
8. Nilsson F, Söderman O, Hansson P, Johansson I (1998) Langmuir 14:4050
9. Nilsson F, Söderman O, Reimer J (1998) Langmuir 14:6396
10. Thiem J, Böcker T (1992) In: Karsa DR (ed) Industrial applications of surfactants III. Royal Society of Chemistry, Cambridge, p 123
11. (a) Rohm and Haas product broschure for Triton BG-10; (b) Kaniecki TJ (1979) US Patent 4,147,652
12. (a) Friberg SE, Brancewics C, Morrison DS (1994) Langmuir 10:2945; (b) Guo R, Compo ME, Friberg SE (1996) J Dispersion Sci Technol 17:493
13. Friberg S (1992) In: Karsa DR (ed) Industrial applications of surfactants III. Royal Society of Chemistry, Cambridge, p 227
14. Matero A, Mattsson Å, Svensson M (1998) J Surfactants Deters 1:485
15. Balzer D (1993) Langmuir 9:3375
16. Maclay WNJ (1956) J Colloid Sci 11:272
17. Schott H, Royce AE, Han SK (1983) J Colloid Interface Sci 98:196
18. (a) Kahlweit M, Busse G, Faulhaber B (1995) Langmuir 11:3382; (b) Kahlweit M, Busse G, Faulhaber B (1996) Langmuir 12:861; (c) Kahlweit M, Busse G, Faulhaber B (1997) Langmuir 13:5249; (d) Kahl H, Kirmse K, Quitzsch K (1996) Tenside Surfactants Deters 1:26
19. Zhang L, Somasundaran P, Maltesh C (1996) Langmuir 12:2371
20. Rendleman JA (1973) In: Advances in Chemistry Series 17. American Chemical Society, Washington, DC, p 51
21. Johansson I, Strandberg C, Karlsson B, Karlsson G, Hammarstrand K In: Karsa DR (ed) Industrial applications of surfactants IV.
22. Thompson L (1994) J Colloid Interface Sci 163:61
23. Malmsten M, Lindman B (1989) Langmuir 5:1105
24. Johansson I, Strandberg C, Karlsson B, Karlsson G, Hammarstrand K (1999) SE Patent 510 989

Progr Colloid Polym Sci (2000) 116 : 33–36
© Springer-Verlag 2000

P. C. A. Barreleiro
G. Olofsson
E. Feitosa

Vesicle-to-micelle transition in dioctadecyldimethylammonium bromide and dioctadecyldimethylammonium chloride dispersions induced by octaethylene glycol *n*-dodecyl monoether. An isothermal titration calorimetry study

P. C. A. Barreleiro · G. Olofsson
Physical Chemistry 1,
Center for Chemistry
and Chemical Engineering
University of Lund
P.O. Box 124, 221 00 Lund, Sweden

E. Feitosa (✉)
Physics Department
IBILCE/UNESP, 15054-000
São José do Rio Preto, SP, Brazil
e-mail: eloi@df.ibilce.unesp.br

This work is part of the results presented at the First Nordic–Baltic Meeting on Surface and Colloid Science, in Vilnius, Lithuania, from 21–25 August 1999

Abstract We have used isothermal titration calorimetry to investigate the vesicle-to-micelle transition in dioctadecyldimethylammonium bromide (DODAB) and chloride (DODAC) vesicle dispersions induced by the nonionic surfactant octaethylene glycol *n*-dodecyl monoether ($C_{12}E_8$) at room temperature. Small and giant unilamellar vesicles were prepared by sonication and without sonication, respectively, of the pure cationic surfactants at low concentrations in water. The titration of 1.0 mM DODAX ($X = Cl^-$ and Br^-) by a concentrated micellar solution of $C_{12}E_8$ shows that the enthalpy of interaction (ΔH_{obs}) of $C_{12}E_8$ in micellar form with DODAX is always endothermic. The titration curves are understood on the basis of superposition of the enthalpies of partitioning of $C_{12}E_8$ into the bilayer, of micelle formation and of vesicle-to-micelle transformation. The enthalpy, ΔH_{obs}, initially increases owing to the incorporation of $C_{12}E_8$ into the vesicle bilayer until the $C_{12}E_8$/DODAX saturation ratio (R_{sat}) is reached, then ΔH_{obs} decreases, in different ways for DODAB and DODAC, owing to degradation of vesicles and formation of mixed micelles and intermediary structures up to the $C_{12}E_8$/DODAX solubilization ratio, R_{sol}. Above R_{sol} only mixed micelles exist. The surfactant solubilization takes place in three stages. All the critical ratios are lower for DODAB than for DODAC, meaning that $C_{12}E_8$ solubilizes more strongly in DODAB; for example, R_{sat} is 0.8 for DODAB and 1.2 for DODAC. Sonication has no significant effect on the transition.

Key words · Cationic vesicles · Dioctadecyldimethylammonium bromide · Dioctadecyldimethylammonium chloride · Isothermal titration calorimetry · Surfactant solubilization

Introduction

Dioctadecyldimethylammonium bromide (DODAB) and chloride (DODAC) have been the most investigated long-chain (C18) vesicle-forming cationic surfactants since the first preparation and characterization of small unilamellar DODAB vesicles in 1977 by Kunitake and Okahata [1]. Since then a number of methods to prepare cationic vesicles from these and homologous surfactants have been developed to give long-term stability and size-controlled structures to be used in specific applications, such as membrane mimetic systems [2], drug and DNA carrier vehicles [3] or as catalysts [4].

The methods commonly used to prepare DODAX (here X denotes Br^- or Cl^-) vesicles are the same as used to prepare phospholipid liposomes [3]. They include tip- or bath-type sonication [1, 5], ethanol- [6], chloroform- [7] or dichloromethane-injection [8], extrusion [8] and the novel nonsonication (nonextrusion or noninjection) method, which gives giant unilamellar vesicles (GUVs)

[9–11]. Small amounts of DODAB or DODAC are simply mixed with the aqueous solvent above the gel-to-liquid-crystalline phase transition temperature (T_m) (around 45 °C for DODAB and 48 °C for DODAC in water) to prepare GUV dispersions, as reported recently by our group [9–11]. The physical properties of these vesicles are described in the literature [1, 5–11].

In our attempt to obtain "high-quality" DODAX vesicles by reducing their structure and size polydispersity as well as to search for new methods to prepare stabilized vesicles with well-controlled size distribution we have been investigating the effects of mixing DODAX with micelle-forming surfactants, such as octaethylene glycol n-dodecyl monoether ($C_{12}E_8$). Isothermal titration calorimetry is suitable for obtaining critical composition and thermodynamic parameters related to the surfactant solubilization [12]. Our results indicate that these surfactants mix together according to the "three-stage" model [13–15] to form stabilized structures like mixed vesicles and micelles or intermediary structures depending on the relative amount of the compounds in the mixture and on the nature of the counterion.

Materials and methods

Chemicals

DODAB with purity better than 99% (Avanti Polar Lipids, Alabaster, Ala.) was used as purchased. DODAC was obtained by counterion exchange from DODAB (Eastman Kodak) and recrystallized as described elsewhere [8]. $C_{12}E_8$ (Nikko Chemicals) was used as purchased. Ultrapure water of Milli-Q-Plus quality was used in the sample preparations.

Vesicle preparation

GUV dispersions were prepared by simply mixing DODAB or DODAC and water at a concentration of 1.0 mM at 60 °C (above the critical transition temperature, T_m) to obtain a homogeneous dispersion [9–11]. Small unilamellar vesicle (SUV) dispersions were prepared by bath-sonicating the GUV dispersion at 60 °C using a bath-type sonicator (Starsonic 90, Liarre) to obtain optically clear dispersions. After preparation the vesicle dispersions were cooled to room temperature and stored for at least 24 h before the calorimetric measurements.

Titration calorimetry measurements

The enthalpies of reaction (ΔH_{obs}) in the $C_{12}E_8$/DODAX/water system were measured in a stainless-steel titration vessel of 3 cm^3 in the prototype of the thermal activity monitor four-channel microcalorimeter at 25 °C [16]. The calorimetric signal was calibrated electrically [16]. Small aliquots of a concentrated $C_{12}E_8$ micellar solution were added to a 2.7-cm^3 aqueous DODAX vesicle dispersion of 1.0 mM placed in the vessel of the calorimeter. The additions were made using a gas-tight Hamilton syringe connected with a computer-operated syringe drive for good control of the injection volume and injection rates. Typically, 10-μl aliquots were injected over 120 s, with a time period between injections of at least 100 min. All measurements were made at least twice to ensure reproducibility.

Results and discussion

The vesicle-to-micelle transition of DODAB and DODAC vesicles in nonsonicated (GUV) and sonicated (SUV) dispersions induced by the nonionic amphiphile $C_{12}E_8$ was studied by isothermal titration calorimetry. The results of the calorimetric titration of a 1 mM DODAB vesicle dispersion with 54 mM $C_{12}E_8$ solution are shown in Fig. 1, where the observed enthalpy changes ΔH_{obs} (calculated per mole of $C_{12}E_8$ added) are plotted against the $C_{12}E_8$/DODAB molar ratio (R) in the final solution. When repeating the titration series but using pure water in the calorimetric vessel, the first injection gave a significant exothermic peak owing to demicellization but after a couple of injections the concentration is well above the critical micelle concentration (cmc) and only a weekly exothermic dilution enthalpy of -1.0 kJ mol^{-1} was seen. The cmc of $C_{12}E_8$ is 9.1×10^{-5} M^{-1} and the enthalpy of micelle formation, ΔH_{mic}, is 16.3 ± 0.4 kJ mol^{-1} at 25 °C [17]. With DODAB vesicles present in the solution, the first injection gave a weekly exothermic enthalpy change but then ΔH_{obs} increased steeply to a peak value of about 17.5 kJ mol^{-1} at $R = 0.6$. This indicates that a critical concentration of $C_{12}E_8$ must be reached before solubi-

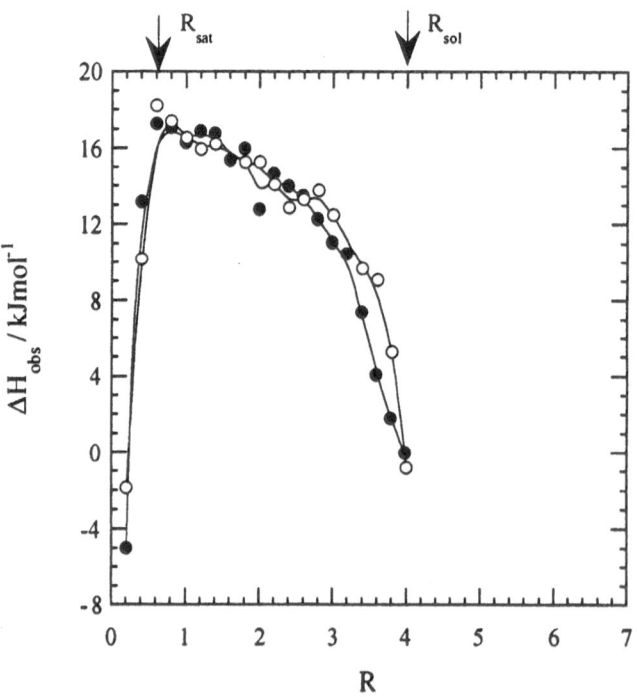

Fig. 1 Enthalpy of interaction as a function of octaethylene glycol n-dodecyl monoether ($C_{12}E_8$)/dioctadecyldimethylammonium bromide (*DODAB*) molar ratio, R, obtained by consecutive additions of aliquots of concentrated (54 mM) $C_{12}E_8$ solution to 1.0 mM DODAB dispersion. Dispersion prepared by sonication (●) and without sonication (○) (25 °C)

lization in DODAB vesicles takes place. After the peak ΔH_{obs} decreased steadily to R about 3.2, after which it dropped to close to zero. As seen from Fig. 1, sonication did not significantly change the titration curves. The results of addition of $C_{12}E_8$ solution to a 1 mM DODAC vesicle dispersion are shown in Fig. 2. After reaching a maximum value of about 6.0 kJ mol^{-1} at $R = 1.2$, ΔH_{obs} stays roughly constant till R is about 3.2, after which it gradually decreases to become close to zero at R around 5.8. Thus, the calorimetric titration curves show the same general features but the changes are more gradual and the enthalpic effects are smaller in the DODAC system than in the DODAB one.

The $C_{12}E_8$/DODAB/water system

When the $C_{12}E_8$ micellar solution is added to the DODAB vesicle dispersion, three stages can be discerned. In the first stage, $C_{12}E_8$ is partitioned between the bulk solution and the vesicle bilayers, resulting in a progressive increase in ΔH_{obs} to around 17.5 kJ mol^{-1} at the $C_{12}E_8$/DODAB mole ratio, R_{sat}, of about 0.6. In the second stage, above R_{sat}, the vesicles become unstable and start to degrade to form mixed micelles with the added $C_{12}E_8$. Dynamic light scattering results to be

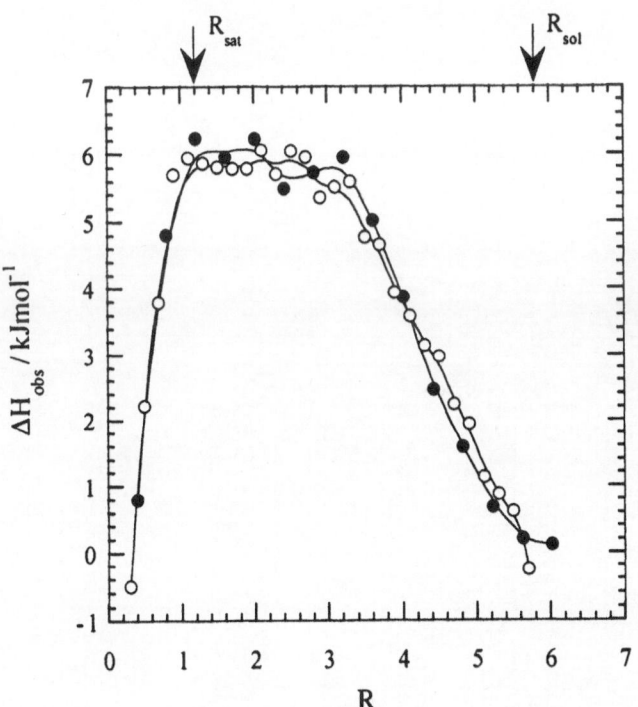

Fig. 2 Enthalpy of interaction as a function of $C_{12}E_8$/dioctadecyl-dimethylammonium chloride (*DODAC*) molar ratio, R, obtained by consecutive additions of aliquots of concentrated (54 mM) $C_{12}E_8$ solution to 1.0 mM DODAC dispersion. Dispersion prepared by sonication (●) and without sonication (○) (25 °C)

reported by our group show the existence in this stage of two populations with characteristic sizes of vesicles and micelles. As R is further increased the vesicle population decreases, whereas the micelle population increases continuously until R_{sol} is reached. Above R_{sol} (in the third stage) the vesicle-to-micelle transition is complete and only mixed micelles exist. Considering the ease with which cationic and nonionic surfactants mix nearly ideally to form mixed micelles [18], the formation of a homogeneous population of mixed micelles is more likely than a mixture of pure $C_{12}E_8$ and mixed micelles. The rapidly vanishing ΔH_{obs} indicates that the enthalpy content of $C_{12}E_8$ is approximately the same in the pure and in the mixed micelles.

The $C_{12}E_8$/DODAC/water system

About the same general behavior was observed for the addition of $C_{12}E_8$ to a DODAC vesicle dispersion, except in the following respects (see Fig. 2):

1. The interaction enthalpy, ΔH_{obs}, is lower for DODAC than for DODAB.
2. Just above R_{sat} the enthalpy change reaches a plateau for DODAC but decreases for DODAB.
3. The critical ratios R_{sat} and R_{sol} are higher for DODAC than for DODAB.
4. The transition from mixed vesicles to mixed micelles occurs more steeply for DODAB.

Such effects are surprisingly large considering that the only difference between DODAB and DODAC is the counterion nature. Br$^-$ binds more strongly to the vesicle interfaces to produce stronger screening effects, which appears to favor the incorporation of $C_{12}E_8$ in the vesicle bilayer. As in DODAB, there is no effect of sonication on the $C_{12}E_8$ solubilization in DODAC vesicles (Fig. 2).

Conclusion

The titration results indicate that the vesicle-to-micelle transition for the $C_{12}E_8$/DODAX/water systems follows the three-stage model: below the $C_{12}E_8$/DODAX saturation ratio $C_{12}E_8$ is partitioned between the bulk solution and the vesicle bilayer, accompanied by a steep increase in the enthalpy of reaction (ΔH_{obs}). After saturation and below the complete solubilization, the $C_{12}E_8$ excess destabilizes the saturated vesicles and mixed micelles are formed while mixed vesicles are degraded continuously. Above the solubilization ratio, all vesicles have been degraded and only mixed micelles remain. For both the DODAB and DODAC systems, there is no significant effect of sonication on the titration curve. A detailed investigation of solubilization phenomena in

$C_{12}E_8$/DODAX/water systems is currently underway by our group using different experimental techniques to providing a better understanding of the vesicle-to-micelle transition mechanism and morphology of the aggregates in each stage of the transition.

Acknowledgements E.F. wishes to thank FUNDUNESP for partially granting his participation in the "First Nordic–Baltic Meeting on Surface and Colloid Science". P.C.A.B. acknowledges the PRAXIS XXI, JNICT for financial support, scholarship BD/ 13788/97. We express our gratitude to I. M. Cuccovia for supplying the DODAC samples.

References

1. Kunitake T, Okahata Y (1977) J Am Chem Soc 99:3860
2. Fendler JH (1982) Membrane mimetic chemistry. Wiley–Interscience, New York
3. Lasic DD (1993) Liposomes. From physics to applications. Elsevier, Amsterdan
4. Bunton CA (1991) In: Rubingh DN, Holland PM (eds) Cationic surfactants. Physical chemistry. Surfactant science series, vol 37. Dekker, New York, p 323
5. Cuccovia IM, Feitosa E, Chaimovich H, Sepulveda L, Reed WF (1990) J Phys Chem 94:3722
6. Cuccovia IM, Aleixo RMV, Mortara RA, Berci Filho P, Bonilha JBS, Quina FH, Chaimovich H (1979) Tetrahedron Lett 3065
7. Carmona-Ribeiro AM (1992) Chem Soc Rev 21:209
8. Cuccovia IM, Sesso A, Abuin E, Okino PF, Tavares PG, Campos JFS, Florenzano FH, Chaimovich H (1997) J Mol Liq 72:323
9. Feitosa E, Brown W (1997) Langmuir 13:4810
10. Benatti CR, Tiera MJ, Feitosa E, Olofsson G (1999) Thermochim Acta 328:137
11. Feitosa E, Barreleiro PCA, Olofsson G (2000) Chem Phys Lipids 105:210
12. Heerklotz H, Lantzcsch G, Binder H, Klose G, Blume A (1995) Chem Phys Lett 235:517
13. (a) Lichtenberg D, Robson RJ, Dennis EA (1983) Biochim Biophys Acta 737:285; (b) Lichtenberg D (1985) Biochim Biophys Acta 821:470
14. Dennis EA (1974) Arch Biochim Biophys 165:764
15. Helenius A, Simons K (1975) Biochim Biophys Acta 415:29
16. Bäckman P, Bastos M, Briggner L-E, Hägg S, Hallén D, Lönnbro P, Nilsson S-O, Olofsson G, Schön A, Suurkuusk J, Teixeira C, Wadsö I (1994) Pure Appl Chem 66:375
17. Olofsson G (1985) J Phys Chem 89:1473
18. Holland PM, Rubingh DN (1991) In: Rubingh DN, Holland PM (ed) Cationic surfactants. Physical chemistry. Surfactant science series, vol 37. Dekker, New York, p 141

Progr Colloid Polym Sci (2000) 116:37–41
© Springer-Verlag 2000

R. Angelico
L. Ambrosone
A. Ceglie
G. Palazzo
K. Mortensen
U. Olsson

Structure and dynamics of polymer-like reverse micelles

R. Angelico (✉) · L. Ambrosone
A. Ceglie
DISTAAM, Università del Molise
v. De Sanctis, 86100 Campobasso, Italy
e-mail: angelico@.unimol.it

G. Palazzo
Dipartimento di Chimica, Università di
Bari, v. Orabona 4, 70126 Bari, Italy

K. Mortensen
Condensed Matter Physics and Chemistry
Department, Risø National Laboratory
4000 Roskilde, Denmark

U. Olsson
Physical Chemistry 1, Center for Chemistry
and Chemical Engineering
Lund University, P.O. Box 124, 22100
Lund, Sweden

Abstract We present pulsed-field-gradient NMR and time-resolved small-angle neutron scattering experimental data to bear out the polymer-like nature of long, flexible micelles in the lecithin–water–cyclohexane system. The molecular transport of lecithin is described in terms of curvilinear lateral diffusion along the contour length of unbranched entangled cylindrical micelles. Moreover, an isotropic-to-nematic phase transition is induced under shear, characterized by a certain degree of alignment of the micelles. As a final result the lecithin micelles appear to be more polymer-like than wormlike micellar systems in general.

Key words Small-angle neutron scattering · Pulsed-field-gradient nuclear magnetic resonance · Curvilinear diffusion · Shear-induced phase transition

Introduction

The polymer-like nature of very long cylindrical micelles has been ascertained in the lecithin–water–cyclohexane system by scattering techniques [1] and NMR self-diffusion measurements [2]. These aggregates are formed by adding small amounts of water to solutions of phosphatidylcholine (lecithin) in cyclohexane, thus producing an isotropic L_2 phase with the peculiarity that the micellar contour length can be tuned, varying opportunely the water-to-lecithin mole ratio, W_0. Above a threshold lecithin volume fraction, ϕ^*, flexible cylindrical micelles start to overlap, leading to the formation of a three-dimensional transient network similar to that of polymer chains in their semidilute or concentrated solutions. A first consequence is that the system turns jelly-like with viscoelastic properties. On the other hand the self-assembly aggregates possess an essential difference from polymeric molecules: they have a finite lifetime, dissociating and recombining with time, and their contour length is also composition- and tempera-

ture-dependent. Owing to these latter properties they are often referred to as living polymers [3]. Further and more peculiar results of the dynamics of the lecithin wormlike micelles have been gained through a lecithin NMR self-diffusion study [4] dominated by the experimental evidence of a lecithin curvilinear diffusion along the micellar contour length of very long (of the order of microns) unbranched aggregates with a lifetime for the lecithin molecules in the micelles which is longer than 1.5 s. That the breaking/recombination process is slow for the present system has also been argued from the rheological behavior [5] characterized by an unusually broad spectrum of relaxation times, where the stress is primarily relaxed by micellar reptation. For the present system, lecithin–water–cyclohexane, W_0 is a crucial parameter. Generally, cylindrical reverse aggregates are found at low W_0 values, $(4 < W_0 < 15)$, while a shape transition has been shown to occur at higher W_0, where the aggregates are now spherical (these are in equilibrium with water at the emulsification failure) [6]. Recently, the peculiar structural properties of the isotropic L_2 phase have been interpreted in terms of a flexible surface model and the excluded-volume effect, in the phase equilibria investigation of the lecithin–water–cyclohexane phase diagram [7]. Indeed, at high lecithin volume fraction and low W_0, where the excluded-volume effect becomes relevant, a reverse nematic birefringent N_2 phase is encountered. Here, the polymer-like micelles orient on a large scale with a degree of ordering quantified in terms of the second rank order parameter $P_2 = \langle 3\cos\theta - 1 \rangle /2$, where θ is the local angle between the tangent of the micellar contour and the director of the phase [8]. As shown later, an anisotropic nematic phase can also be induced by shearing samples with a high lecithin volume fraction belonging to the isotropic L_2 micellar phase. The presence of aligned aggregates under the influence of a steady-shear rate, $\dot{\gamma}$, creates anisotropic small-angle neutron scattering (SANS) patterns along the two orthogonal directions of the scattering vector $Q_\perp$ and $Q_{||}$, i.e. perpendicular and parallel to the flow direction. A first SANS study of shear-aligned viscoelastic reverse micellar solutions of lecithin in isooctane at low volume fractions was carried out at high $\dot{\gamma}$ and direct evidence for the presence of cylindrical reverse micelles was obtained [9]. Here, we report some results of a preliminary SANS study of the shear-induced anisotropy for a sample in the liquid isotropic L_2 phase, at $W_0 = 10$ and a volume fraction of the micelles (sum of water and lecithin) ϕ of 0.28. More details on this study are reported elsewhere [10]. To emphasize the strong analogy of the lecithin reverse micelles to polymer solutions, we also report further NMR self-diffusion data concerning the lecithin diffusion for a sample in the liquid isotropic L_2 phase with micellar volume fraction $\phi = 0.16$ and $W_0 = 12$, in very close agreement with the results obtained for samples at $W_0 = 10$ [4].

Materials and methods

The samples were made up of pure soybean phosphatidylcholine (lecithin) with the trade name Epikuron 200 obtained with a purity of 95% and an average mass of 772 Daltom from a generous gift of Lucas Meyer (Hamburg, Germany). The lecithin was used as received. For the purpose of neutron contrast for the SANS experiments and to minimize the intensity from the solvent protons for the ^1H NMR self-diffusion measurements, we used perdeuterated cyclohexane purchased from Dr. Glaser, which was also used as received. Water was Millipore-filtered. Samples were prepared as previously described [11]. To probe the liquid-to-nematic phase transition, SANS measurements under shear were performed on a lecithin–water–C_6D_{12} sample with micellar volume fraction $\phi = 0.28$ and $W_0 = 10$. The sample was placed in a 0.5-mm-gap Couette shear cell (especially conceived for SANS measurements [12]), placed at the small-angle instrument at the DR3 reactor at Risø National Laboratory (Denmark), which allows the recording of SANS experiments under shear with shear rate $\dot{\gamma}$ in the range 1–3000 s^{-1}. In a system with x-, y- and z-axes the primary neutron beam is defined in the y direction, which also defines the direction of the velocity gradient in the Couette shear geometry. The velocity vector is parallel to the x-axis, while the z-axis defines the

cylindrical symmetry axis of the Couette cell. The temperature was adjusted to 16.7 °C. The sample, isotropic and highly viscous at rest, was exposed to a steady shear of $\dot{\gamma} = 10$ s^{-1} in the Couette cell. This was sufficient to transform the whole sample into a nematic state where the polymer-like micelles preferentially oriented parallel to the velocity direction (x-axis). The shear-induced anisotropy of the sample was well recognized by recording the scattered intensity in the form of two symmetric correlation peaks, with essentially a Lorentzian angular dependence, along the direction perpendicular to the primary beam (z-axis).

Lecithin self-diffusion was measured for a lecithin–water–C_6D_{12} sample with micellar volume fraction $\phi = 0.16$ and $W_0 = 12$ by pulsed-field-gradient (PFG) ^1H NMR experiments on a Bruker DMX 200 instrument. A stimulated echo sequence [13] was used rather than the classical Hahn echo owing to the rapid transverse relaxation of lecithin protons. The time between the first two 90° pulses in the stimulated echo sequence was kept short and constant at 6 ms, while the time between the second and the third 90° pulses was varied between 44 and 1494 ms, corresponding to a variation of the gradient pulse separation between 50 and 1500 ms. The measurements were performed at a fixed gradient pulse duration of $\delta = 5$ ms by varying the gradient magnitude, G, up to approximately 8 T/m. The echo attenuation was measured by recording the intensity of the trimethylammonium and terminal methyl protons in the spectra obtained by Fourier transform of the second half of the echo. Experiments were performed for different values of t, the experimental timescale in the self-diffusion experiments, from 0.050 to 1.5 s.

Results and discussion

Small-angle neutron scattering

The particular flow behavior of solutions of long cylindrical wormlike micelles has been observed in a number of systems, characterized by banded flow over a certain range of shear rates [14, 15]. This phenomenon has been interpreted, for some systems, as a shear-induced phase transition from a disordered liquid isotropic solution, L, to a nematic phase, N, with the aggregates aligned in the direction of shear. As a consequence, the phase sequence L → L + N → N is observed with increasing shear rate [14]. For the sample under investigation the N phase can be induced at low shear rate ($\dot{\gamma} = 10$ s^{-1}), giving the typical SANS pattern shown in Fig. 1 (curve A), where the angular dependence of the scattered intensity within a narrow Q range around the peak position (0.06 Å^{-1} ≤ Q ≤ 0.09 Å^{-1}) is plotted. In Fig. 1, curve B represents the scattering pattern recorded 15 min after the shear has been turned off and curve C is the scattering pattern after approximately 45 min, where no more signs of anisotropy are detectable. The process of relaxation from the oriented to the isotropic state occurs gradually. From a careful analysis of the SANS data [10] it has been found that, after the shear has been turned off, the decay of the two peaks is characterized by a continuous increase in the angular width, producing at the end of the relaxation process a ring on the two-dimensional detector. During the whole process no experimental evidence of a growth in the circular

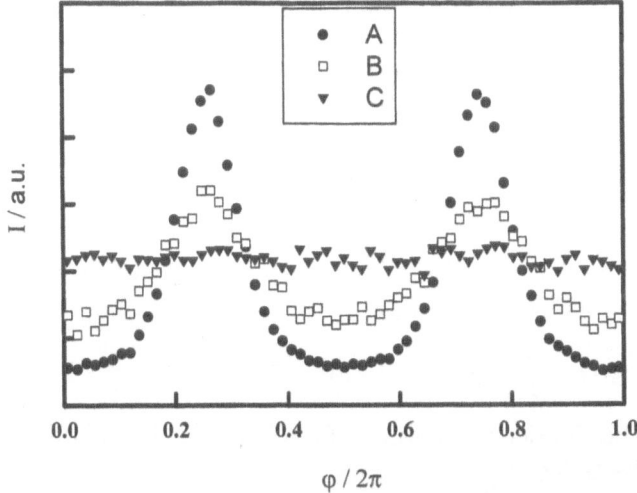

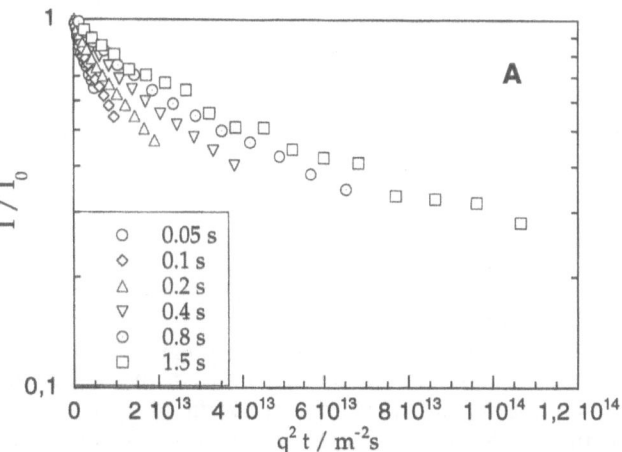

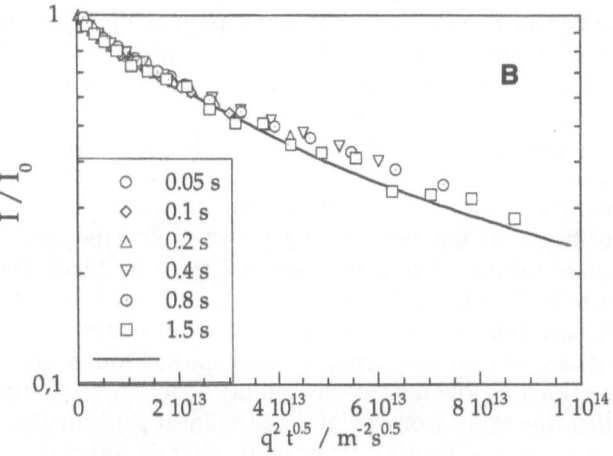

Fig. 1 Angular dependence of the scattered intensity (not normalized) in the Q range 0.06–0.09 Å^{-1}; sample composition $\phi = 0.28$ and $W_0 = 10$. *A* During shear $\dot{\gamma} = 10$ s^{-1}; *B* 15 min after cessation of the shear; *C* 45 min after cessation of the shear

scattering intensity was recorded, which means that we can exclude the nucleation and growth mechanism for the relaxation of the N phase back to the L phase. From the analysis of the anisotropic structure factor peak [8] it is also possible to estimate the orientational order P_2 extrapolated to $t = 0$. For the present system it has been found [10] that $P_2 = 0.41$ represents a useful and valuable experimental datum to estimate, in turn, the micellar persistence length, l_p. Indeed, following the theoretical relation of van der Schoot and Cates [16] we obtain $l_p \approx 500$ Å, in very close agreement with the value calculated by analyzing the self-diffusion data as shown later.

Pulsed-field-gradient NMR

Within the short gradient pulse approximation ($\delta \rightarrow 0$, $G \rightarrow \infty$ keeping the product δG finite), the normalized NMR echo attenuation $E(q,t)$ corresponds to the Fourier transform of the one-dimensional diffusion average propagator parallel to the applied magnetic field gradient (here z direction). For a homogeneous and isotropic solution we may write [17]

$$E(q,t) = \int_{-\infty}^{+\infty} \overline{P}(z,t) \exp(iqz)\mathrm{d}z \ . \tag{1}$$

Here, $q = \gamma g \delta / 2\pi$, where γ is the magnetogyric ratio of the nucleus (here ^{1}H). The average diffusion propagator $\overline{P}(z,t)$ describes the probability that the molecule has diffused a distance z during a time t. In the PFG echo experiment the diffusion time, t, is given by the time separation between the two gradient pulses. In a simple

liquid characterized by a self-diffusion coefficient, D, $\overline{P}(z,t)$ is Gaussian

$$\overline{P}(z,t) = \frac{\exp(-z^2/4Dt)}{(4\pi Dt)^{1/2}} \ , \tag{2}$$

resulting in a Gaussian echo attenuation:

$$E(q,t) = \exp(-4\pi^2 q^2 Dt) \ . \tag{3}$$

Experimental echo decays recorded on different time-scales are plotted in Fig. 2A, as commonly done, on a semilogarithmic scale as $E(q,t)$ versus tq^2 (Stejskal–Tanner plot) for a sample at $W_0 = 12$ and $\phi = 0.16$. As is seen, the data deviate significantly from simple Gaussian diffusion behavior. In contrast to the prediction of Eq. (3), a straight line is not obtained in the Stejskal–Tanner plot and, moreover, there is a significant dependence on t. In order to rationalize this behavior we have to outline a treatment for curvilinear diffusion of lecithin along a wormlike micelle.

Fig. 2 Normalized echo attenuation from the sample at $W_0 = 12$ and $\phi = 0.16$ measured at 25 °C plotted **A** on a semilogarithmic scale versus tq^2 and **B** on a semilogarithmic scale as a function of $q^2t^{1/2}$. The solid line is the best fit of Eq. (7) to the data

Let us start by considering a continuous flexible chain, Γ, characterized by a contour length, L, with $L \gg d$, where d is the chain diameter. Let us define a laboratory-fixed frame xyz; any particle moving onto the chain can be identified using a local curvilinear abscissa, s, starting from an arbitrary point $s_0 = 0$, (s_0 is the origin of the curvilinear abscissa on the chain). The flexible chain may be more conveniently treated in terms of the freely jointed chain model, also referred to as the discrete random-walk model, where now the chain is modeled by a set of links as shown in Fig. 3.

The direction of one link of length l_p (also called the persistence length in the isotropic limit) in any part of the chain is independent of the direction in space to its neighbors. Small molecular species (say, a surfactant molecule) can freely diffuse with local self-diffusion coefficient D_s along the curvilinear path of the chain on the experimental timescale t. The net displacement, $R(t)$, in the laboratory frame, is given by $R(t) = r_t - r_0$. Assuming any external force to be negligible or absent on the chain, $\langle R(t) \rangle$ must be zero, where the angled brackets mean an average over a large number of particles distributed over different chains and with different initial positions on the same chain; however, for the mean-square displacement, we have $\langle R^2(t) \rangle \neq 0$. Since the particle is forced to move along the curve Γ, the displacement $R(t)$ can be decomposed into two contributions: one coming from the curvilinear motion on Γ, described by the density probability function $f(s,t)$, the other due to the isotropy of the curvilinear geometry and given by the density probability $\psi(R,s)$. What it is measured in a PFG–NMR experiment is the mean-square displacement along the magnetic gradient axis, say z-axis, in the laboratory frame, i.e. $\langle z^2(t) \rangle$. Therefore,

we have to consider the function $\psi(z,s)$ in order to describe the projection of $R(t)$ along the z-axis. For the Brownian characteristic of the diffusional process, both functions must satisfy the Fokker–Planck equation, i.e.

$$\frac{\partial f}{\partial t} = D_s \frac{\partial^2 f}{\partial s^2} + \delta(s)\delta(t) \quad \text{and} \quad \frac{\partial \psi}{\partial s} = l_p \frac{\partial^2 \psi}{\partial z^2} + \delta(s)\delta(z) \ , \tag{4}$$

where δ is the Dirac function. In this case, the average propagator is

$$\overline{P}(z,t) = \int_{-\infty}^{+\infty} f(s,t)\psi(z,s)\,\mathrm{d}s \ . \tag{5}$$

The Fourier transform of Eq. (5) is the observed echo decay $E(q,t)$:

$$E(q,t) = \int_{-\infty}^{+\infty} f(s,t)\overline{\psi}(q,s)\,\mathrm{d}s \ , \tag{6}$$

where $\overline{\psi}(q,s) = \int_{-\infty}^{+\infty} \psi(z,s)\exp(iqz)\,\mathrm{d}z$. Then, we calculate the temporal derivative of Eq. (6) and by transforming Eq. (4) we finally have $\frac{\mathrm{d}E}{\mathrm{d}t} = D_s l_p^2 q^4 E - D_s l_p q^2 f(0,t)$.

On solving the differential equation we have the following solution:

$$E(q,t) = \exp(\rho x)^2 \operatorname{erfc}(\rho x) \ , \tag{7}$$

where $\rho = (4/3\sqrt{\pi})D_s^{1/2} l_p$, $x = q^2 t^{1/2}$ and erfc is the complementary error function.

The form of the echo attenuation in Eq. (7), which has also been given by Fatkullin and Kimmich [18] in terms of the mean-square displacement, is not Gaussian and suggests that if the data are plotted as a function of $x = q^2 t^{1/2}$ all data points should fall on the same master curve. That this is indeed the case is shown in Fig. 2B, where the experimental data are replotted against $q^2 t^{1/2}$.

Fits to the data, shown as solid lines, gave the value 6.66×10^{-14} m^2 s$^{-1/2}$ for the fitting parameter $D_s^{1/2} l_p$. In bilayers formed in the lamellar phase of the binary water–lecithin system the lateral diffusion coefficient of lecithin is approximately 2×10^{-12} m^2 s^{-1} at room temperature [19].

If we calculate l_p using this value for D_s we obtain $l_p = 470$ Å.

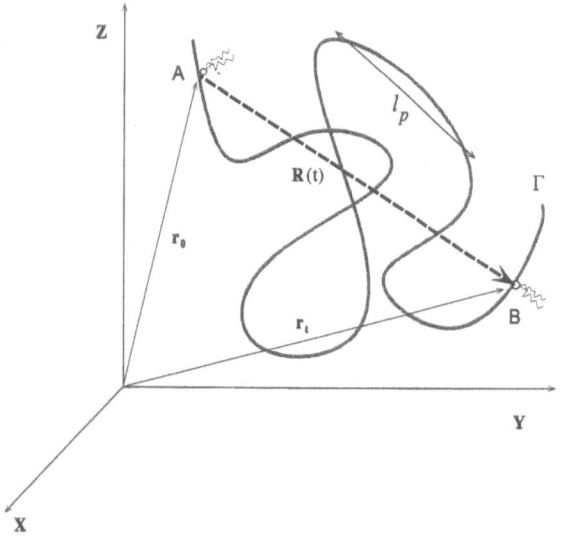

Fig. 3 Schematic representation of the diffusion of lecithin along the contour of a giant wormlike micelle (see text for details)

Conclusions

From the data reported here a number of conclusions can be drawn.

1. The micellar contour lengths have to be very long. In the analysis of the experimental results we see no signs of end effects. This implies that the micellar contour length must be much longer than $\langle s^2 \rangle^{1/2}$ for the longest observation time of 1.5 s. With this observation time and $D_s = 2 \times 10^{-12}$ m^2 s^{-1}, we obtain $\langle s^2 \rangle^{1/2} = 2.4$ μm. Thus,

the micellar contour lengths have to be much longer than micrometers.

2. The characteristic time for the breaking of the micelles also has to be very long. This is consistent with shear experiments where a shear-induced nematic phase needs several minutes to relax back to the liquid phase.

3. The micelles do not form branches. Had they done so we would have observed simple Gaussian surfactant diffusion.

4. For high ϕ values both SANS and PFG–NMR techniques suggest a persistence length of about 500 Å, a value larger than the previously reported estimate (120 Å at low ϕ) [1].

Acknowledgements This work was supported by the Swedish Natural Science Research Council (NFR), the Consorzio Interuniversitario per lo Sviluppo dei Sistemi a Grande Interfase (CSGI-Florence) and MURST of Italy (Prog. Cofinanz. 1998).

References

1. Schurtenberger P, Jerke G, Cavaco C, Pedersen JS (1996) Langmuir 12:2433, and references therein
2. Angelico R, Balinov B, Ceglie A, Olsson U, Palazzo G, Söderman O (1999) Langmuir 15:1679, and references therein
3. Lequeux F (1996) Curr Opin Colloid Interface Sci 1:341
4. Angelico R, Olsson U, Palazzo G, Ceglie A (1998) Phys Rev Lett 81:2823
5. Cavaco MC PhD thesis (1994) ETH Zürich
6. Angelico R, Cirkel PA, Colafemmina G, Palazzo G, Giustini M, Ceglie A (1998) J Phys Chem B 102:2883
7. Angelico R, Ceglie A, Olsson U, Palazzo G (2000) Langmuir 16:2124
8. Deutsch M (1991) Phys Rev A 44:8264
9. Schurtenberger P, Magid LJ, Penfold J, Heenan R (1990) Langmuir 6:1800
10. Angelico R, Olsson U, Mortensen K, Palazzo G, Ceglie A (manuscript in preparation)
11. Angelico R, Colafemmina G, Della Monica M, Palazzo G, Giustini M, Ceglie A (1997) Prog Colloid Polym Sci 105:1184
12. Mortensen K, Almdal K, Bates FS, Koppi K, Tirell M, Nordén B (1995) Physica B 213:682
13. Stilbs P (1987) Prog Nucl Magn Reson Spectrosc 19:1
14. (a) Berret J-F, Roux DC, Porte G (1994) J Phys II 4:1261; (b) Cappelaere E, Berret J-F, Decruppe JP, Cressely R, Lindner P (1997) Phys Rev E 56:1869
15. (a) Britton MM, Callaghan PT (1999) Eur Phys J B 7:237; (b) Porte G, Berret J-F, Harden JL (1997) J Phys II 7:459
16. van der Schoot P, Cates ME (1994) Europhys Lett 25:515
17. Callaghan PT (1991) Principles of nuclear magnetic resonance microscopy. Clarendon, Oxford
18. Fatkullin N, Kimmich R (1995) Phys Rev E 52:3273
19. Lindblom G, Orädd G (1994) Prog Nucl Magn Reson Spectrosc 26:483

Progr Colloid Polym Sci (2000) 116:42–47
© Springer-Verlag 2000

A. Lopes
B. Lindman

Gelation of steroid-bearing dextrans by surfactants

A. Lopes (✉)
Instituto de Tecnologia Química
e Biológica-ITQB
Universidade Nova de Lisboa
Apartado 127, 2781-901 Oeiras, Portugal
e-mail: alopes@itqb.unl.pt
Tel.: +351-21-4469725
Fax: +351-21-4411277

A. Lopes · B. Lindman
Physical Chemistry 1, Lund University
Lund, Sweden

Abstract Gelation of hydrophobically modified dextrans, prepared by covalent attachment of deoxycholic acid (DCA) to dextran (Dex), with several surfactant categories was investigated with viscosity and rheological techniques. Freshly prepared Dex–DCA solutions do not gel over a large range of polymer and surfactant concentration for any surfactant categories, anionic, cationic, nonionic, and zwitterionic, as indicated by only minor change in the solution's apparent viscosity (capillary viscometers). For 1% (w/w) Dex–DCA solutions at intermediate sodium deoxycholate (DCA) concentrations (0.008–0.03 M) a stiff macroscopic transparent gel is formed when the samples mature for more than 10 days at room temperature (independently of temperature and sonication during preparation). The time required for gel formation varies with the surfactant concentration and shows a minimum of 10 days around about 10 mM DCA and increases up to about 30 days near the gel phase boundaries. Continuous-flow measurements were carried out for the Dex–DCA/DCA gel mixture. The rheological measurements indicate a high yield stress but an easy breakdown of the gel into small particles when stressed. The extrapolated zero-shear viscosity shows a transition from a gel to a fluid solution at about 33 °C (heat scan) and an opposite gelation around 20 °C (cool scan). In these gelation–fluidization cycles the gel recovers practically its initial mechanic properties.

Key words Physical gel · Hydrophobically modified polymer · Dextran · Bile salts · Rheology

Introduction

Water-soluble polymers where hydrophobic or amphiphilic side groups such as alkyl, arylalkyl, and cycloalkyl [1–5] or steroids such as cholesterol [6–10] or bile acids [11, 12] are grafted, known as hydrophobically modified polymers (HMP), have been extensively studied for theoretical reasons and for their wide range of applications in medicine, biotechnology, and food, paint, or pharmaceutical formulation technologies [13–16]. When dissolved in water a HMP can self-associate due to intra- and/or intermolecular hydrophobic interactions between the hydrophobic chains [12, 13], and depending on the degree of substitution (DS) and molecular weight (M_w) the hydrophobe domains may interconnect the polymer chains, which leads to an increase in the viscosity of the solution with polymer concentration. However, for HMP with two or more hydrophobes per polymer in the semidilute regime, the addition of a surfactant, in general, changes the viscosity of the solution dramatically, with a maximum as a function of the surfactant concentration [17–23]. This behavior is explained by considering an efficient transitory network development congruent with the mixed micelle formation between the hydrophobic grafts of the polymer and the surfactant molecules. The formation of such mixed aggregates favors, relative to the absence of surfactant case, the interpolymer chain cross-linking; hence, the steep viscosity enhancement around the surfactant critical micelle concentration (cmc). However, when the molar ratio between the hydrophobic grafts and the mixed micelles in which they participate becomes small (≤ 1), the system looses the connectivity; hence, the viscosity decreases.

In some cases the transient network formed may lead to the formation of gels with special macroscopic mechanical properties. Several of these gels have been studied using hydrophobized glucopyranosidic polymers such as cellulose [18, 22, 24–27], pullulans [6–10], and chitosan [28] or synthetic polymers such as poly(sodium acrylate) [19, 20] or poly(acrylamide) [21, 29]. In this work a glucopyranosidic polymer – dextran (Dex) with $M_w = 30$–500 kDa – is used as a backbone and deoxycholic acid (DCA) is used as the hydrophobizer with a DS of 3 mol% (Scheme 1). These HMPs were recently fully characterized with fluorescence and light scattering [12] and spherical monodisperse nanosized particles of 2–3 physically cross-linked polymer units per aggregate were found above 0.2% (w/w), with structural hydrophobic domains of four steroid units per aggregate, which roughly corresponds to the aggregation number (N_{agg}) of DCA in water [30]. Unlike the other types of HMP mentioned earlier, the phase behavior and the rheology of these HMP in the presence of anionic alkylsulfates and bile salts, cationic alkyl trimethylammonium bromide, nonionic poly(ethyleneglycol monoalkyl ether) and alkyl

glucopyranoside, and zwitterionic egg lecithin surfactants seems quite uninteresting as indicated by the minor increase in the solution's apparent viscosity with surfactant concentration. However, when the samples are left to mature at room temperature a macroscopic gel develops in an intermediate concentration range for DCA. The parameters which favor the process of gelation are analyzed here with particular emphasis on the rheological and thermotropic properties of the gels.

Experimental

Materials

The Dex–DCA used in this study has the same characteristics as the hydrophobically modified Dexs used in previous work with different DS [12]. Dex with weight-average molecular weights ranging from 30 to 500 kD reacts with DCA in the presence of N,N-dicyclohexylcarbodiimide as a coupling agent and 4-(N,N-dimethyl)aminopyridine as a catalyst. A DS of 3 mol% (moles of DCA bound/100 glucopyranosidic structural units) was used. Sodium octyl sulfate (purity above 98%) from Merck, sodium dodecyl sulfate ($SC_{12}S$) (specially pure) from BDH, sodium hexadecyl sulfate (purity above 98%) from Merck, sodium cholate (CA) and sodium deoxycholate (DCA) from Merck (purity above 99%), sodium taurocholate (TCA), sodium taurodeoxycholate (TDCA), and sodium glycocholate (GCA) from Sigma (approximately 98% pure), pentaethyleneglycol monododecyl ether and decaethyleneglycol monododecyl ether (from Nikko Chemicals (purity above 99%), dodeyltrimethylammonium bromide and hexadecyltrimethylammonium bromide ($C_{16}TAB$) from TCI Tokyo Kasei (purity above 99%), n-octyl-α-D-glucopyranoside from Sigma (purity above 98%), and egg L-α-phosphatidyl choline (eLec) from Sigma (purity above 85%) were used as supplied. Water was purified using a Milli-Q system from Millipore.

Methods

Sample preparation

Solutions were made by dilution of a 4% (w/w) polymer stock solution in water and adding the required amount of surfactant from a surfactant stock solution of 0.15 M (except for $C_{16}TAB$, where the stock solution was 0.02 M). In the case of eLec the surfactant was directly dispersed into the polymer aqueous solution. In order to ascertain the effect of sonication and temperature during preparation on the sample properties some samples were bath-sonicated for 10–15 min and/or heated at temperatures ranging from room temperature up to 70 °C for up to 2 h. All the samples were stirred with a magnetic stirrer and kept at 25 °C. For long-time studies platforms with a smooth rocking movement were used to continuously homogenize the samples. The samples were inspected visually on a daily basis to study the kinetics of the gel formation and once the gel formation started the stirrer/platform was turned off.

Viscosity/rheology measurements

The relative viscosity of the solutions was measured using capillary viscometers (Schott Geräte) with the appropriate capillary immersed in a controlled temperature bath (± 0.1 °C). For the gelled samples, rheological measurements were performed in the continuous-flow mode using a Carri-Med CSL100 rheometer equipped

Degree of substitution,

$$DS = \frac{m}{m+n} \, 100 \, mol\%$$

with automatic gap setting and a broad variety of cone and plate geometries. The temperature in all runs was kept within ± 0.1 °C of the desired value. Owing to the breakdown of the gel when stress was applied all measurements on gelled samples were made on fresh samples, one for each temperature.

In the fluidization/gelation cycles each sample was left to equilibrate for 1 h at the selected temperature to allow equilibration before stress was applied (longer periods of time did not result in any change in the measured parameters).

Results and discussion

The relative viscosity of the 1% Dex–DCA (30 kD) solution when freshly mixed with several surfactant categories is shown in Fig. 1 as a function of surfactant concentration. It can be noticed that for all surfactant categories – anionic (alkyl, steroid), cationic, nonionic, zwitterionic – there are only minor changes in the apparent viscosity up to a surfactant concentration greater than 5 times the cmc (similar results were obtained for all the values of M_w up to 500 kD). This is a different situation from that observed with other hydrophobically modified glucopyranosidic-based polymers, such as cellulose [18, 22, 24–27] or pullulans [6–10], and deserves some discussion. In the general case a HMP/surfactant mixture dramatically changes its viscosity as a function of the surfactant concentration around its cmc [17–23, 26, 27]. This behavior is explained

by considering an efficient transitory network which appears with the formation of mixed aggregates between the hydrophobic grafts of the polymer and the surfactant molecules. The formation of such mixed aggregate favors interpolymer chain cross-linking; hence, the steep viscosity enhancement around the surfactant cmc. However, when the molar ratio between the hydrophobic tails and the mixed micelles in which they participate becomes small, we can conceive the grafts as being "solubilized" individually by the increasing number of micelles, and consequently the connectivity of the system decreases, hence the lowering of the viscosity. This picture seems to provide a description of the system when both the grafts and the hydrophobic tails of the surfactant are alkyl or arylalkyl groups, which leads to more or less compact spherical micellar aggregates with tens of hydrophobes per aggregate. In the case of a different nature of the two hydrophobe constituents the change in the viscosity even if noticeable is not dramatic anymore: for example, in the case of hydrophobically modified pullulan where cholesterol was grafted with a DS = 1.5% [10] the low solubility of cholesterol in $SC_{12}S$ micelles shifts the maximum of the curve from around the $SC_{12}S$ cmc to a value 6 times higher, and the viscosity enhancement is only one order of magnitude.

In the present case the hydrophobe grafts on the polymer are also of steroid nature, from which we can anticipate a not so large compatibility with the alkyl-type surfactants. In spite of that, the total absence of a perceptible enhancement of the viscosity up to 5 times the cmc for the different alkyl surfactants was not expected, for example, there is extensive miscibility of $SC_{12}S$ with DCA surfactants although the compactness of the resulting aggregates is highly dependent on the molar ratio between the constituents. Nevertheless, for the case of steroidlike surfactants, CA, DCA, TCA, TDCA and GCA, because they form small micellar aggregates in water with $N_{agg} = 4$–20 [30] the expected connectivity when in the presence of Dex–DCA should not be very high because the simultaneous occurrence of two grafts of different polymer structural units in the same micellar aggregate composed of only two to three more surfactant monomers should be ephemeral. This fact could explain the almost constancy of the viscosity for these mixtures (for the sake of clarity only CA and DCA are included in Fig. 1 but the viscosity values are comparable for the other members of the steroid family). In the case of the Dex–DCA/eLec mixture the phase separation that occurs around 0.75% (w/w) eLec does not allow us to draw a conclusion on the enhancement of the viscosity for the lowest surfactant concentrations. Nevertheless it is worth noting that at the phase boundary the ratio between the hydrophobe graft and eLec surfactant is similar to previously published values for the phase boundaries of free bile salt/eLec mixtures [31], an indication that for this kind of backbone (Dex) the free/grafted state of the

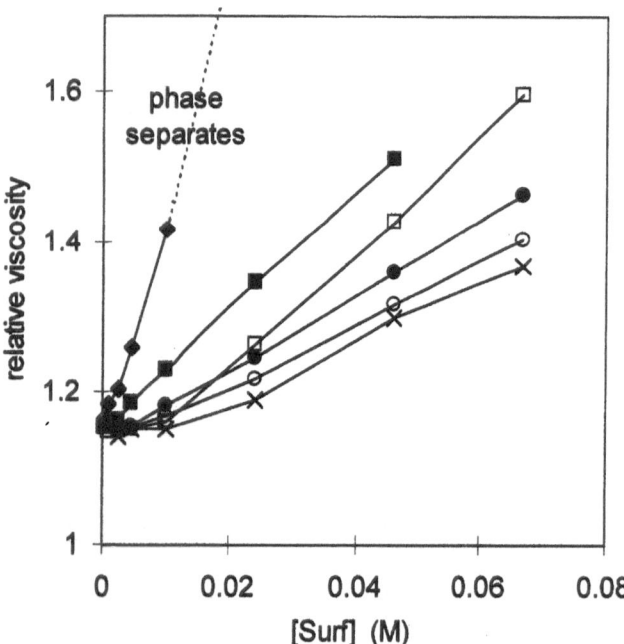

Fig. 1 Relative viscosity of the 1% dextran (*Dex*)–deoxycholic acid (*DCA*) (30 kD) solution, when freshly mixed with surfactant, as a function of surfactant concentration for several surfactant categories: sodium dodecyl sulfate (□), hexadecyltrimethylammonium bromide (■), sodium cholate (○), DCA (●), egg L-α-phosphatidyl choline (◆), *n*-octyl-α-D-glucopyranoside (×). The *lines* are only guides for the eye and do not represent a fit to any model

steroid hydrophobe seems to be of minor importance for the phase behavior of the system.

If the samples are left to equilibrate at room temperature under stirring all remain as fluid solutions (viscosity does not change) except the Dex–DCA/DCA system, where a stiff physical (macroscopic) gel is formed at intermediate DCA concentrations as depicted in Fig. 2a. Moreover, the gel is transparent over almost all the surfactant concentration range (a peculiarity with maybe some technological interest). The gelation process is only observed for the narrow range of polymer concentrations between 0.6 and 1.5% (w/w) for the same DCA concentration range (data not shown). This gelation process (as well as the gel properties) is independent of the sample preparation for temperatures – up to 70 °C for up to 2 h – and sonication – up to 10–15 min. The time required for gel formation varies with the surfactant concentration (Fig. 2b): it shows a minimum of 10 days around about 10 mM DCA and increases up to about 30 days near the gel phase boundaries. Both these results can be related with the general trend of apparent viscosity versus surfactant concentration in the gel-forming systems discussed in the Introduction. From Fig. 2a we can infer that the viscosity goes trough a maximum around the surfactant cmc – the stiff macroscopic gel phase whose apparent viscosity we will analyze in more detail. The results depicted in Fig. 2b can be viewed as the inverse curve of the viscosity versus surfactant concentration mentioned earlier, which seems to indicate how the two parameters are related – the first gel to appear corresponds to the peak in the viscosity.

The apparent viscosity versus shear rate is presented in Fig. 3a as a function of the temperature for the system 1% Dex–DCA/DCA (0.02 M). As can be noticed, the apparent viscosity decreases with shear up to 30 °C and presents a more Newtonian behavior above 40 °C.

Moreover, the apparent viscosity is very high for the low shear values up to 30 °C, which is consistent with the fact that the gel is stiff at these temperatures. Nevertheless, while the apparent viscosity is highly dependent on the shear the gel does not show thixotropy, which is related to the fact that there is gel breakdown into small particles (observable with a transparent geometry in the rheometer) when stressed.

On heating, a gel-to-fluid transition is observed around 33–34 °C and is accompanied by an abrupt reduction in the extrapolated zero-shear viscosity as a function of temperature (heat scan) (Fig. 3b). Conversely, if this fluidized solution is cooled gelation is observed around 20–21 °C, where the gel practically recovers its initial mechanic properties (also depicted in Fig. 3b). This hysteresis is not significantly reduced if the samples are analyzed after a longer equilibration time. Moreover, if a gel destroyed by shearing is heated and allowed to cool it behaves in the same way as if no destruction had occurred.

However, if about 0.02 M DCA is added to the fluidized gel ($T > 40$ °C) the gel phase is not recovered when the samples are cooled. Furthermore, this effect is also observed for the other surfactant categories. This can be explained as the addition of the excess surfactant brings the system beyond the gel phase boundary (Fig. 2a).

Conclusions

Freshly prepared mixtures of semidilute Dex–DCA with several surfactant categories do not gel at intermediate surfactant concentrations – a situation different from what is observed with other hydrophobically modified polysaccharides, such as pullulans, hydrophobized with similar steroids. However, for a large range of molecular

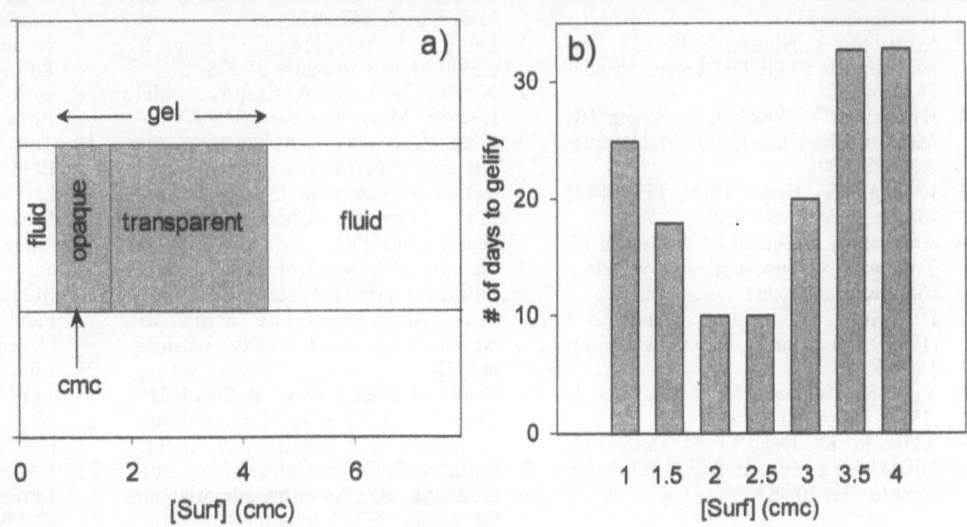

Fig. 2 a Phase behavior of the system Dex–DCA 1% (w/w) as a function of DCA concentration (in units of critical micelle concentration, *cmc*) when the samples equilibrate for long periods at room temperature. **b** Kinetics of gel formation for the system 1% Dex–DCA ($M_w = 30$ kD) (number of days to start to develop a gel) as a function of DCA concentration (in units of cmc)

Fig. 3 a Viscosity versus shear rate for the 1% Dex–DCA/ DCA (0.02 M) gel system as a function of temperature obtained from continuous-flow experiments: −5 °C (□), −10 °C (◇), −20 °C (○), −30 °C (△), −40 °C (×). **b** Fluidization–gelation cycles of the system 1% Dex–DCA/DCA (0.02 M): heat scan (◆), cool scan (■), cool scan when 0.02 M DCA is added to the fluidized gel ($T > 40$ °C) (●). The lines are only guides for the eye and do not represent a fit to any model

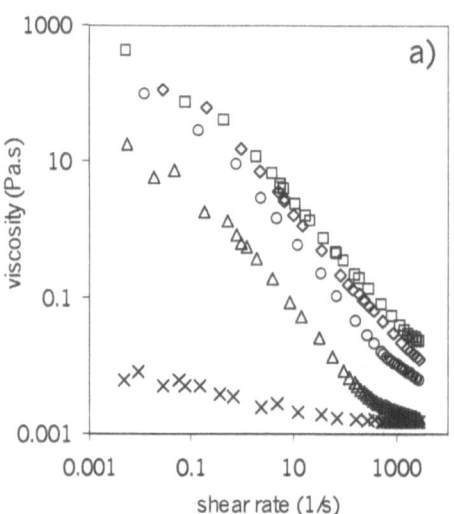

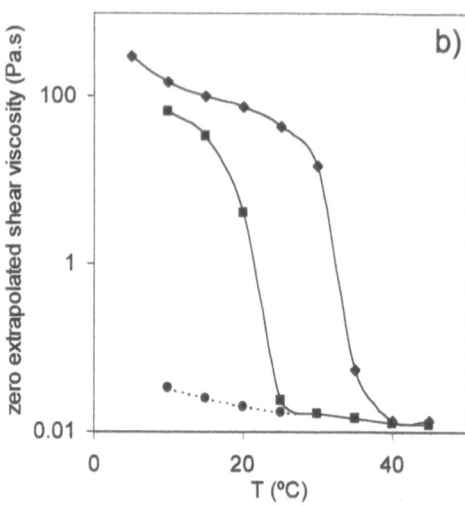

weights (30–500 kD), semidilute Dex–DCA mixed with DCA as a surfactant forms a stiff, but soft macroscopic, transparent gel when samples are left for more than 10 days at room temperature. This gelation process is independent of the sample preparation temperature and of sample sonication. The kinetics of the gel formation as a function of the surfactant concentration shows a minimum of 10 days around about 10 mM DCA and increases up to about 30 days near the gel phase boundaries.

Continuous-flow measurements were carried out for the Dex–DCA/DCA gel mixture. The rheological measurements indicate a high yield stress but an easy breakdown of the gel into small particles when stressed. The extrapolated zero-shear viscosity shows a transition from gel to fluid solution at about 33 °C (heat scan) and an opposite gelation around 20 °C (cool scan). In these gelation–fluidization cycles the gel recovers practically its initial mechanic properties; however, gelation is suppressed if a surfactant is added to the fluidized gel ($T > 40$ °C).

Acknowledgement The authors are grateful to M. Tsianou, Physical Chemistry 1, Lund University, for many helpful discussions concerning the interpretation of rheological data. A.L. is indebted to Foundation for Science and Technology (FCT) of Portugal for the grant PRAXIS XXI/BPD/16373/98.

References

1. Winnik (1998) J Phys Chem 93:7452
2. Biggs S, Selb J, Candau F (1992) Langmuir 8:838
3. Morishima Y, Nomura S, Ikeda T, Seki M, Kamachi M (1995) Macromolecules 28:2874
4. Kramer MC, Welch CG, Steger JR, McCormick CL (1995) Macromolecules 28:5248
5. Hwang FS, Hogen-Esch TE (1995) Macromolecules 28:3328
6. Akiyoshi K, Deguchi S, Moriguchi N, Yamaguchi S, Sunamoto J (1993) Macromolecules 26:3062
7. Deguchi S, Akiyoshi K, Sunamoto J (1994) Macromol Rapid Commun 15:705
8. Yusa S, Kamachi M, Morishima Y (1998) Langmuir 14:6059
9. Akiyoshi K, Deguchi S, Tajima H, Nishikawa T, Sunamoto J (1997) Macromolecules 30:857
10. Deguchi S, Kuroda K, Akiyoshi K, Lindman B, Sunamoto J (1999) Colloids Surf A 147:203
11. Lee KY, Jo WH, Kwon IC, Jeong SY (1998) Macromolecules 31:378
12. Nichifor M, Lopes A, Carpov A, Melo E (1999) Macromolecules 32:7078
13. Glass JE (ed) (1989) Polymers in aqueous media: performance through association, Advances in Chemistry Series 223. American Chemical Society, Washington, DC
14. Shalaby SW, McCormick CL, Butler GB (eds) (1991) Water-soluble polymers, ACS Symposium Series 467. American Chemical Society, Washington, DC
15. Dubin PJ, Bock J, Davis R, Schulz DN, Thies C (eds) (1994) Macromolecular complexes in chemistry and biology. Springer, Berlin Heidelberg New York
16. Goddard ED, Ananthapadmanabham KP (eds) (1993) Interactions of surfactants with polymers and proteins. CRC, Boca Raton
17. Lindman B, Thalberg K (1993) In: Goddard ED, Ananthapadmanabham KP (eds) Interactions of surfactants with polymers and proteins. CRC, Boca Raton, pp 203–276
18. Tanaka R, Meadows J, Williams P, Phillips G (1992) Macromolecules 25:1304
19. Loyen K, Iliopoulos I, Olsson U, Audebert R (1995) Prog Colloid Polym Sci 98:42
20. Magny B, Iliopoulos I, Audebert R, Picullel L, Lindman B (1992) Prog Colloid Polym Sci 89:118
21. Effing J, McLennan I, van Os N, Kwak J (1994) J Phys Chem 98:12397
22. Nyström B, Thuresson K, Lindman B (1995) Langmuir 11:1994
23. Kästner U, Hoffman H, Dönges R, Ehrler R (1994) Colloids Surf A 82:279

24. Kästner U, Hoffman H, Dönges R, Ehrler R (1995) Prog Colloid Polym Sci 98:57
25. Goddard E, Leung P (1992) Colloids Surf 65:211
26. Guillemet F, Picullel L, Nilsson S, Djabourov M, Lindman B (1995) Prog Colloid Polym Sci 98:47
27. Carlsson A, Karlström G, Lindman B (1990) Colloids Surf 47:147
28. Kjøniksen A-L, Nyström B, Nakken T, Palmgren O, Tande T (1997) Polym Bull 38:71
29. Bigs S, Selb J, Candau F (1992) Langmuir 8:838
30. Small DM (1971) In: Nair PP, Kritchevsky D (eds) The bile acids. Chemistry, physiology, and metabolism, vol 1. Plenum, New York, pp 249–355
31. Söderman O, Nyden M, in preparation
32. Schurtenberger P, Lindman B (1985) Biochemistry 24:7161

Progr Colloid Polym Sci (2000) 116:48–56
© Springer-Verlag 2000

DISPERSE SYSTEMS

A. Ruplis

Sorption and catalytic properties of Latvian clay powders

A. Ruplis
Riga Technical University
Chemical Technology Faculty
14/20, Azenes Street
Riga, 1048, Latvia
e-mail: auruplis@acad.latnet.lv
Tel.: +371-708-9275

Abstract The characteristics of the most important Latvian clay deposits are considered. The locations of the deposits, the thickness of the cover and the industrial bed, demonstrated and inferred reserves of clays as well as the content of their main fractions and chemical composition are reported. A survey on the surface properties and sorption phenomena of Latvian clay is given. Adsorption and desorption isotherms of nitrogen, argon (at −195 °C), water, methanol, benzene, n-hexane and carbon tetrachloride (at ambient temperature) have been measured. It was established that the values of the specific surface area were between 10 and 120 m^2/g. Two types of pore size distribution curves are given. The samples from Devonian and Quaternary deposits (main component illite) are characterised by pore size distribution curves without remarkable maxima. The samples from Triassic clay (main component smectite) ex-

hibit a characteristic shoulder at the closure of the hysteresis loop of the sorption isotherms. This corresponds to the maximum on the curve of pore size distribution which was due to the presence of slit-shaped pores. Increasing the treatment temperature decreases the magnitude of the specific surface area of the samples as well as their total pore volume. The treatment of natural samples with acids (sulphuric, hydrochloric, phosphoric) changes the specific surface area. The cation-exchange capacity was evaluated. The samples of Latvian clay might be successfully used for the bleaching of the local rapeseed oil. Recently, it was found that Latvian clay samples may be used as catalysts for 1,4-butanediol dehydration into tetrahydrofuran.

Key words Latvian clay deposits · Sorption isotherm · Carbon tetrachloride · Catalysis · 1,4-Butanediol

General part

Characteristics of Latvian clay deposits

Clay is one of the most widespread sedimentary rocks in the Earth's crust. Latvia is very rich in clays of different mineral composition [1–5]. Clay in Latvia is found both in separate beds of deposits and as a mixture together with other types of deposits.

Nowadays the clays are widely used in ceramics and building material production. The most important clay

deposits in Latvia occur as Upper Devonian and Quaternary [3]. The largest and practically most important clay deposits are the Burtnieki, Lode and Katleši formations. Reserves, grain size and chemical compositions of the clay deposits are shown in Tables 1 and 2.

Triassic clays have not been used yet; however, these will probably be good raw materials for the production of sorbents in the future. Different parameters of Triassic clays are shown in Tables 3 and 4.

Quaternary clay is very widely spread in Latvia and is extensively used. The characteristics of the most impor-

Table 1 Deposits of Devonian clay

Deposit	Location of civil parish	Thickness (m)		Reserves (10^6 m^3)		Content of main fractions (%)				Coarse inclusions (%) > 0.5 mm
		Cover	Industrial bed	Demonstrated	Inferred	Sand > 0.05 mm	Silt 0.05–0.005 mm	Clay < 0.005 mm	< 0.001 mm	
Burtnieks formation (D$_2$br)										
Tuja	Liepupe	1.6	3.9	0.86	0.90	35.2	34.2	30.6	17.1	1.8
Vitrupe	Vilkene	6.5	8.4	2.04	21.34	20.8	41.2	38.0	26.5	0.4
Austrumi	Ramata	0.4	1.8	–	0.10	27.0	31.0	42.0	30.1	1.1
Elkškene	Targale	1.9	6.3	–	3.90	24.6	41.9	33.5	19.6	1.8
Jesperi	Svetupe	1.6	7.1	–	0.18	31.6	32.4	36.0	23.2	3.0
Pale	Pale	1.6	4.2	0.63	–	22.7	36.9	40.4	29.0	1.2
Lodes formation (D$_3$ld)										
Liepa	Liepa	5.0	15.2	7.9	35.6	16.9	37.5	45.6	26.4	0.2
Red						18.6	38.9	42.5	24.5	0.2
Grey				0.5		6.0	28.5	65.5	38.6	0.2
Garšas	Raiskums	5.5	14.5	2.36	21.2	21.5	24.05	54.4	34.4	0.1
Cesu Gluda	Cesis Town	6.3	11.7	0.37	6.1	20.6	72.7	Not determ.	0.2	
Pavari	Kauguri	1.1	4.0	0.42	–	25.6	33.3	41.1	33.1	0.3
Turaida	Krimulda	3.0	8.9	0.24	0.12	26	34	40	30	0.4
Katlešu formation (D$_3$ktl)										
Kuprava	Susaji	2.6	12.0	14.99	1.90	8.6	29.8	61.6	53.9	0.2
Kastrane	Suntazi	1.6	4.2	0.70	–	27.7	31.1	41.2	34.6	0.2

tant and interesting deposits of the clays are given in Table 5.

Sorption studies of Latvian clay

The sorption properties of Latvian clay became the subject of notable research only after the renewal of Latvia's independence in the 1990s.

The main attention has been paid to the measurement of the adsorption and desorption isotherms of nitrogen, argon, n-hexane, carbon tetrachloride, benzene, methanol, and water vapours [6–17]. The recommendations of IUPAC's Commission on Colloid and Surface Chemistry (1985) have been followed concerning the methodology, experimental procedures, evaluation of adsorption data, determination of surface area and assessment of meso- and microporosity [18].

Table 2 Chemical composition of Devonian clay (w/w%)

Deposit	CO$_2$	SiO$_2$	Al$_2$O$_3$	Fe$_2$O$_3$	CaO	MgO	Na$_2$O	K$_2$O
Burtnieks formation (D$_2$br)								
Tuja	1.4	66.25	12.60	5.40	2.56	2.61	0.25	3.18
Vitrupe	2.4	67.50	12.30	4.76	1.97	2.83	0.18	3.60
Austrumi	0.1	65.81	12.72	8.81	0.81	2.51	Not determined	
Elkškene	1.2	68.83	13.06	6.48	2.41	2.33	Not determined	
Jesperi	2.5	68.02	11.16	6.10	1.94	2.99	Not determined	
Pale	0.7	67.62	13.56	6.76	0.73	2.10	Not determined	
Lodes formation (D$_3$ld)								
Liepa								
Red	< 0.1	68.41	14.01	5.55	0.91	1.41	0.29	3.48
Grey	< 0.1	68.53	16.57	4.94	0.95	1.65	0.25	3.79
Garšas	< 0.1	68.49	15.49	5.22	1.02	1.52	Not determined	
Cesu Gluda	0.7	58.20	20.45	6.38	1.23	2.15	Not determined	
Pavari	0.4	66.23	14.23	7.08	0.94	2.00	Not determined	
Turaida	0.1	72.17	12.46	5.37	0.35	1.56	Not determined	
Katlešu formation (D$_3$ktl)								
Kuprava	3.6	56.65	15.58	7.34	2.73	3.23	5.14	
Kastrane	9.7	52.18	11.56	5.15	7.64	6.69	Not determined	

Table 3 Deposits of Triassic clay

Deposit	Industrial bed (m)	Cover (m)	Reserves (10^6 m³)	Content of main fraction (%)			
				> 0.05 mm	0.05–0.005 mm	< 0.005 mm	< 0.001 mm
Pampali	3.6	7.5	22.7	41.0	34.2	24.8	8.3
Zana	11.0	15.0	33.0	11.4	44.0	44.6	19.7
Jaunauce	5.0	14.2	90.0	38.8	37.2	24.7	10.4
Vadakste	6.0	10.0	37.2	15.4	43.3	43.4	23.5

Table 4 Chemical composition of Triassic clay (w/w%)

Deposit	CO_2	SiO_2	Al_2O_3	Fe_2O_3	CaO	MgO	Na_2O	K_2O
Pampali	7.4	61.5	8.3	3.0	10.0	1.9	2.2	0.3
Zana	7.8	51.2	11.7	6.8	9.3	3.1	2.6	0.3
Jaunauce	8.5	54.9	9.3	4.3	10.4	3.1	2.6	0.4
Vadakste	9.7	59.7	10.7	5.3	8.2	3.0	2.2	0.6

After the results of different experiments had been gathered together, it was concluded that sorption isotherms may be divided in two groups depending on the forms of the isotherms and hysteresis loops (Figs. 1, 2). In this aspect the crystalline structure of clay minerals is important. The samples of nonswelling structures, such as illites or hydromices, are included in the first group. A narrow and smooth hysteresis loop was observed for the isotherms of this group. The samples with a smectite structure are included in the second group. The orientation of the hysteresis loop parallel to the pressure axis, and a shoulder on the desorption isotherm at $P/P_o \approx 0.3$ is significant for the isotherms of this group. Accordingly there are two types of pore size distribution curves. The samples from Devonian and from Quaternary deposits (main component illite) are characterised by a pore size distribution curve without a definite remarkable maximum. The samples from Triassic clay deposits (main component smectite) exhibit a maximum on the pore size distribution curve. The maximum corresponds to the characteristic shoulder at the closure of the hysteresis loop and is due to the presence of the slit-shaped pores.

The Brunauer–Emmett–Teller (BET) method was used to calculate the specific surface area. An overview of the magnitude of the specific surface area of Latvian clay powders is given in Table 6.

On the basis of many experiments it was concluded that the treatment with mineral acids is one of the most important ways to change the surface and porosity of clays.

In general, the following trends in specific surface area and pore structure changes with increasing acid concentration were established:

1. The values of the specific surface area pass through a maximum.
2. There is a continuous increase in the characteristic pore radius.
3. There is an increase in the total pore volume for the samples of smectite-type structure.
4. Initially the total pore volume increased, but then it decreased for the samples of illite-type structure.

Some clay samples (from deposits Liepas, Usmas and others) were found to give homogeneous adsorbents as the products of acidic treatment. The potential barriers between the adjacent adsorption positions on the surface of such adsorbents are low. They can be used in gas-adsorption chromatography as adsorbents or in gas–liquid chromatography as support for the nonvolatile phase.

The use of the clay samples studied for bleaching of vegetal oils may be considered as the most important contribution to the recent history of the utilisation of

Table 5 Deposits of quaternary clay

Deposit	Location of civil parish	Reserves (10^6 m³)	Content of main fraction (%)			Carbonates CO_2 (%)
			> 0.05 mm	0.05–0.005 mm	< 0.005 mm	
Apriki	Laza	5.1	2.4	11.7	85.9	4.5
Laza	Laza	0.93	5.8	17.9	76.3	5.2
Liberti	Cena	4.2 (10^6 t)	3.2	36.2	60.1	8.6
Nicgale	Nicgale	2.02	4.5	22.2	73.3	5.6
Priekule	Priekule	0.17	5.4	34.8	59.8	8.2
Rolava	Grobinas	0.35	15.2	42.4	42.4	7.5
Ugale	Ugale	0.48	1.6	9.7	88.7	6.3
Usma	Usma	5.52	8.2	22.3	69.5	9.1

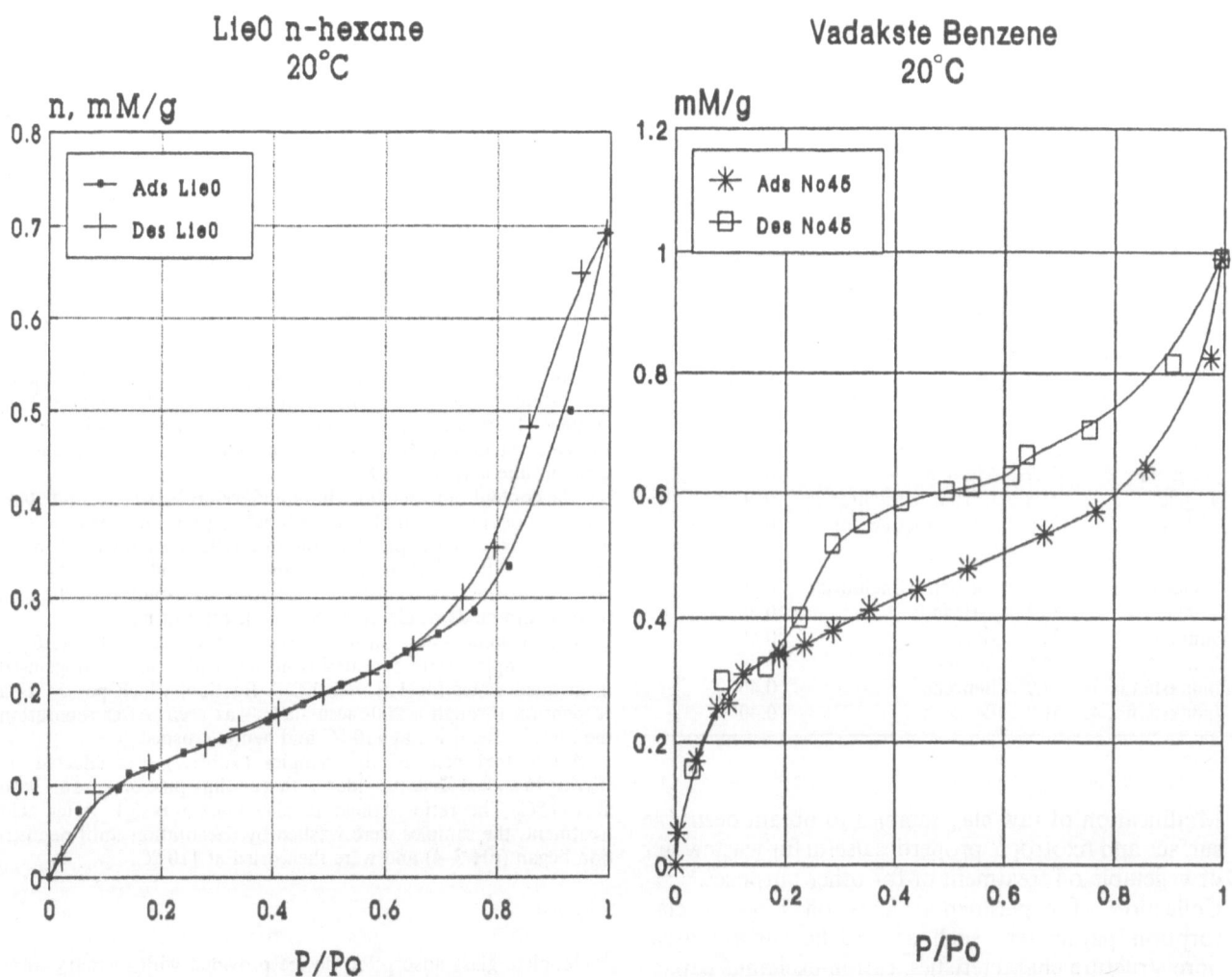

Fig. 1 Adsorption (*Ads*) and desorption (*Des*) isotherms of clay sample from the Liepa deposit (illite + kaolinite)

Fig. 2 Adsorption and desorption isotherms of clay sample from the Vadakste deposit (smectite + kaolinite + chlorite)

Latvian clays [19–22]. It was shown that the clays might be successfully used for bleaching of vegetable oils. New types of sorbents obtained as by-products of rapeseed oil bleaching may be of interest. It should be considered as an example of waste-free technology. After bleaching the used sorbents contained more than 40% rapeseed oil; this is very difficult to separate. The sorbents were thermally treated (coked) at several temperatures up to 600 °C in air. A series of new types of sorbents was obtained. The particles of the sorbents have two parts: a core and a surface layer. The core was aluminosilicate (matrix), which was covered with a layer of coke (similar to active charcoal). The new sorbents have a good capacity for organic compound adsorption from the gas phase and a high thermal resistance, which is important in regeneration for repeated use.

The research on the adsorption of cation-active dyes from water solution (methylene blue, methyl violet) made it possible to show that Latvian clays may be used as cheap and efficient sorbents in paper, textile and similar industries for wastewater treatment [23–27]. Great attention has been paid to the improvement of experimental measurements to obtain reproducible results. It was established that the time for attaining the adsorption equilibrium is sometimes longer then previously observed [22]. Measured adsorption isotherms were of the Langmuir type. The ion-exchange capacity was calculated using the values of maximal adsorption of dyes. It was established that the ion-exchange capacity depends on the concentration of the mineral acid (sulphuric, hydrochloric) used for activation. Increasing the acid concentration decreases the ion-exchange capacity.

Today the study of Latvian clay properties is in the following areas:

Table 6 Specific surface area, A, (m^2/g) of some Latvian clays

Deposit	A	Sorbat	Molecular cross-sectional area (nm^2)
Devonian			
Liepa red	35.2	Argon	0.166
Liepa grey	66.1	Argon	0.166
Liepa grey	55.6	Methanol	0.25
Liepa grey	45.0	Carbon tetrachloride	0.30
Kuprava	75.5	Argon	0.166
Kuprava	49.7	Carbon tetrachloride	0.30
Quaternary			
Usma	44.3	Argon	0.166
Usma	26.9	Carbon tetrachloride	0.30
Laza	31.2	Carbon tetrachloride	0.30
Rolava	20.7	Carbon tetrachloride	0.30
Ozolnieki	43.9	Methanol	0.25
Nicgale	23.8	Carbon tetrachloride	0.30
Priekule	10.0	Carbon tetrachloride	0.30
Triassic			
Akmens	37.0	Carbon tetrachloride	0.30
Akmens	33.2	n-Hexane	0.40
Pampali	58.6	Benzene	0.40
Zana	54.4	Benzene	0.40
Vadakste no. 14	72.0	Benzene	0.40
Vadakste no. 45	61.9	Benzene	0.40

1. Modification of raw clay samples to obtain desirable surface and tixotropic properties useful for wastewater or vegetable oil treatment or for other purposes.
2. Collection of experimental facts on Latvian clay sorption parameters such as specific surface area, pore structure characteristics, cation-exchange capacity, etc.

It seems that in the future Latvian clay used as a catalyst or support will find varied application in organic chemistry. The study of the sorption properties of recently found Latvian paligorskite and vermiculite could also be interesting.

Special part

Treatment of natural clay samples from the Kuprava deposit by phosphoric acid

Acidic treatment is one of the simplest ways to change the surface and sorption parameters of natural clay samples. The use of phosphoric acid was investigated.

Experimental

Materials

The Kuprava deposit is used industrially for production of bricks, ceramics, expanded clay globules and other materials. The industrial bed is composed of two or three red 2–12-m-thick clay layers separated by two 6-m-thick grey-green layers of sand or sand–clay. Clay samples are characterised by the prevailing (40–70%) clay fraction (particle size less than 0.005 mm). Usually the content of the silt fraction is not larger than 20–30%, and the sand fraction is 15%. Unlike other clay deposits in Latvia the clay fraction in clay samples from the Kuprava deposit are monomineralic and composed of illite. However, X-ray analysis shows that sometimes the distance between the reflecting planes is a little larger then 10 Å, which is characteristic of an illite structure. It was supposed that illite/smectite (montmorillonite) structure minerals represent the observed mixed layered minerals. These observations as well as a large fraction of fine particles present in samples indicate that it is advantageous to use the Kuprava clay as a sorbent. It is well known that smectites are better sorbents than illites. The content of kaolinite is not larger than 5%.

The chemical composition of Kuprava clay also has some interesting peculiarities. The samples exhibit the highest K_2O content (4–5% and more) found in Latvian clays. This infers that the clay fraction of these samples contain only illite, because in general illite is richest in K_2O.

The content of iron oxides in boundaries is between 2 and 10%, but for most of the samples the content of Fe_2O_3 ranges from 6 to 7%. For red clay samples half the iron oxides are included in the crystalline lattice, but the rest are present as oxide and hydroxide coatings of the clay particle. The uniform distribution of the iron oxide compounds results in a red colour after firing.

The carbonates content in Kuprava clay samples is low (3.6%).

The samples were separated from impurities and homogenised in distilled water for 1–2 months. A fraction, which passed as a suspension through a 200-mesh sieve, was used. After separation the sample was dried at 110 °C and gently crushed.

Acidic treatment of the samples (Table 7) was effected by stirring in a glass flask provided with a cooling pipe on a water bath at 100 °C. The ratio of acid to clay mass was 5:1. After acid treatment, the samples were washed by decantation until peptisation began (pH 3–4) and were then dried at 110 °C.

Sorption

A complete glass adsorption device provided with mercury traps and a McBain–Bakr quartz spring balance was employed [28]. The sensitivity of the quartz springs was about 2–3 mg/mm. A Katetometer KM-2 (former USSR) was used to measure the elongation and the shortening of the quartz spring during the adsorption and desorption of carbon tetrachloride vapour and to measure the mercury level of the manometer. A U-tube filled with Au powder separated the samples from other part of the device to eliminate the sorption of mercury vapour. Measurements of vapour sorption were made at 23.7 ± 0.1 °C.

The samples were outgassed at temperatures not higher than 70 °C to constant weight at a pressure 1×10^{-3} mmHg.

Sorption equilibrium was quickly attained. Sorption measurements were made 1 h after increasing or decreasing the vapour pressure in the region of low and intermediate relative pressure, but in the region near saturated pressure the measurements were made after 3–4 h.

Table 7 Acidic treatment conditions of Kuprava clay samples

Concentration of phosphoric acid (w/w%)	0	3	10	25
Temperature (°C)	100	100	100	100
Duration (h)	6	6	6	6
Samples	K0	K3P	K10P	K25P

Table 8 X-ray analysis data of Kuprava clay samples

Sample	Reflection angle (θ)	Distance between the planes (D_x)	Mineral
K0	4.48	9.911	Illite
	6.25	7.070	Kaolinite
	13.44	3.3222	Illite
K3P	4.55	9.718	Illite
	6.25	7.081	Kaolinite
	13.55	3.2957	Illite
K10P	4.55	9.718	Illite
	6.24	7.092	Kaolinite
	13.4	3.3319	Illite
K25P	4.24	9.415	Illite
	6.24	7.092	Kaolinite
	12.63	3.5314	Undefined
	13.4	3.3319	Illite

X-ray analysis

X-ray analysis was performed using the DRON-3.0 X-ray diffractometer (made in the USSR) with Cu Kα radiation, a voltage of 18 kV, and a current of 8 mA (Table 8).

It was estimated that all the samples studied were composed of illite (dominant mineral) and kaolinite (5%).

Differential thermal analysis and thermogravimetric analysis

Thermogravimetric analysis was performed using equipment made in Hungary (F. Paulic, J. Paulic, L. Erdey). The temperature interval was 20–1000 °C; the rate of heating was 10 °C/min. The differential thermal analysis and the thermogravimetric analysis data indicated the presence of illite in the samples.

Results and discussion

The sorption isotherms of carbon tetrachloride vapour are shown in Fig. 3. The isotherms demonstrate the influence of the acid concentration used for activation. The shape of the adsorption isotherms may be described as -S type (type II according to the IUPAC classification). The isotherms are convenient for the calculation of the specific surface area following BET theory. The isotherms exhibit a closed reproducible hysteresis loop located at a relative equilibrium pressure $P/P_o = 0.2$–0.99.

The BET parameters for the isotherms are listed in Table 10. The monolayer capacity, n_m, increases as the concentration of the acid used for activation increases. The value of the specific surface increases as the acid concentration increases. The values of the specific surface

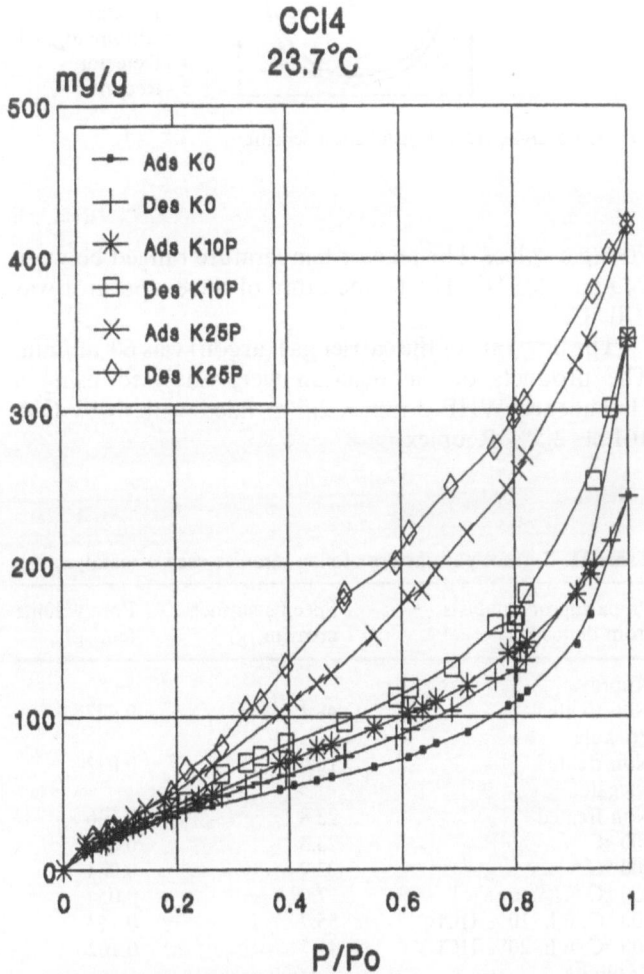

Fig. 3 Adsorption and desorption isotherms of clay samples from the Kuprava deposit (illite)

Table 9 Conversion of 1,4-butanediol on clay samples from the Nicgale deposit

Temperature of reaction (°C)	Conversion of 1,4-butanediol (%)	Selectivity of tetrahydrofuran formation (%)
	Not treated	
245	89	97
275	92	93
305	96	97
	Thermal treatment 7 h at 750 °C	
245	15	80
275	13	77
305	20	75
	Treated with hydrochloric acid	
245	100	99
	Thermal treatment 7 h at 750 °C + treated with hydrochloric acid	
245	56	77
275	57	95
305	68	93

Table 10 Brunauer–Emmett–Teller equation parameters and total pore volume

Samples	n_m (mg/g)	C	Range of linearity	A (m²/g)	V mg/g	cm³/g
K0	39.3	9.7	0.038–0.262	46.1	200	0.126
K3P	40.7	8.1	0.038–0.262	47.7	220	0.138
K10P	49.4	7.9	0.038–0.262	57.9	320	0.201
K25P	76.9	4.3	0.038–0.262	90.2	405	0.260

area were calculated using the value of n_m for carbon tetrachloride, assuming the molecular cross-sectional area to be 0.30 nm² [28]. The results of the pore structure analysis indicate that the pore size distribution is uniform. The pore size distribution does not exhibit a maximum. The total pore volume was estimated as the adsorption capacity of the samples at $P/P_o = 0.95$. The total pore volume increases as the concentration of the acid increases (Table 10). The results obtained testify that phosphoric acid is a very powerful agent capable of changing the surface parameters of clay similar to the influence of sulphuric and hydrochloric acids. The small values of the BET constant, C, indicate that the samples have negligible microporosity.

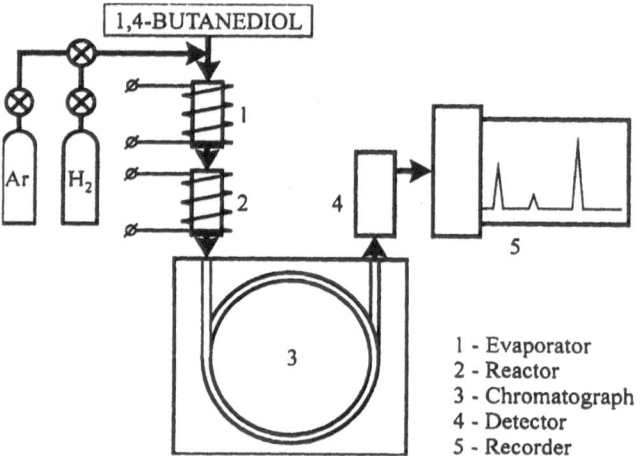

Fig. 4 Catalytic reactor installation scheme

1 - Evaporator
2 - Reactor
3 - Chromatograph
4 - Detector
5 - Recorder

Catalytic activity of Latvian clays [17]

Clay as a catalyst and as a support is widely used for organic reactions [29]. The catalytic activity of Latvian clay samples was studied. Clay from the Kuprava, Priekule, and Nicgale deposits and kaolinite from the Prosjanovo (Russia) deposit were used for 1,4-butanediol dehydration. The target compound was tetrahydrofuran.

Experimental

Preparing of catalysts

The natural clay samples were dried at 105–110 °C to constant weight, crushed in a mortar, and sieved through sieve no. 065. The prepared material was treated thermally or was activated by hydrochloric acid. The ratio of acid to clay mass was 7:1. Water was added to the powder obtained (1 ml/g). The samples were dried and crushed. The fraction of 0.25–1.00 mm was used.

Carbon tetrachloride sorption isotherms were measured. The samples were kept in carbon tetrachloride vapour and than vacuated. The reversion to the initial weight indicated that physical adsorption has occurred.

Study of the catalytic properties

An impulse microcatalytic device was used (Fig. 4). The volume of the reactor was 0.32 ml, corresponding to 60–

70 mg catalyst. The reactor temperature ranged between 240 and 305 °C. The temperature of the evaporator was 270 °C.

The flow rate of the carrier gas (argon) was 60 ml/min. The products of the reaction were analysed using a Hromosorb WHP 3 mm × 2.5 m filled with 10% OV-101 + 2.5% Reoplex.

Table 11 Surface parameters of clay samples used as catalysts

Preparing of catalysts from deposit	Specific surface area (m²/g)	Pore volume (cm³/g)
Kuprava		
Non treated	48.1	0.472
Priekule		
Non treated	10.0	0.072
Nicgale		
Non treated	23.8	0.126
300 °C	23.3	0.072
500 °C	22.7	0.084
750 °C	7.0	0.054
100 °C, 6 h, 10% HCl	55.7	0.153
100 °C, 6 h, 20% HCl	41.7	0.162
Kaolinite		
Non treated	13.7	0.062
750 °C + 1 h 18% HCl	12.1	0.092

Results and discussion

Some results obtained are shown in Table 9.

The surface parameters of the clay samples used as catalysts are given in Table 11.

In general, the following observations may be made. The clays and the kaolinite show high activity in the reaction studied. The preheating at 750 °C decreases the clay activity but increases the activity of kaolinite. Treatment with 18% HCl increases the activity of all the catalysts. It was found that the catalytic activity of the catalysts depends on their specific surface area.

Conclusions

1. The characteristics of the most important Latvian clay deposits have been considered. The locations of the deposit, the thickness of the cover and the industrial bed, demonstrated and inferred reserves as well as content of their main fractions and chemical composition have been reported.
2. The investigations in field of Latvian clay surface properties and sorption phenomena have been given.
3. The influence of acid treatment on the sorption characteristics (specific surface area, total pore volume) of samples from the Kuprava deposit has been reported.
4. It is shown that Latvian clays can be successfully used as catalysts for the conversion of 1,4-butanediol in tetrahydrofuran.

Acknowledgements I am grateful for the technical assistance and helpful discussion of my colleagues and collaborators A. Stinkule, G. Stinkulis, V. Lakevichs, L. Leite, A. Lebedevs, R. Bümans and M. Veidis.

References

1. Stinkule A (1999) In: Lectures on the use of Latvian mineral raw materials in national economics. Theses. Latvian Council of Sciences, Riga, pp 11–12 (in Latvian)
2. Grosvalds I (1970) Treasures of the Earth's crust in Latvia. Zinatne, Riga, p 72 (in Latvian)
3. Kuršs V, Stinkule A (1997) Mineral deposits of Latvia. University of Latvia, Riga (in Latvian)
4. Kuršs V, Stinkule A (1972) Clay in the Earth's crust and industry of Latvia. Liesma, Riga (in Latvian)
5. Birger AJ, Karpov VN, et al (1977) Mineral sources of Latvia for production of building materials. Zinatne, Riga (in Russian)
6. Ruplis A, Ramans A, Eiduks J (1973) In: Lukjanova T (ed) Studies in chemistry and chemical technology, Riga Polytechnic Institute, Riga, p 67 (in Russian)
7. Ruplis A, Višs R (1991) In: Congress of Latvian Scientists. Theses. Section of Chemistry, Riga, p 130 (in Latvian)
8. Ruplis A, Bumans R (1993) In: IUPAC Symposium on the Characterization of Porous Solids, COPS III, Marseille, France, May 1993, Book of Abstracts. IUPAC, p 47
9. Ruplis A, Bumans R (1993) Adsorption Sci Technol 10:137
10. Ruplis A, Bumans R, Saveljeva I, Tjumina (1994) In: Scientific Meeting on the Problems of Latvian Environment Protection, Riga, 20–21 September 1994. Summaries reproduced from manuscripts submitted by authors, Riga Technical University, Riga, p 61
11. Ruplis A, Ramans A (1994) Latv J Chem 3:286 (in Russian)
12. Ruplis A, Bumans R, Martcin I, Tjumina A, Višs R (1995) Latv J Chem 5/6:36
13. Bumans R, Ruplis A, Tjumina A, Višs R (1995) In: Braga I, Cavallini S, Di Cesare GF (eds) Fourth Euro Ceramics Society Symposium (ECERS IV). Proceedings, vol 12. Bricks and roofing tiles. Gruppo Editoriale Faenca Editrice, Italy, pp 139–146
14. Ruplis A, Mezinskis G, Chaghuri M (1998) International Conference Eco-Balt'98, Riga, 22–23 May 1998. Summaries reproduced from manuscripts submitted by authors. Riga Technical University, Riga, p54
15. Ruplis A (1998) 13th International Congress of Chemical and Process Engineering CHISA'98, 23–28 August 1998, Prague, Czech Republic, Summaries 4. Summaries and full texts of papers reproduced from manuscripts submitted by authors. Novosad, Prague, p 180
16. Lakevich V, Ruplis A (1999) In: ECerS Topical Meeting Sedimentary Rocks in Ceramics Technology, April 29–30, Riga, Latvia. Riga Technical University, Riga, p 30
17. Lebedevs A, Lakevichs V, Leite L, Ruplis A (1999) In: Beresnevièius Z (ed) International Conference on Organic Chemistry. Kaunas Technical University, Kaunas, pp128–132 (in Russian)
18. Sing KSW, Everett DH, Haul RA, Moscou L, Pierotti RA, Rouquerol J, Siemieniewska T (1985) Pure Appl Chem 57:603
19. Serzane R, Gudriniece E, Šantare D, Strele M, Ruplis A, Kalniņš R (1993) Latv J Chem 6:738 (in Latvian)
20. Ruplis A, Gudriniece E, Bumans R, Serzane R, Strele M, Ramans A, Saveljeva I (1994) Latv J Chem 4:497
21. Serzane R, Strele M, Ruplis A, Gudriniece E (1995) Latv J Chem 5/6:111 (in Latvian)
22. Ruplis A, Serzane R, Strele M, Gudriniece E (1998) In: International Conference on Organic Chemistry. Kaunas Technical University, Kaunas, pp 84–85 (in Russian)
23. Švinka R, Švinka V, Petersone E (1994) Latv J Chem 3:280 (in Latvian)
24. Ruplis A, Saveljeva I, Buholca L, Kugure K, Taurina D (1995) In: Colloquium on Problems of Environment Pollution. Riga, 22 September. Summaries reproduced from manuscripts submitted by authors. Riga Technical University, Riga, p 41 (in Latvian)
25. Ruplis A, Saveljeva I, Tjumina A (1995) In: Colloquia on Problems of Environment Pollution, Riga, 22 September. Riga Technical University, Riga, p 43 (in Latvian)
26. Ruplis A, Saveljeva I, Denisenko D (1997) In: Bortzmeyer D, Boussuge M, Chartier Th, Fantozzi G, Lozes G, Rousset A (eds) Fifth Euro Ceramics

Society Symposium (ECERS V). Proceedings. Key Engineering Materials, vols 132–136, Part I (1997). Trans Tech Publications, Switzerland, pp 256–259

27. Lakevich V, Ruplis A (1999) In: International conference EcoBalt'99, Riga, 14–15 May 1999. Summaries reproduced from manuscripts submitted by authors. Riga Technical university, Riga, pp 78–79

28. Lepin l, Ruplis A, Zemturis M (1968) Latv PSR Zin Akad Vestis Kim Ser 525 (in Russian)

29. Balogh M, Laszlo P (1993) Organic chemistry using clays. Springer, Berlin Heidelberg New York

Progr Colloid Polym Sci (2000) 116:57–66
© Springer-Verlag 2000

J. Kekkonen
P. Stenius

Deposition of wood resin emulsion droplets on ceramic oxides

J. Kekkonen · P. Stenius (✉)
Laboratory of Forest Products Chemistry
Helsinki University of Technology
P.O. Box 6300, 02015 HUT, Finland

Abstract The kinetics of deposition of wood resin emulsion droplets from aqueous dispersions onto Al_2O_3, ZrO_2, Cr_2O_3 and TiO_2 surfaces in stagnation point flow was studied using reflectometry. A semiquantitative measure for the amount deposited was obtained from theoretical mass transfer to the collector surfaces. Parameters investigated were electrolyte concentration and valence and pH (4.5 and 6.5). Electron spectroscopy for chemical analysis experiments showed that all oxide surfaces contained a very thin layer of hydrophobic impurities that covered a fraction of the surfaces. Taking the different isoelectric points into account, no qualitative differences between the oxides were observed. It is suggested that surface contamination affects the deposition process but its importance depends on the type of electrolyte used. For stable dispersions the initial rate of deposition was rate-determining, the rate remaining constant up to 50% or more of total deposition, depending on the electrolyte. The initial rate was a function of electrolyte concentration and decreased when the dispersion was made unstable by addition of electrolyte. Deposition could not be described in terms of the model for diffusion-controlled particle deposition developed by Dabros and van de Ven. The final amount deposited was a function of emulsion concentration, which is interpreted as a surface kinetic effect resulting from re-formation of droplets upon deposition.

Key words Wood pitch emulsion ·
Optical reflectometry ·
Ceramic oxides · Deposition
kinetics

Introduction

Deposition of particles on solid surfaces has been studied extensively, and several theories describing deposition have been developed [1–14]. However, when it comes to deposition of emulsions in general, and especially deposition kinetics, the picture is far more obscure. The only investigations of deposition kinetics that we know of are ellipsometric studies of fat emulsions [15]. Hence, this study was designed to compare the deposition of emulsion droplets from a dilute dispersion on solid surfaces with descriptions of particle deposition found in the literature.

The system chosen is of considerable practical importance since it is well known that wood pitch (resin) emulsions occurring in paper machine process waters cause troublesome deposits on process equipment [16, 17]. Some studies of the deposition tendency of wood resin have been reported [18], but the kinetics of wood resin emulsion deposition on solid surfaces has not been studied extensively. In a previous article [19] we reported the effects of short-chain cationic polymers on resin deposition on silica. In this work these studies are extended to deposition of model wood resin emulsion on Al_2O_3, ZrO_2, Cr_2O_3, and TiO_2 surfaces. These oxide surfaces are believed to serve as adequate although simplified models of ceramic surfaces used in, for example, press roll covers in press section of paper machines.

Materials and methods

Reflectometry

The rate of deposition of emulsion droplets was studied by reflectometry. The instrument and its calibration for silica surfaces have been described previously [19]. The same experimental setup, including alignment of the optics, as used previously for silica was used for all the oxides. The instrument is essentially similar to that described by Dijt et al. [20]. The adhesion of any material with a refractive index different from that of the collector surface changes the polarisation, S_0, of the beam reflected from the collector surface by δS. The relative change, $\Delta S/S_0$, was monitored on-line by photodiodes coupled to a computer. The value of the signal was recorded every 2 s. Using a stagnation point flow cell [1] the hydrodynamic conditions could be controlled.

Interpretation of reflectometric signal

An approximate value of the amount deposited can be found by considering the mass transfer to the collector surface. In the pH range investigated in this study, resin droplets and collector surfaces are oppositely charged and, consequently no repulsive energetic barriers between droplet and collector will occur. For such surfaces the deposition rate can usually be predicted theoretically under well-defined hydrodynamic conditions [4].

Thus, a conversion factor relating $\Delta S/S_0$ to the amount adsorbed can be obtained by calculating the theoretical flux towards the collector [1]. This can be done by calculating the Péclet number (Pe) for the geometry of the stagnation point flow cell (the radius of the inlet tube was 1 mm, the ratio between the wall-to-collector distance and the inlet tube radius was 1.7, Reynolds number approximately 30).

$$Pe = \frac{2\alpha a^3}{D} \quad , \tag{1}$$

where α is the strength of the stagnation point flow, a is the droplet radius, and D is the diffusion coefficient. The values for α can be calculated from plotted values shown in Ref. [1]. The Sherwood number (Sh) is then calculated from the Péclet number such that

$$Sh = 0.616 Pe^{1/3} \quad . \tag{2}$$

The flux of droplets towards the surface can then be calculated from

$$j = \frac{Sh D c}{a} \quad , \tag{3}$$

where j is the droplet flux towards the collector, D is the diffusion coefficient, c is the number concentration of droplets, and a is the droplet radius.

For instance, for droplets with a diameter 205 nm the theoretical flux is 8.2 $\mu g/m^2s$, whereas the initial rate measured in $\Delta S/S_0$ was 0.00064 s^{-1}. This gives a conversion factor of 12.8. The quantitative results for droplets with diameter 205 nm are calculated on the basis of this factor and so the deposited amount, Γ (in milligrams per square meter), was obtained from

$$\Gamma = 12.8 \frac{\Delta S}{S_0} \quad . \tag{4}$$

note that the mass transfer also depends on the droplet size distribution. The conversion factor of 12.8 is thus a mean value and

can be considered to yield only an approximate value of the amount adsorbed. Nevertheless, it should be useful for indicating the order of magnitude for the final amount deposited in various experiments. Two examples of the conversion factors determined for experiments carried out with Al_2O_3 and annealed Al_2O_3 surfaces as well as the properties of the respective dispersions are presented in Table 1. The conversion factor for unstable dispersions was calculated from the initial diffusion constant of the droplets.

Collector surfaces

The Al_2O_3, TiO_2, and ZrO_2 surfaces were prepared by Microchemistry (Espoo, Finland) on polished silicon wafers (Okmetic OY, Finland) using atomic layer epitaxy (ALE) technique. The thicknesses of the Al_2O_3, TiO_2, and ZrO_2 layers were 90, 40, and 40 nm, respectively. The isoelectric points (IEP) of the oxides used are listed in Table 2.

Cr_2O_3 surfaces were prepared by radio-frequency magnetron sputtering at the Department of Materials Science of Tampere University of Technology. The thickness of the oxide layer was optimised such that the reflectometric signal increased upon deposition of resin. The oxide layer thickness was less than 0.5 μm.

Examples of compositions of the oxides as studied by electron spectroscopy for chemical analysis (ESCA) are shown in Table 3. The atomic concentration of carbon in the oxides prepared by the ALE technique was 20% ± 5%. ESCA also showed that the carbon, which is only weakly oxidised, lies on top of the surface, i.e. the surfaces are partially covered by hydrocarbons.

A slowly decreasing reflectometric signal indicated that some of the Al_2O_3 surfaces – which are amorphous – were unstable in the electrolytes used. Unstable surfaces were rejected and only initially stable surfaces were used in this study. The stability of the initially stable surfaces was tested in water by means of reflectometry, and if the surface was initially stable, no indication of instability appeared within 5000 s (i.e. the baseline remained stable). The reason for the stability seems to be the high atomic concentration of carbon in the surface [23]. Despite the problems with stability, hydrocarbon-contaminated Al_2O_3 turned out to be the most suitable substrate for deposition studies (availability, signal-to-noise ratio, drift in the signal), and most of the systematic work was carried out using this substrate.

Table 1 Two examples of the properties of resin dispersions (standard deviation, SD, diffusion coefficient, D, and the theoretical rate of mass transfer, J_0) as well as the conversion factors determined for oxide surfaces. Dispersions 1 and 2 were used in studies of Al_2O_3 and annealed Al_2O_3, respectively

	Size (nm)	SD (nm)	D (10^{-8} cm^2/s)	J_0 ($\mu g/m^2s$)	Conversion factor
Dispersion 1	205	24	1.74	8.2	12.8
Dispersion 2	140	36	2.71	11.0	11.3

Table 2 Isoelectric points (IEP) of the oxides

Oxide	pH at IEP	Reference
Al_2O_3	7	21
ZrO_2	7.6	22
Cr_2O_3	7	21
TiO_2	6	21

Table 3 Composition of oxide surfaces, determined by electron spectroscopy for chemical analysis. M denotes the respective metal in each oxide. The values are atomic concentrations in percent

Oxide	M	O	C	N	Si	Na
Al_2O_3	23.5	53.8	21.5	1.2	–	–
ZrO_2	22.4	53.0	22.7	–	1.9	–
TiO_2	22.1	55.8	20.2	2.0	–	–
Cr_2O_3	23.2	45.2	30.3	–	1.2	0.2

Annealing the amorphous Al_2O_3 surfaces in air at 900 °C for 30 min stabilises the Al_2O_3 surfaces in water. This treatment also imparts changes in the optical properties of the oxide layer owing to some crystallisation [24]. Nevertheless, such annealed surfaces are a viable option for studies with Al_2O_3. Since most of the Al_2O_3 surfaces were stable as supplied, however, annealed surfaces were not used extensively.

Cleaning of the surfaces

In order to find a method for cleaning the surfaces, we studied the effects of gas and ethanol flames on the metal oxide films used in deposition experiments. ESCA analyses showed that flame treatment (5–30 s) removed most of the organic contaminants from the surfaces and so the residual atomic concentration of carbon in the surface was roughly 5%. However, both ESCA and reflectometry studies showed that the flame treatment also imparted changes in the structures of the oxides (chemical shift of Ti^{4+} peak in TiO_2), instability of the oxide layer in water (annealed Al_2O_3, ZrO_2), and changes in the optical response (Cr_2O_3).

The effectiveness of acetone extraction in cleaning the surfaces after deposition experiments was also investigated; however, it was observed that reproducible results could not be obtained. The final amount deposited on acetone-extracted surfaces was only about 30% of the final amount deposited found on the surfaces studied as supplied by the manufacturers. Thus, acetone extraction probably does not effectively clean the surface after deposition experiments. Finally, we did not attempt to clean the surfaces in surfactant solutions since there is a possibility that adsorbed surfactants change the surface properties of the collectors.

For these reasons all surfaces were used as supplied without further cleaning and each surface was used only once.

Wood resin

Emulsions of wood resin were prepared according to the method described in Ref. [25]. Unbleached thermomechanical pulp (TMP) at 4% consistency was obtained from a Finnish paper mill and stored in a freezer at −24 °C until used. Freeze-dried TMP was extracted in a Soxhlet apparatus with hexane. After extraction the hexane was evaporated and the resin was dissolved in acetone. The solution was injected in water under stirring. Finally, the acetone was removed by dialysis.

The resin droplets are negatively charged in the pH range used in this study. A study of the ζ potential of the droplets in the electrolytes used was reported in Ref. [19]. The mean size of the resin droplets in the various dispersions varied between 140 and 230 nm as determined by dynamic light scattering (Coulter N4). When referring to details of stability of the dispersion in various electrolytes, data from Ref. [25] have been used.

The concentration of resin in the emulsion was determined by gravimetry. It was found that successive determinations by means of gas chromatography as in Ref. [26] varied between 50 and 100% of the concentration determined by gravimetry.

Other reagents

All reagents used in preparing the electrolyte solutions were of analytical grade. Distilled, degassed water was used in reflectometry.

Results

Deposition of wood resin on Al_2O_3, ZrO_2, and Cr_2O_3

All these oxides have an IEP at pH $\approx$ 7 (Table 2). Thus, they are expected to be cationic in the pH range used in papermaking (i.e. roughly 4.5–7). We studied the deposition of wood resin droplets on these surfaces at pH 6.5 in solutions of NaCl and $CaCl_2$ (resin concentration 10 mg/l). The results are presented in Figs. 1–5.

The deposition behaviour of wood resin in a given electrolyte is qualitatively similar for all the different oxides studied. The main features are

1. The kinetic curves reach a plateau if the dispersion is colloidally stable.
2. No plateau is reached if the dispersion is colloidally unstable.

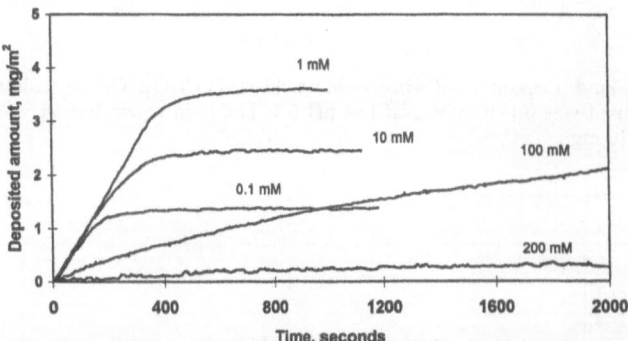

Fig. 1 Deposition of wood resin emulsion on Al_2O_3. The electrolyte used was 0.1–200 mM NaCl at pH 6.5. The resin concentration was 10 mg/l

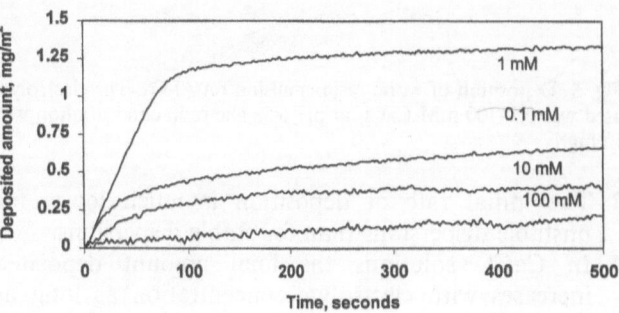

Fig. 2 Deposition of wood resin emulsion on annealed Al_2O_3. The electrolyte used was 0.1–100 mM NaCl at pH 6.5. The resin concentration was 10 mg/l

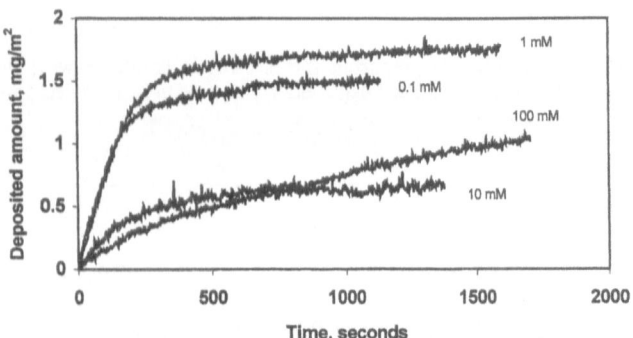

Fig. 3 Deposition of wood resin emulsion on ZrO_2. The electrolyte used was 0.1–100 mM NaCl at pH 6.5. The resin concentration was 10 mg/l

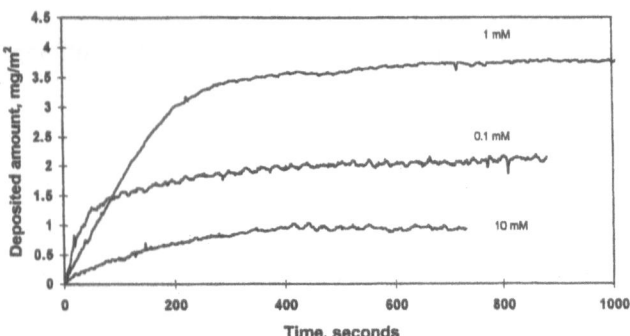

Fig. 4 Deposition of wood resin emulsion on Cr_2O_3. The electrolyte used was 0.1–10 mM NaCl at pH 6.5. The resin concentration was 10 mg/l

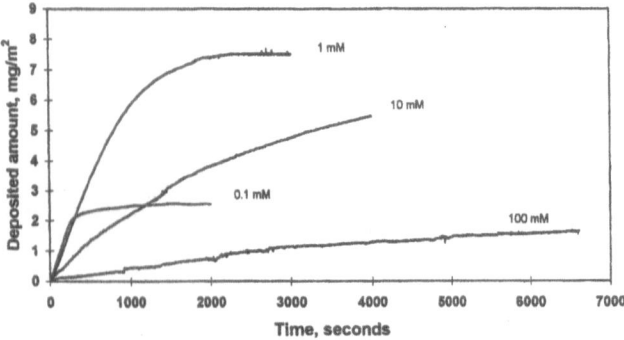

Fig. 5 Deposition of wood resin emulsion on Al_2O_3. The electrolyte used was 0.1–100 mM $CaCl_2$ at pH 6.5. The resin concentration was 10 mg/l

3. The initial rate of deposition is much lower for unstable dispersions than for stable dispersions.
4. In $CaCl_2$ solutions the final amount deposited increases with electrolyte concentration as long as the dispersion is stable.

These observations corroborate previous research on particle deposition [4].

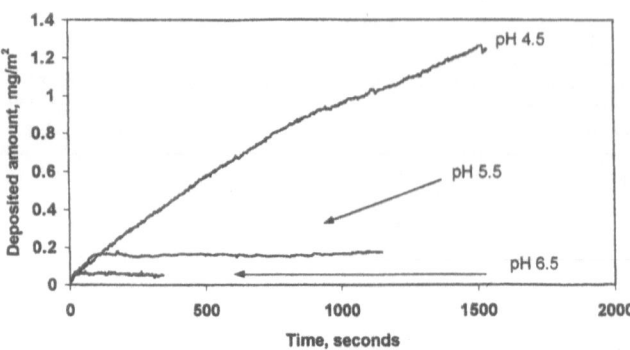

Fig. 6 Deposition of wood resin emulsion on TiO_2 as a function of pH. The electrolyte used was 1 mM NaCl at the pHs indicated in the figure. The resin concentration was 10 mg/l

More detailed inspection of the kinetic curves reveals some features which are unexpected on the basis of previous research:

5. In NaCl solutions the final amount deposited goes through a maximum as a function of salt concentration (Figs. 1–4).
6. In most cases the initial rate of deposition from stable dispersions is about the same; however, in 10 mM NaCl the initial rate of deposition decreases although the dispersion is stable (Figs. 2–4).

Also, the final amount deposited varied somewhat on the semiquantitative scale used. Sometimes there was a slowly creeping deposition in the flat part of the curve, and in some cases also the time needed to reach a plateau varied considerably (e.g. Al_2O_3 versus annealed Al_2O_3). Owing to the uncertainties in the experimental system (optics, contaminated surfaces), these factors were not investigated more accurately in this study.

Resin deposition on Cr_2O_3 or on TiO_2 in 100 mM NaCl (see next section) could not be analysed reliably by reflectometry because the experiments resulted in anomalies, i.e. the kinetics curve started to decrease after an initial increase. While this effect could be attributed to weak deposition it is also possible that it is due to strong light scattering from coagulating resin droplets in the laser beam, which would render determination of the actual amount deposited impossible.

Deposition of wood resin on TiO_2 compared with Al_2O_3

TiO_2 differs from the other oxides used in this study in that its IEP (pH 6) is located in the pH region used in papermaking. Resin deposition on TiO_2 at pH 4.5, 5.5, and 6.5 is shown in Fig. 6.

The final amount of resin deposited on TiO_2 at pH 6.5 is very low and if the pH is increased to 6.7 it becomes too low to be detected by reflectometry. Significant deposition occurs at lower pH values. This is clearly due to the surface charge of TiO₂. At pH 6.5–7 the surface is weakly anionic, while at pH 4.5 and 5.5 the surface is cationic.

Thus, deposition of resin on TiO₂ was compared with that on Al₂O₃ at pH 4.5 rather than at pH 6.5 as for the other oxides, using a higher resin concentration (40 mg/l) in order to shorten the time needed to reach a plateau value. The kinetics of deposition of resin on Al₂O₃ and TiO₂ surfaces from NaCl solutions and on Al₂O₃ from CaCl₂ solutions is shown in Figs. 7–9.

Deposition on TiO₂ and Al₂O₃ surfaces at pH 4.5 is qualitatively similar. In NaCl solutions the main features of deposition are the same as those at pH 6.5. However, when the NaCl concentration increases, the final amount deposited from stable emulsions does not go through a maximum but decreases monotonously (Figs. 7–8). Note that at pH 4.5 only emulsions in 0.1 and 1 mM NaCl are stable.

Initial rate of deposition

The initial rate of deposition from different electrolytes on Al₂O₃ surfaces at pH 6.5 is shown in Fig. 10.

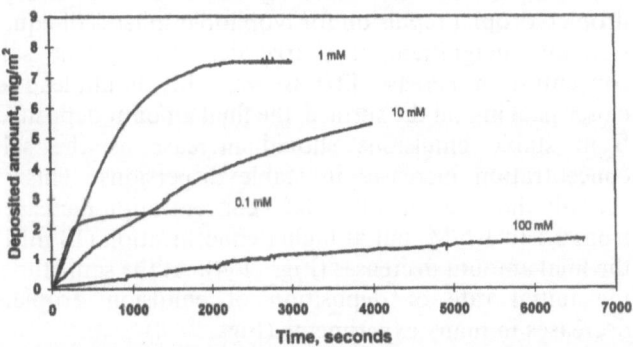

Fig. 7 Deposition of wood resin emulsion on TiO₂. The electrolyte used was 0.1–10 mM NaCl at pH 4.5. The resin concentration was 40 mg/l

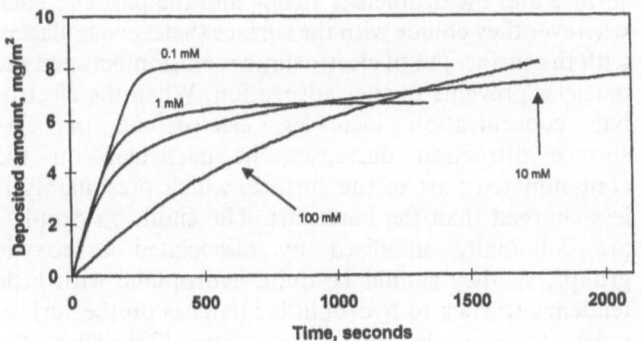

Fig. 8 Deposition of wood resin emulsion on Al₂O₃. The electrolyte used was 0.1–100 mM NaCl at pH 4.5. The resin concentration was 40 mg/l

Deposition from NaCl solutions

In this case, the initial rate is independent of salt concentration as long as the emulsions are colloidally stable, i.e. when the critical coagulation concentration (ccc) is not exceeded. From a colloidal point of view, deposition of resin droplets can be considered as heterocoagulation between the cationic collector surface and the anionic droplets. According to the Derjaguin–Landau–Verwey–Overbeek (DLVO) theory, the droplet–surface interaction should be attractive due to combined van der Waals and double-layer forces. The double-layer force can even be repulsive at short separations, but the combination should nevertheless result in attraction.

When the concentration of NaCl increases the electrostatic attraction decreases, which would decrease the rate of deposition if colloidal interactions were rate-determining; however, the initial rate of deposition from stable emulsions does not depend on electrolyte concentration (Fig. 10). Hence, it seems that the Al₂O₃ surface

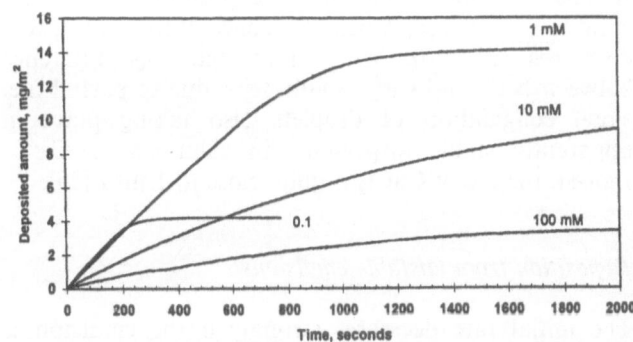

Fig. 9 Deposition of wood resin emulsion on Al₂O₃. The electrolyte used was 0.1–100 mM CaCl₂ at pH 4.5. The resin concentration was 40 mg/l

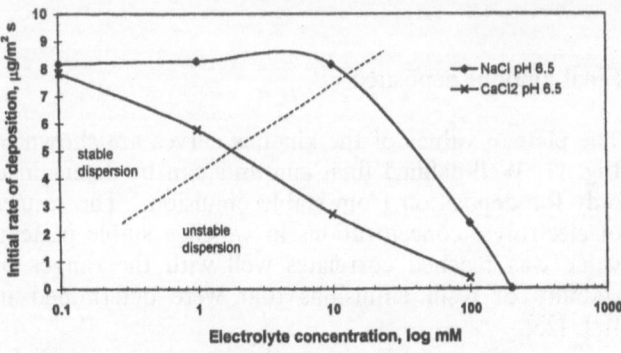

Fig. 10 The initial rate of deposition on Al₂O₃ surfaces at pH 6.5 as a function of electrolyte concentration. The *dashed line* indicates the experiments in which a plateau is reached (stable dispersions). Note that the dashed line does not indicate the exact location of the critical coagulation concentration. The error in the initial rate due to noise in the reflectometric signal is 10%

is initially a perfect sink for the droplets, so the deposition from stable emulsions is transfer-limited, i.e. the initial rate is determined by the supply rate and not by surface forces. There are, however, some exceptions to this, which will be discussed later.

The initial rate of deposition from 10 mM NaCl on annealed Al_2O_3 as well as on ZrO_2 and Cr_2O_3 is lower than from lower NaCl concentrations although the emulsion is still stable (Figs. 2–4). This indicates that the deposition from 10 mM NaCl is attachment-limited and not transfer-limited. It seems that this is a result of the same interactions that cause the final amount deposited to go through a maximum as a function of salt concentration. The interactions that we believe to be significant in this case – mainly electrostatic surface–droplet attraction – are discussed in the following section.

Deposition from stable CaCl₂ solutions

In this case, the initial rate of deposition from stable emulsions decreases when the concentration of $CaCl_2$ decreases (Fig. 10). We believe that this difference between Na^+ and Ca^{2+} solutions is due to partial (i.e. slow) coagulation of droplets also taking place in apparently stable dispersions in solutions of Ca^{2+}. Indeed, the ccc of $CaCl_2$ is quite close to 1 mM [25].

Deposition from unstable emulsions

The initial rate decreases strongly if the emulsion is rendered unstable by addition of electrolyte. This is consistent with previous studies of resin deposition on silica surfaces coated with cationic polymers [19]. The low rate of deposition from unstable emulsions can be attributed to coagulation of droplets, which results in a low rate of mass transfer [4].

Final amount deposited

The plateau values of the kinetics curves are shown in Fig. 11. Well-defined final amounts can be determined only for deposition from stable emulsions. The ranges of electrolyte concentrations in which a stable plateau value was reached correlates well with the ranges of stability of resin emulsions that were determined in Ref. [25].

Current theories describing the deposition of particles predict that the final amount deposited should be a function of droplet–droplet repulsion [2, 13]. In our experiments, this probably applies when deposition takes place from $CaCl_2$ solutions; however, for NaCl solutions the situation is more complex. In this case, the amount

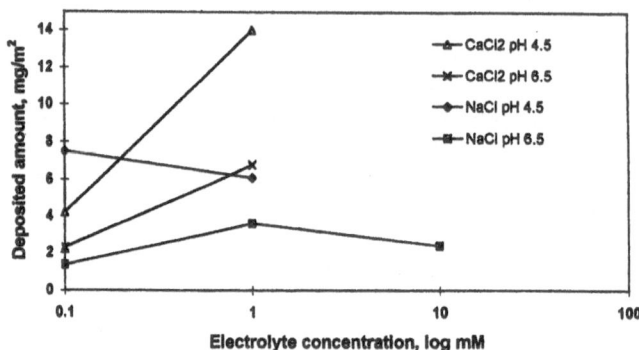

Fig. 11 Final amount deposited for stable resin dispersions as a function of electrolyte concentration, type of cation, and pH. Al_2O_3 collector surface, resin concentration 10 mg/l at pH 6.5 and 40 mg/l at pH 4.5

deposited either decreases monotonously or goes through a maximum when the electrolyte concentration increases.

The long-range interactions between particles and the surface should be the sum of van der Waals attraction, electrostatic repulsion between anionic droplets, and electrostatic attraction between the anionic droplets and the cationic surface. According to the DLVO theory, droplet–droplet repulsion for lyophobic spheres of equal sign and magnitude of charge decreases as the salt concentration increases [27]. Because this should lead to closer packing on the surface, the final amount deposited from stable emulsions should increase as the salt concentration increases in stable dispersions. This is actually the case when the NaCl concentration increases from 0.1 to 1 mM, but at higher concentration (10 mM) the final amount decreases (Figs. 1–4). At the same time, the initial rate of deposition of emulsion droplets decreases in many experiments (Figs. 2– 4).

We suggest that this effect can also be understood in terms of the screening effect of electrolyte on electrostatic interactions, combined with the fact that about 20% of the surface is covered by hydrophobic contaminants. At low ionic strength electrostatic attraction between the surface and the droplets is strong and the particles stick wherever they collide with the surface (heterocoagulation with the surface) until electrostatic repulsion between the particles prevents further adsorption. When the electrolyte concentration increases, electrostatic particle–surface attraction decreases, in particular on the contaminated part of the surface, which presumably is less charged than the bare part. The emulsion droplets are colloidally stabilised by dissociated carboxylic groups, so they should be quite hydrophilic with little tendency to stick to hydrophobic patches on the surface unless there is electrostatic attraction [28]. Thus, the probability that particles will stick on collision decreases, so the rate of deposition decreases. Because there is little tendency for the particles to stick to the contaminated part of the surface, adsorption ceases to increase when the surface is only partially covered ($\Gamma < \Gamma_{max}$ in lower salt concentration).

Thus, the results suggest that diminishing droplet–surface attraction becomes important at a lower degree of coverage than would be required for the droplet–droplet repulsion to prevent further deposition, if the surface were homogenous. Note that it is the droplet–droplet repulsion which also prevents deposition in this case; increasing the salt concentration only decreases the available "sites" for deposition. This effect is attributed to hydrocarbon contamination, which partially covers the surfaces.

The phenomenon described here, i.e. decreasing electrostatic droplet–surface attraction, is also believed to cause the decrease in the initial rate of deposition in some 10 mM NaCl solutions (cf. Discussion on the initial rate of deposition).

On the other hand when the electrolyte is $CaCl_2$, the droplet–droplet interactions determine the final amount deposited as predicted by the DLVO theory. However, since the divalent cations should be more effective in screening the electrostatic surface–droplet attraction than monovalent cations, there is no reason why the final amount deposited should increase in solutions of divalent cations. Consequently, it seems that some other interaction must be effective in calcium solutions.

It is well known that fatty and resin acids form soaps of very low solubility with Ca^{2+} ions and that the resulting metal soaps promote formation of deposits [29–32]. It seems very reasonable that formation of Ca^{2+} soaps in the droplet surface should make the droplets stickier, and, consequently, increase their affinity for the parts of the collector surface largely covered by hydrocarbons. Thus, deposition in calcium solutions follows the predictions of DLVO theory because the importance of the surface contamination is less.

If the semiquantitative final amounts deposited are analysed in terms of the area occupied by a single droplet it can be seen that the area taken by one droplet on Al_2O_3 – after the deposition has reached saturation – varies from 0.29 to 3.45 μm^2 (in 1 mM $CaCl_2$ at pH 4.5 and 0.1 mM NaCl at pH 6.5, respectively). As the area occupied by a projection of a sphere with a diameter of 205 nm is 0.033 μm^2 it can be suggested that each droplet occupies at least 10 times its projected area and in some cases even 100 times the projected area. This is mainly attributed to electrostatic droplet–droplet repulsion. It is also noticeable that the projection of a sphere is probably not the best estimate of the area on the surface occupied sterically by a single droplet since the droplets are expected to spread out somewhat. Nevertheless, these figures are in fairly good agreement with previous studies which suggest that the final coverage of deposited particles is in the range 2–5% [5].

Deposition as a function of resin concentration

Stable dispersions

For stable emulsions, both the initial rate of deposition and the final amount deposited are a function of resin solution concentration (Fig. 12). This contradicts earlier studies of particle deposition as well as current theories of deposition, according to which deposition of particles continues until a jamming limit is reached [3, 9, 13].

A possible explanation for the dependence on resin concentration is that the deposition of resin droplets is a combination of two processes: mass transfer to the surface and reconformation of droplets at the surface. This type of kinetics as well as the dependence between concentration and amount deposited have been suggested for adsorption of enzymes [33]. Re-formation is made possible by the liquid nature of the droplets. Possible mechanisms are, for example, spreading or coalescence of droplets. Spreading of the droplets would result in coverage of a larger surface area, which increases the blocking of the surface from droplets approaching from stable dispersions. A detailed picture of the interface is, however, needed until definitive conclusions can be drawn. The working hypothesis is that the droplets spread out somewhat upon deposition on the surface.

Thus, deposition of resin at one concentration only does not give a full picture of resin deposition. It seems that the final amount deposited could be a result of competition between mass transfer and surface kinetics. Note that this affects the results shown in Fig. 11 such that the experiments carried out at different pH are not comparable owing to the different resin concentrations used.

Model for stable dispersions

The kinetics curves that reached a plateau were analysed in terms of the model presented in Refs. [3, 5, 34]. The

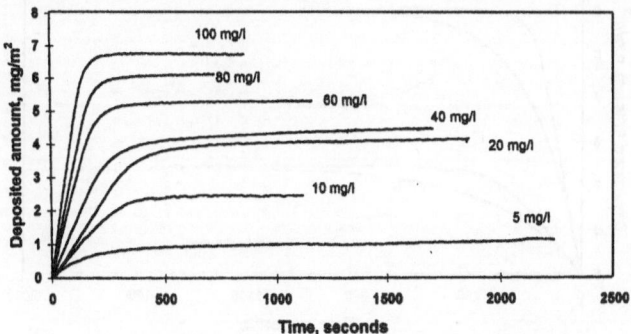

Fig. 12 Final amount deposited for stable dispersions as a function of resin concentration. The experiments were carried out at pH 6.5 in 10 mM NaCl in different concentrations of resin using an Al_2O_3 collector surface

analysis was applied to stable emulsions only. It is assumed that the droplets are irreversibly deposited, that deposition is diffusion-controlled, and that the rate of attachment decreases with time because the surface becomes blocked by already deposited droplets.

That deposition is indeed irreversible has not been verified systematically but in several test runs (detailed results not shown here) we found no indication of desorption of resin from the Al_2O_3 surfaces. Thus, deposition can be described according to

$$\Gamma = \left(\frac{d\Gamma}{dt}\right)_0 \tau_{bl}\left(1 - e^{-t/\tau_{bl}}\right) , \tag{5}$$

where Γ is the amount deposited, $(d\Gamma/dt)_0$ is the initial rate of deposition, τ_{bl} is the blocking time, and t is the time elapsed from the beginning of the experiment. The blocking time can be obtained from

$$\Gamma_{final} = \left(\frac{d\Gamma}{dt}\right)_0 \tau_{bl} . \tag{6}$$

Curves calculated from Eq. (5) are compared with experiment in Fig. 13.

The agreement is not satisfactory. The theoretically calculated curves reach the same level (final amount deposited) as the experimentally determined curves, but the main reason for the deviation between the experimental and theoretical curves (Fig. 13) is that the initial linear increase is not accounted for in Eq. (5). Typical features of the experimental curves are the linear initial increase and the flattening out after 50–90% of the final amount deposited has been reached.

Experimental curves like ours have also been recorded in studies of adsorption of nanoparticles and polymers [9, 20]. The random sequential adsorption (RSA) model [13] was tested for such curves in Ref. [9] and it was found that the model does not adequately describe the experimental results.

However, if Eq. (5) is combined with an initial linear dependence of Γ on t,

$$\Gamma = \left(\frac{d\Gamma}{dt}\right)_0 t , \tag{7}$$

a good fit between the model and experimental results is achieved. This is shown in Fig. 14. After the beginning of the nonlinear part of the curve the amount deposited can be obtained by combination of Eqs. (5) and (7)

$$\Gamma = \left(\frac{d\Gamma}{dt}\right)_0 t_1 + \left(\frac{d\Gamma}{dt}\right)_0 \tau_{bl}\left(1 - e^{(-t/\tau_{bl})}\right) , \tag{8}$$

where t_1 is a constant representing the time elapsed at the beginning of the nonlinear part of the curve. The blocking time was calculated by considering the nonlinear part of the curve only.

It has been found that deposition of anionic resin droplets leads to the reversal of ζ potential of the collector [19]. Hence, the flattening out of the kinetics curve and a plateau can be interpreted as a result of electrostatic interface–droplet repulsion. Since the amount deposited increases at a constant rate in the early stages, it seems that the electrostatic repulsion does not become effective in blocking the approaching droplets until a certain surface coverage is reached. Intuitively, the point where the linear part of the kinetics curve ends can be seen as the point where the ζ potential of the collector is neutralised.

The kinetics curves obtained by the surface diffusion RSA model seem to resemble ours [35]. Unfortunately, no analytical expression of deposition kinetics according to this model exists. According to the model, the droplets may reenter the surface at another site owing to surface diffusion after having been blocked by a deposited droplet. This seems possible since near the actual stagnation point there is a circular region (diameter 35 μm, see Appendix) where the deposition is strongest. This region, where the mass transfer apparently occurs

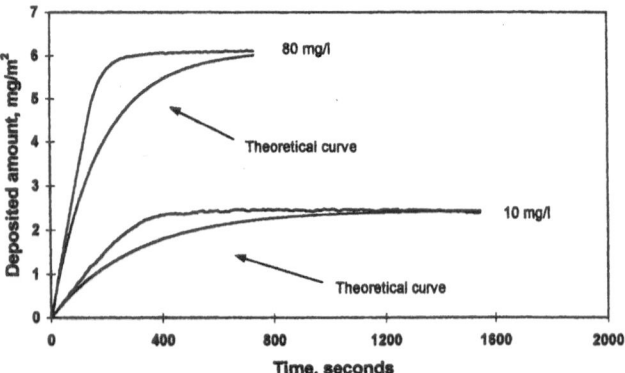

Fig. 13 Comparison of experimentally obtained curves with theoretical curves calculated using Eq. (5). The resin concentrations are indicated in the figure. The electrolyte used was 10 mM NaCl and the pH was 6.5

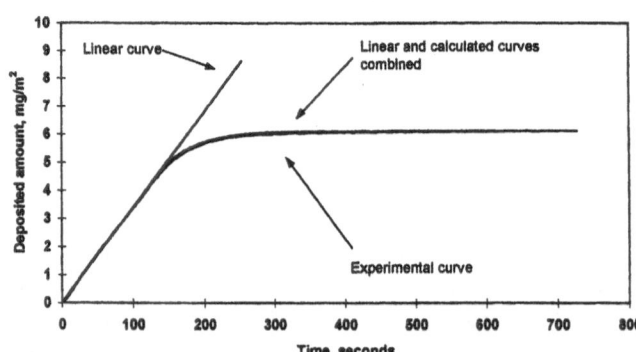

Fig. 14 Comparison of an experimentally obtained curve with one calculated from Eq. (8). τ_{bl} is obtained by considering the nonlinear part of the curve only. The resin concentration was 80 mg/l. The electrolyte used was 10 mM NaCl and the pH was 6.5

by diffusion, is large compared with the size of the droplets (diameter roughly 200 nm). Thus, lateral surface diffusion might play a part in the deposition. Such surface diffusion could explain the initial linear rise of the kinetics curves.

The good fit of Eq. (8) shows that an exponential dependence can be used to model the nonlinear part of the curve. The last term on the right-hand side of Eq. (8), however, is based on well-defined parameters describing the interactions of the particles after deposition. It is clear that these parameters are not necessarily directly applicable to the deposition of droplets. For instance, the parameters may not remain constant during the experiment owing to possible spreading of the droplets. The predictive power of Eq. (8) depends on how t_1 and τ_{bl} can be linked with surface parameters such as the ζ potential and the kinetics effects of the droplets after deposition.

Unstable emulsions

As can be seen in Fig. 15 the initial rate of deposition increases with increasing resin concentration, as expected. For unstable dispersions no adsorption plateau is reached since droplet–droplet repulsion is not sufficient to prevent droplets in solution from approaching the interface. Thus, the highest amounts are deposited from high salt concentrations and high concentrations of resin. The rate of deposition decreases with time, but a slow increase in the amount deposited continued for at least 10,000 s.

The decreasing rate of deposition with time for unstable dispersions can be attributed to coagulation of the droplets [4]; however, other factors cannot be totally excluded. For instance, some (electrostatic) droplet–interface repulsion could be effective even.

Also, the optical response might not be linear at high coverages of resin. The theory of reflectometry predicts that the reflectometric signal should start to decrease when the amount deposited grows to a high value [20, 36]. The experiment for which the results are shown in Fig. 15 (60 mg resin/l) was continued for 10 000 s but, surprisingly, we found no evidence of a decreasing signal even though the final amount deposited was very high on the semiquantitative scale used. Optical microscopy showed that the surface was indeed covered by a multilayer of coalesced resin droplets (see Appendix).

Concluding remarks

No qualitative differences were observed for deposition of resin emulsion droplets on the different oxides studied. In most aspects the deposition behaviour could be understood on the basis of previous research on particle deposition. There were two exceptions to this:

1. The final amount deposited increased as a function of supply rate. The mechanism suggested for this is re-formation of the droplets upon deposition.
2. When deposition takes place from NaCl solutions there is a maximum in the final amount deposited as a function of electrolyte concentration. It seems that electrostatic surface–droplet attraction and surface contamination play a major role in deposition on heterogeneous surfaces from NaCl solutions.

For calcium solutions the importance of surface heterogeneity is less marked owing to the increasing stickiness of the resin droplets and, consequently, the higher affinity of the droplets for surfaces partly covered by hydrocarbons. The stickiness of the droplets is probably increased by specific interactions of the Ca^{2+} ions with the resin droplets. Nevertheless, since the maximum in NaCl solutions is attributed to the properties of the collector surface, the only phenomenon observed in this study that seems to be characteristic of the deposition of emulsion droplets compared with solid particles is the effect of supply rate in stable emulsions.

It was found that the theory for deposition kinetics by Dabros and van de Ven does not fully explain the deposition kinetics of stable resin droplets. The main reason for this is that that the initial rate of deposition stays constant until 50–90% of the final amount deposited is reached.

Acknowledgements Leena-Sisko Johansson at the Center for Chemical Analysis, Helsinki University of Technology, is thanked for the ESCA analyses. Minnamari Vippola at the Department of Materials Science of Tampere University of Technology is thanked for preparation of the Cr_2O_3 surfaces.

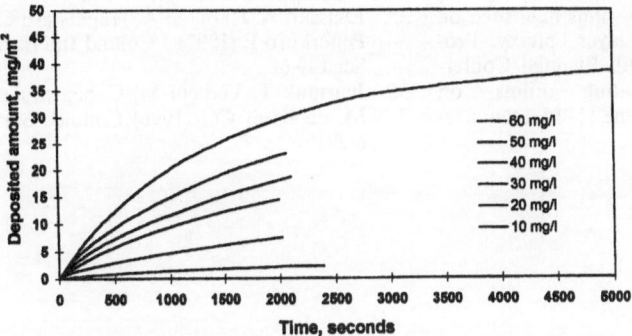

Fig. 15 The effect of resin concentration on deposition of resin from an unstable dispersion on an Al_2O_3 collector surface. The electrolyte was 100 mM NaCl at pH 6.5

Appendix

An image obtained by an optical microscope connected to a video screen is shown in Fig. 1A. The long side of the

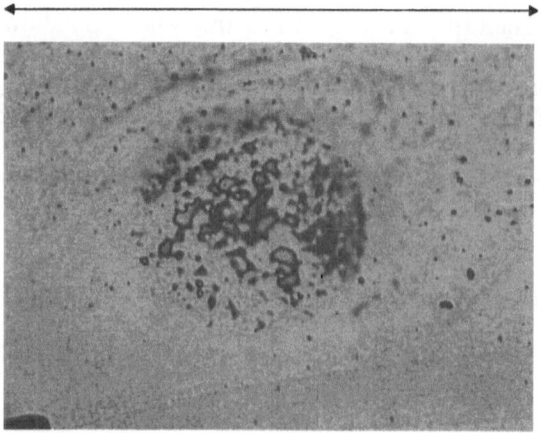

80 µm

Fig. 1A Resin deposited in stagnation point flow on Al_2O_3 in 100 mM NaCl at pH 6.5 (resin concentration 60 mg/l) as seen be means of optical microscopy. The magnification was × 10

figure is 80 µm. This specimen was obtained from the experiment presented in Fig. 15, i.e. deposition of resin on Al_2O_3 at pH 6.5 in 100 mM NaCl. The resin concentration in the solution was 60 mg/l. Owing to the high electrolyte concentration the dispersion was unstable.

In the middle there seems to be a circular region – with a diameter of roughly 35 µm – where the deposition is strongest (coalesced resin droplets). This is believed to be the region where the actual stagnation point is situated. The shades of grey in the (originally coloured) figure indicate that the area on the outside of the circular region around the stagnation point is largely covered by a "film" formed by deposited resin. In the foreground (right corner of the figure) there are darker areas, where the deposition of resin is weaker and the Al_2O_3 surface can be seen.

References

1. Dabros T, van de Ven TGM (1983) Colloid Polym. Sci 261:694
2. Dabros T, van de Ven TGM (1993) Colloids Surf 75:95
3. van de Ven TGM (1998) Colloids Surf 138:207
4. van de Ven TGM (1989) Colloidal hydrodynamics, Academic, London
5. Boluk MY, van de Ven TGM (1990) Colloids Surf 46:157
6. Schenkel JH, Kitchener JA (1960) Trans Faraday Soc 56:161
7. Marshall JK, Kitchener JA (1966) J Colloid Interface Sci 22:342
8. Lüthi Y, Ricka J, Borkovec M (1998) J Colloid Interface Sci 206:314
9. Böhmer MR, van der Zeeuw EA, Koper GJM (1998) J Colloid Interface Sci 197:242
10. Böhmer MR (1998) J Colloid Interface Sci 197:251
11. Harley S, Thompson DW, Vincent B (1992) Colloids Surf 62:163
12. Oberholzer MR, Stankovich JM, Carnie SL, Chan DYC, Lenhoff AM (1997) J Colloid Interface Sci 194:138
13. Adamczyk Z, Warszynski P (1996) Adv Colloid Interface Sci 63:41
14. Adamczyk Z, Siwek B, Zembala MJ (1997) Colloid Interface Sci 185:236
15. Malmsten M, Lindström A-L, Wärnheim T (1996) J Colloid Interface Sci 179:537
16. Allen LH (1975) Pulp Pap Can 76:70
17. Allen LH (1980) Tappi J. 62:81
18. Sihvonen A-L, Sundberg K, Sundberg A, Holmbom B (1998) Nord Pulp Pap Res J 13:64
19. Kekkonen J, Stenius P (1999) Colloids Surf A 156:357
20. Dijt JC, Cohen Stuart MA, Hofman JE, Fleer GJ (1990) Colloids Surf 51:141
21. Parks GA, de Bruyn PL (1962) J Phys Chem 66:967
22. Solomon MJ, Saeki T, Wan M, Scales PJ, Boger DV, Usui H (1999) Langmuir 15:20
23. Kekkonen J, Stenius P (1999) Nord Pulp Pap Res J 14:300
24. Ericsson P, Bengtsson S, Skarp J (1997) Properties of Al_2O_3 films deposited on silicon by atomic layer epitaxy. Proceedings of the 10th Biennial Conference on Insulating Films on Semiconductors, June 11–14, Stenungsund, Sweden
25. Sundberg K, Pettersson C, Eckerman C, Holmbom B (1996) J Pulp Pap Sci J 22:248
26. Ekman R, Holmbom B (1989) Nord Pulp Pap Res J 4:16
27. Derjaguin BV, Churaev NV, Muller VM (1987) Surface forces. Consultants Bureau, New York
28. Israelachvili JN (1985) Intermolecular and surface forces. With applications to colloidal and biological systems. Academic London
29. Gustafsson C, Tammela V, Lindh T (1954) Pap Puu 36:269
30. Bergmann BE, Rying S (1975) Tappi J 58:147
31. Douek M, Allen LH (1980) Pulp Pap Can 81:T 318
32. Allen LH (1988) Tappi J 71:61
33. van Eijk MCP, Cohen Stuart MA (1997) Langmuir 13:5447
34. Dabros T, van de Ven TGM (1982) J Colloid Interface Sci 89:232
35. Elaissari A, Haouam A, Huguenard C, Pefferkorn E (1992) J Colloid Interface Sci 149:68
36. Jeurnink T, Verheul M, Cohen Stuart M, de Kruif CG (1996) Colloids Surf 6:291

Progr Colloid Polym Sci (2000) 116:67–73
© Springer-Verlag 2000

E. Dingsøyr
A. A. Christy

Effect of reaction variables on the formation of silica particles by hydrolysis of tetraethyl orthosilicate using sodium hydroxide as a basic catalyst

E. Dingsøyr (✉) · A. A. Christy
Faculty of Mathematics and Natural
Sciences, Agder College,
Tordenskjoldsgate 65, Sevicebox 422
4604 Kristiansand, Norway
e-mail: eldar.dingsoyr@hia.no
Tel.: 47-38-141560; Fax: 47-38-141011

Abstract The formation of silica particles by hydrolysis of tetraethyl orthosilicate (TEOS) in ethanolic solutions was studied by multivariate analysis using a 2^{4-1} reduced experimental design. The variables (factors) in the analysis were the starting concentrations of TEOS (0.1–0.5 M), NaOH (1–5 mM) and water (1–6 M) and the temperature (23–40 °C). The target parameters were the size and the polydispersity of the silica particles and the amount of silica formed. The size and the polydispersity of the particles were determined using photon correlation spectroscopy, while the amount of silica formed was determined gravimetrically after centrifugation. Analysis of the data by multiple regression analysis indicated that the variable of greatest importance for the amount of silica formed, the particle size and the polydispersity was the starting concentration of NaOH. In the concentration range studied, the conversion of TEOS to silica was low; however, the amount of silica formed increased in proportion to the concentration of NaOH. Our interpretation is that hydroxyl ions are consumed during the hydrolysis reaction, leading to the formation of charged intermediates. When most of the hydroxyl ions have been consumed the hydrolysis will stop. A higher concentration of NaOH resulted in the formation of smaller and more uniform silica particles and these were formed in greater number. In addition, both the particle size and the polydispersity index showed a strong negative correlation with the interaction term between water and NaOH, indicating that the ionisation of NaOH increases with increasing water content.

Key words Monodisperse silica ·
Particle size · Polydispersity index ·
Silica sols · Multivariate analysis

Introduction

In the presence of basic catalysts, hydrolysis of tetraalkyl orthosilicates followed by condensation reactions leads to the formation of silica particles, with the size and the size distribution depending on the experimental conditions. The chemistry of the reactions can briefly be described as the hydrolysis (Eq. 1) and the condensation reactions (Eq. 2).

$$Si(OR)_4 + 4H_2O \rightarrow Si(OH)_4 + 4ROH \qquad (1)$$

$$nSi(OH)_4 \rightarrow nSiO_2 + 2nH_2O \qquad (2)$$

Since tetraalkyl orthosilicates are not soluble in water, the reactions are carried out in alcoholic solutions. Water, for the hydrolysis, is added in controlled quantities and ammonia is usually used as a catalyst.

Kolbe [1], in 1956, was the first to discover that monodisperse silica particles could be prepared as described previously. The early studies by Stöber et al. [2] and Bogush et al. [3] focused on finding the experimental conditions leading to the formation of uniform,

68

spherical silica particles of controlled size. Tailor-made silica particles are of interest in different fields, including chromatography, catalysis, ceramics and speciality applications, such as polishing of silicon wafers.

Later more attention was given to the mechanism of the formation of the monodisperse silica particles [4–12]. Understanding the mechanism may help to control the properties of the silica particles as well as be useful in the preparation of other types of monodisperse particles. As briefly mentioned later, several possible mechanisms have been suggested; however, no general agreement on the processes leading to the formation of the monodisperse particles seems to have emerged. Matsoukas and Gulari [4, 5] studied tetraethyl orthosilicate (TEOS) and suggested that the growth of the silica particles was controlled by the hydrolysis of TEOS. They found that nucleation was limited to the early stage of the process and suggested that the particles grew only by monomer addition. Bogush and Zukoski [6, 7] found that the particle size of the silica prepared by hydrolysis of TEOS also depended on parameters other than the rates of hydrolysis and condensation. Addition of NaCl, for example, increased the particle size dramatically. They suggested that particle growth is rate-limited by some step in the condensation pathway and that the growth of particles occurs through an aggregation mechanism where subparticles in the 1–10-nm range aggregate to form the final particles.

Bailey and Mecartney [8] observed the growth of the silica particles by using cryogenic transmission electron microscopy techniques. They suggested a particle growth model where hydrolysed monomers react to form microgels of branched polymers, which collapse upon reaching a certain size and cross-linking density. Further growth and consolidation of these seed particles occur by monomer addition or addition of polymer species to the particle surface. Van Blaaderen and Kentgens [9] studied the morphology and microstructure of silica particles prepared by hydrolysis of TEOS. They suggested that particles with smooth and spherical shape are the result of the growth by monomers or small oligomers, while growth by larger silicon structures results in rougher surfaces and particles of irregular shapes. The particles were not fully consolidated, but the particle microstructure did not depend on the reaction conditions. Van Blaaderen et al. [10] have reviewed most of the work in this field until 1992. They emphasise that further investigation is needed and in greater detail owing to the complexity of the system.

The aim of this study was to identify the factors (variables) of greatest importance in the formation of uniform silica particles. In due time this may be helpful in elucidating the mechanism of the formation of the monodisperse particles. To identify these factors we chose to use experimental design and multivariate analysis.

In nearly all previous studies in this field, ammonia has been used as the catalyst. Hydroxyl ions (OH$^-$) then form through the reaction between ammonia and water. The concentration of hydroxyl ions is expected to be an important variable, since it is believed [9, 10] that the hydrolysis of the tetraalkyl orthosilicate occurs by a nucleophilic substitution reaction (SN$_2$ mechanism) where the nucleophilic hydroxyl ion attacks the silicon atom. If this is the case, it will be difficult to distinguish between the effects of hydroxyl and water when ammonia is used as the catalyst. We therefore chose to use NaOH as a catalyst in this study. If NaOH is used, hydroxyl ions form from a direct, partial ionisation of NaOH.

Experimental design and multivariate analysis has previously been used by Lindberg et al. [12] to study the formation of silica by ammonia-based catalysis. They used six variables: temperature, time of reaction and concentrations of TEOS, water, ammonia and sodium chloride. The influence of these variables on the target parameter, the diameter of the silica particles, was studied.

In this study the system is less complex since NaOH was chosen as the catalyst. The number of variables has been reduced to four; temperature and concentrations of TEOS, water and sodium hydroxide. To keep the number of variables low, the reaction time was kept constant in all experiments. Furthermore, sodium chloride was added only to give a constant concentration of sodium in the system and is therefore not regarded as an independent variable. In this investigation, the number of target parameters has been increased. In addition to particle size as studied by Lindberg et al., the polydispersity index of the particles and the amount of silica formed have been included. The last parameter, of great interest both scientifically and technically, seems not to have been given much attention in the previous studies in the field.

Experimental

The formation of silica particles during the hydrolysis of TEOS was investigated in two stages. In the first stage, multivariate analysis was used to study the influence of the variables temperature, concentrations of TEOS, NaOH and water on the target parameters, the diameter of the silica particles, the polydispersity index of the particles and the amount of silica formed. In the second stage, the amount of silica formed was studied by changing "one variable at a time" over larger ranges than in the experiments for multivariate analysis.

Multivariate analysis

A comprehensive treatment of the theory of multivariate analysis has been given by Box et al. [13]. To investigate the effect of four variables on a parameter by multivariate analysis, $2^4 = 16$ experiments are required; however, if a 2^{4-1} reduced fractional design is used, as in this case, the number of experiments can be reduced to eight. Multivariate analysis provides information on how the target parameter is affected when the variables are within the limits used in the experiments. The influence of the variables on the target parameters is estimated with and without interaction terms. The interaction terms provide information on whether the variables have a simultaneous effect on the target parameter or not. The multivariate analysis was carried out using Sirius 6.0 software [14].

Chemicals

TEOS (98%) from Aldrich, absolute ethanol (99.9 vol%) from Arcus, NaOH (extra-pure food grade) from Merck and NaCl from Norsk Medisinaldepot were used as received. The water used was deionised (Milli-Q) and had been boiled to remove CO_2. A 1.0 M ethanolic solution of NaOH was prepared by dissolving NaOH in absolute ethanol. Immediately before use the solution was filtered through a 0.45-μm cellulose acetate filter and the exact concentration of NaOH was determined by titration with 0.100 M HCl (Titrisol). A 1.00 M aqueous solution of NaCl was prepared by dissolving NaCl in degassed water. The solution was stored under nitrogen.

Preparation of silica particles

In each experiment the required amount of ethanol, water and NaOH was measured into 50- or 100-ml Erlenmeyer bottles with rubber stoppers. When the starting concentration of NaOH was less than 0.005 M, 1.00 M NaCl was added to give a total concentration of sodium of 0.005 M. The bottles were flushed with nitrogen and kept at room temperature or in a water bath to reach the specified temperature. The reactions were started by adding the required amount of TEOS. As quickly as possible the bottles were flushed with nitrogen and shaken by hand for mixing. Flushing with nitrogen was done to avoid formation of sodium carbonate from dissolution of CO_2 in alkaline solutions. During the reaction the bottles were kept at room temperature or in the water bath. The bottles were shaken by hand every 30 min to avoid settling of the solutions. The reaction time was 180 min in all experiments.

Particle size and polydispersity index

The diameters of the silica particles and their polydispersity index were determined by photon correlation spectroscopy using a Coulter Nanosizer from Coulter Electronics, UK. The measuring time was 2 min. The polydispersity index was given as a number from 0–10. An increasing number indicates increasing polydispersity.

Amount of silica formed

Exactly 50.0 ml of the silica dispersions were pipetted into Erlenmeyer bottles and an amount of HCl equal to the amount of NaOH in 50.0 ml of the original reaction mixture was added. This was done to stop further hydrolysis reactions. The silica particles were removed from the dispersion by centrifugation at 6000 rpm for 15 min. (4500 g) in a Labofuge 6000 centrifuge from

Heraeus Instrument, Germany. This procedure was not used for particles smaller than 200 nm. The supernatant was removed by decanting. The sediment was washed by redispersing it in 10 ml 96% ethanol and then centrifuged for 15 min. at 6000 rpm. The sediment was then redispersed in 5 ml 96% ethanol, transferred to a porcelain crucible and dried at 105 °C for 20 h. The amount of silica was determined by weighing.

Results

Multiple regression models

The four variables chosen for the multivariate analyses, their higher and lower levels and their centre points are shown in Table 1. The effect of the four variables on the three different response parameters, the diameter of the silica particles, their polidispersity index and the amount of silica formed, was studied using 2^{4-1} reduced factorial designs. In such designs there will be overlap between several interaction terms. The main variables x_1, x_2, x_3 and x_4 will contain contributions from the three-factor interactions $x_2x_3x_4$, $x_1x_3x_4$, $x_1x_2x_4$ and $x_1x_2x_3$. The three-factor interactions are, however, assumed to be small and may be neglected. The 2^{4-1} reduced factorial designs also gives information about the two-factor interactions x_1x_2, x_1x_3 and x_1x_4. These terms, however, contain contributions from the other two-factor interactions x_3x_4 ($=x_1x_2$), x_2x_4 ($=x_1x_3$) and x_2x_3 ($=x_1x_4$).

Models for diameter and polydispersity index of the silica particles

Multiple regression models were built for the target parameters, the diameter of silica particles and the polydispersity index of the particles, using the computer program Sirius 6.0 [14].

The regression curve for the model based on the particle diameter is shown in Fig. 1a. The diameter measured for the centre point is only slightly less than that predicted by the model. This indicates that the model is almost linear. The full model with interaction terms explains 99.6% of the data, while only 75.8% of the data are explained when interaction terms are excluded.

The regression curve for the model based on the polydispersity index is shown in Fig. 1b. In this case the measured polydispersity index of the centre point is

Table 1 The experimental variables, their higher and lower values and their centre points

Variable		Low level −	High level +	Centre point
Concentration of TEOS (M)	x_1	0.1	0.5	0.3
Concentration of water (M)	x_2	1.0	6.0	3.5
Concentration of NaOH (M)	x_3	0.001	0.005	0.003
Temperature (°C)	x_4	23.0	40.0	31.5

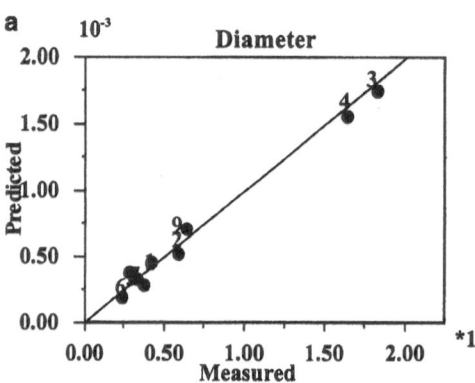
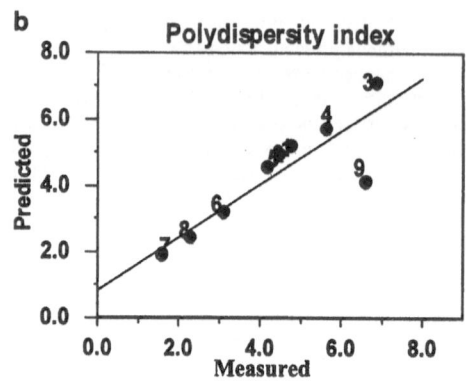

Fig. 1 Measured versus predicted by models **a** particle diameter (nm) and **b** polidispersity index. The *numbers* indicate the numbering of the experiments (1–8) and the centre points (9). 99.6 and 80.2% of the variance is explained when the centre points are included in the datasets

higher than predicted by the model. This shows that at least one of the variables is related to the polydispersity index in a nonlinear manner. The full model explains 80.2% of the data, while 71.9% is explained when interaction terms are excluded.

The regression coefficients of the main variables and the interaction terms in the full models of the particle diameters and the polydispersity indices are shown in Fig. 2a and b, respectively. When the regression coefficients are positive, the response parameter increases with the variable concerned, while negative regression coefficients mean that the response parameter decreases.

In the range studied, the particle diameter increases with the concentration of water and decreases with increasing concentration of NaOH. The other main variables seem to have little influence on the particle size. It is seen that the starting concentration of TEOS has very little influence on the particle diameters. The coefficients for the interaction terms are also small except for the $x_1 x_4 = x_2 x_3$. This can be explained as an interaction between the concentration of TEOS and the temperature or alternatively as an interaction between the concentrations of water and NaOH.

The polydispersity index decreases (meaning that the particles become more monodisperse) with increasing concentration of NaOH. The polydispersity index does not seem to be influenced very much by the starting concentrations of water, TEOS and temperature. The interaction terms are relatively small except for the $x_1 x_4 = x_2 x_3$, which could be an interaction between the concentration of TEOS and the temperature or alternatively an interaction between the concentrations of water and NaOH.

Models for the amount of silica particles formed

The regression curve for the model based on the amount of silica formed is shown in Fig. 3. The amount of silica measured at the centre point is only slightly less than that predicted by the model. This indicates that this model is almost linear. The full model with interaction terms explains 99.4% of the data, while 85.9% of the data are explained when interaction terms are excluded.

The regression coefficients for the variables of the full model for the amount of silica formed are shown in

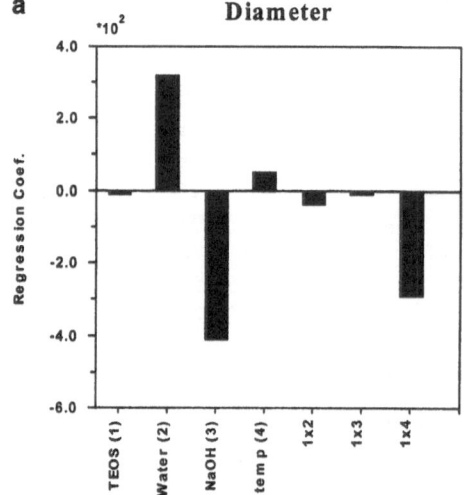
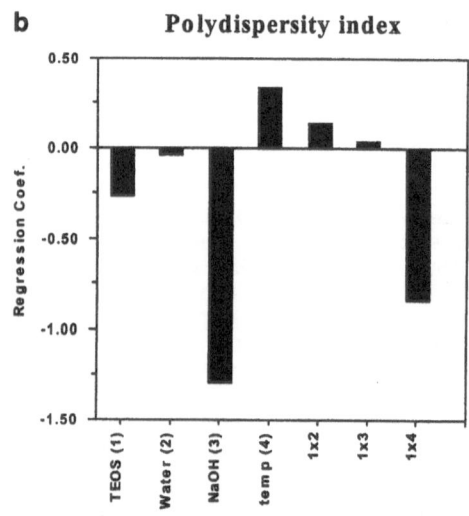

Fig. 2 Regression coefficients for models based on **a** particle diameter (nm) and **b** polydispersity index

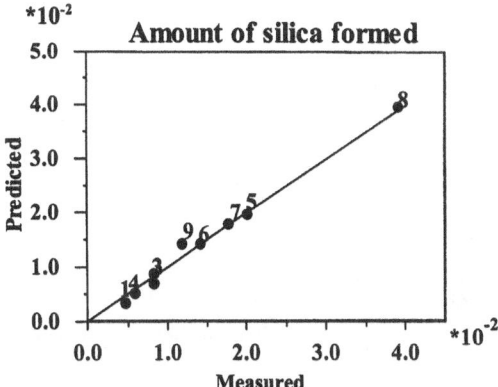

Fig. 3 Measured versus predicted amount of silica formed (mol/l). The *numbers* indicate the numbering of the experiments (1–8) and the centre point (9). The model explains 99.4% of the variance when the centre point is included in the dataset

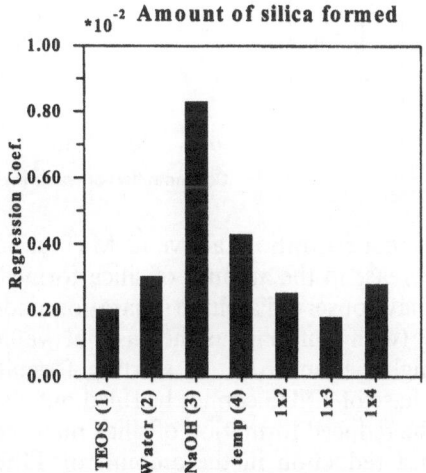

Fig. 4 Regression coefficients for the model based on the amount of silica formed (mol/l)

Fig. 4. All the main variables and all the interaction terms tend to increase the formation of silica. The most important main variable is the concentration of NaOH, followed by temperature. As long as there is enough TEOS in the mixture for the hydrolysis, the reaction will take place; however, the concentration of TEOS does not effect the amount of silica formed to a great extent. All two-factor interactions seem to have some effect.

"One-variable-at-a-time" investigation of amount of silica formed

The amount of silica formed was also studied by varying one variable at a time and over a larger range than in the multivariate analysis. The variables were as before the temperature and the concentrations of TEOS, water and NaOH. The results are shown in Fig. 5a–d.

For concentrations of TEOS higher than 0.3 M, the amount of silica increases only very slowly with increasing concentration of TEOS (Fig. 5a). The results indicate that 100% conversion of TEOS in this system can only be achieved at very low concentrations of TEOS. In the TEOS concentration range 0.1–2.0 M, the molar conversion of TEOS to silica is in the range 18–1%.

In the multivariate analysis the concentration of NaOH was found to be the variable affecting the amount of silica formed to the greatest extent. It is seen (Fig. 5b) that the amount of silica increases linearly with increasing concentration of NaOH up to 0.015 M.

The influence of temperature on the amount of silica formed does not seem to be very significant (Fig. 5c).

Water seems to affect the formation of silica in a complex manner (Fig. 5d). The amount of silica increases with increasing concentration of water up to a maximum at about 12 M water. At higher water concentrations the amount of silica formed was reduced.

Discussion

Amount of silica formed

The multivariate analysis (Fig. 4) shows that a concentration of NaOH is the starting variable influencing the amount of silica formed to the greatest extent. Furthermore, the amount of silica seems to be proportional to the starting concentration of NaOH (Fig. 5b). It is also found (Fig. 5a) that the amount of silica formed is nearly independent of the concentration of TEOS for concentrations greater than 0.3 M. A possible explanation of these results is that NaOH is consumed in the reactions and is involved in the formation of ionic species. The presence of charged intermediates has been indicated by conductivity measurements before [6, 11].

The hydrolysis of TEOS is believed to be a nucleophilic substitution reaction, where OH^- replaces the ethoxy groups [9–11]. Furthermore, it is expected that the rate of hydrolysis will increase as more ethoxy groups are removed. Fully hydrolysed silicic acid compounds should, therefore, dominate. Since the acid dissociation constants for silicic acid in ethanolic solutions are not known, it is difficult to tell exactly which species are formed. Assuming that the consumption of NaOH is the stoichiometrically lowest possible, the hydrolysis reaction may be written as

$$Si(OC_2H_5)_4 + 3H_2O + OH^- \rightarrow Si(OH)_3O^- + 4C_2H_5OH$$

$$(3)$$

Here charged $Si(OH)_3O^-$ may coexist with $Si(OH)_4$ and/or $Si(OH)_2O_2^{2-}$ depending on the acid dissociation constants of silicic acid.

Fig. 5 Amount of silica formed (mol/l) when the following parameters are varied: **a** concentration of tetraethyl orthosilicate (*TEOS*), **b** concentration of NaOH, **c** temperature and **d** concentration of water. When the parameters are not varied, they have the following values: concentration of TEOS = 0.3 M, concentration of NaOH = 0.005 M, concentration of water = 2.0 M and temperature = 23 °C

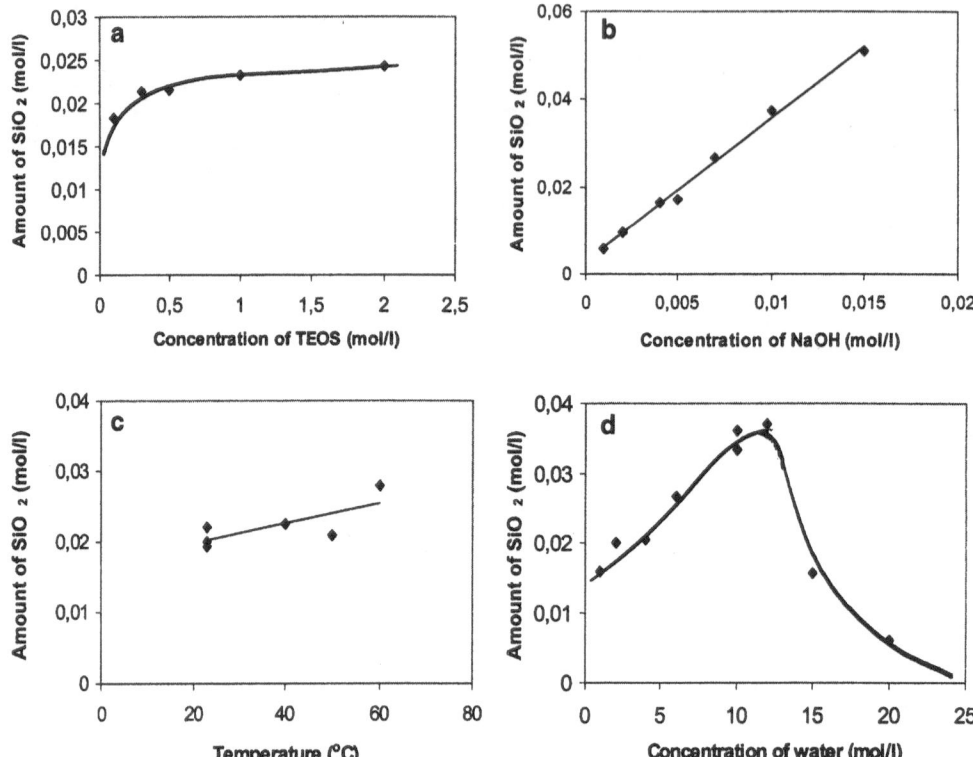

It was not the purpose of this investigation to study the intermediates formed. In one case, however, the final concentrations of TEOS, SiO_2 and nonvolatile soluble material were determined. The starting concentrations of TEOS, water and NaOH were 0.3, 2.0 and 0.005 M, respectively, and the reaction time was 180 min at 23 °C as before. The final concentrations of SiO_2, nonvolatile soluble material (calculated as SiO_2) and TEOS were then estimated to be 0.02, 0.04 and more than 0.2 M, respectively. Here SiO_2 was estimated gravimetrically after centrifugation as before. Nonvolatile soluble material was estimated by drying the supernatant at 105 °C and TEOS was estimated volumetrically after dilution 10 times in water to induce phase separation. The presence of TEOS indicates that the hydrolysis has reached equilibrium owing to the formation of intermediate products or has stopped owing to a lack of catalyst. Since the amount of silica formed is nearly independent of the concentration of TEOS, the last explanation is preferred.

Few authors studying the ammonia-catalysed growth of silica seem to have focused on the amount of silica formed (seed growth techniques excluded). Matsoukas and Gulari [4] measured the particle mass by light scattering, but used much lower concentrations of TEOS than in this study. Furthermore, in the ammonia-catalysed systems hydroxyl ions are less likely to be a limiting factor than in the NaOH-catalysed case. Ammonia is typically present in the 0.1–3.0 M range. When hydroxyl ions are consumed, more hydroxyl ions will be formed from the reaction between ammonia and water.

At water concentrations above 12 M (Fig. 5d) there is a steep decrease in the amount of silica formed. Bridger et al. [15] have observed a phase separation under similar conditions (with ammonia as the base) at water concentrations higher than 15 M. A starting formation of a separate phase of TEOS cannot be ruled out also for this system. The reduced formation of silica may accordingly be due to a reduction in the amount of TEOS in the ethanolic phase. Our results indicate, however, a low correlation between the concentration of TEOS and the amount of silica formed. It is therefore more likely that the decrease in the formation of silica is due to an increase in the solubility of silica. It is to be expected that an acid oxide such as silica will dissolve when the solvent becomes more polar at this high pH.

Particle size and polydispersity

The effects of the different variables on the diameter of the silica particles and their polydispersity indices are shown in Fig. 2a and b, respectively. The starting concentration of NaOH is the variable influencing the particle diameters and the polydispersity index to the greatest extent. Within the concentration range studied, an increase in the starting concentration of NaOH resulted in a reduced particle size and more monodisperse particles.

Both the particle size and the polydispersity index are negatively correlated to the interaction term $x_1 x_4 = x_2 x_3$.

This interaction has two possible interpretations. Since the main effects of the concentration of TEOS and the temperature (x_1 and x_4) are both small, their combined effect would be expected to be small. The interaction between water and NaOH (x_2 and x_3) is, therefore, believed to be the important one. This means that a simultaneous increase in the concentrations of water and NaOH will result in smaller and more monodisperse particles. To understand this, the ionisation of NaOH in ethanolic solutions must be considered. The ionic activity of NaOH in ethanol is probably much lower than unity [6]. If water is added, the solvent become more polar and the ionisation is expected to increase. This explains the interaction terms (x_2 and x_3) between water and NaOH. Also Lindberg et al. [12] found a strong negative interaction term, which could be interpreted as an interaction between ammonia and water; however, they were not able to distinguish this term from other possible interaction terms.

In the concentration range studied here, the particle size decreased with increasing concentration of NaOH. This may seem to be in contradiction with the results from the ammonia-catalysed systems where particle size is usually reported to increase with ammonia concentration up to at least 2 M [2, 3]. This discrepancy may, however, be explained if the chemical differences in the two systems are considered more carefully. If it is assumed that hydroxyl ions are consumed during the reactions, the reactions in the NaOH-catalysed systems will eventually stop owing to a lack of catalyst. In the ammonia-catalysed systems, ammonia is primarily present as unprotonated ammonia [11]. If hydroxyl ions are consumed, they will reform and thereby give higher concentrations of ammonium ions as mentioned previously. It is to be expected that a higher concentration of cations will reduce the thickness of the diffuse double layer [10] and will also affect the hydrolysis and condensation reactions if charged species are involved. It has been observed that the particle size of silica increases dramatically when salts such as NaCl are added [6]. The higher concentration of ammonium cations as a result of consumption of hydroxyl ions may, therefore, explain the increasing size with increasing ammonia concentration.

The results in Fig. 2a and b show that the particle size increases with the concentration of water, while the polydispersity index is not affected very much by this variable. The effect of water on the particle size may be complicated. In some experiments there may have been a lack of water since the concentration ratio of water to TEOS was as low as 2.0.

It is found that the size of the silica particles decreases with increasing starting concentration of NaOH, while the amount of silica formed increases in proportion to the concentration of NaOH. The number of particles or nuclei formed must therefore be higher when the concentrations of NaOH is high. This may be due to a rapider hydrolysis and a higher supersaturation before nucleation starts.

References

1. Kolbe G (1956) Dissertation. Jena
2. Stöber W, Fink A, Bohn E (1968) J Colloid Interface Sci 26:62
3. Bogush GH, Tracy MA, Zukoski IVCF (1988) J Non-Cryst Solids 104:95
4. Matsoukas T, Gulari E (1988) J Colloid Interface Sci 124:252
5. Matsoukas T, Gulari E (1988) J Colloid Interface Sci 145:557
6. Bogush GH, Zukoski IVCF (1991) J Colloid Interface Sci 142:1
7. Bogush GH, Zukoski IVCF (1991) J Colloid Interface Sci 142:19
8. Bailey JK, Mecartney ML (1992) Colloids Surf 63:151
9. Van Blaaderen A, Kentgens APM (1992) J Non-Cryst Solids 149:161
10. Van Blaaderen A, Van Geest J, Vrij A (1992) J Colloid Interface Sci 154:481
11. Chen SL, Dong P, Yang GH, Yang JJ (1996) Ind Eng Chem Res 35:4487
12. Lindberg R, Sundholm G, Pettersen B, Sjöblom J, Friberg SE (1997) Colloid Surf A 123–124:549
13. Box GEP, Hunter WG, Hunter JS (1978) Statistics for Experimenters. Wiley, New York
14. Sirius 6.0. Pattern Recognition Systems AS, Bergen, Norway
15. Bridger K, Fairhurst D, Vincent B (1979) J Colloid Interface Sci 68:190

Progr Colloid Polym Sci (2000) 116:74–78
© Springer-Verlag 2000

F. Joabsson
B. Lindman

Interfacial interaction between ethyl(hydroxy-ethyl)cellulose and sodium dodecyl sulphate

F. Joabsson (✉) · B. Lindman
Department of Physical Chemistry 1
Center for Chemistry
and Chemical Engineering
Lund University
P.O. Box 124, 221 00 Lund, Sweden
e-mail: fredrik.joabsson@fkem1.lu.se
Tel.: +46-46-2220134
Fax: +46-46-2224413

Abstract The effect of surfactants on the adsorption properties of ethyl(hydroxyethyl)cellulose (EHEC) and its hydrophobically modified analogue (HM-EHEC) at the solid–liquid interface has been studied by ellipsometry. On the polar silica surface, a small addition of the anionic surfactant sodium dodecyl sulphate (SDS) caused a threefold to fivefold expansion of a preadsorbed HM-EHEC layer, while the amount adsorbed was influenced less. On the hydrophobised silica surface, SDS could replace EHEC (above 10 mM SDS), while some adsorbed HM-EHEC could still be detected at 14 mM SDS. The nonionic surfactant octa(ethylene oxide) dodecyl ether did not affect the adsorbed layer structure on silica. It is proposed that the layer structure is governed by polymer–surfactant interactions and/or competitive adsorption, depending on the adsorption characteristics of the individual cosolutes.

Key words Mixed adsorption · Polymer · Surfactant · Ellipsometry

Introduction

Mixed polymer–surfactant systems are extensively used for surface modification, for example, for the stabilisation of suspensions of solid particles in aqueous systems, one major application being water-borne paints. The study of adsorption from mixed polymer–surfactant solutions is, therefore, of considerable practical interest, with an obvious need for understanding the behaviour of polymer–surfactant solutions at both polar and nonpolar surfaces. However, any deeper understanding of such complex systems presupposes a thorough knowledge of the behaviour of the individual cosolutes. This has only started to develop during the last few years. Most of the previous work has focused on polymer–surfactant interactions in solution, such as phase behaviour and rheology [1–7]; however, only a few investigations have dealt with polymer–surfactant associations at interfaces [8–13]. With the present investigation, we want to illustrate the importance of interfacial interactions in associating polymer–surfactant systems and try to gain deeper understanding of such interactions.

Ethyl(hydroxyethyl)cellulose (EHEC) is known to form mixed aggregates with ionic surfactants in aqueous solution [3, 6, 14]. EHEC interacts especially strongly with anionic surfactants, and it is generally found that anionic surfactants interact more strongly than cationic surfactants with nonionic polymers. The polymer acts as a nucleation centre for surfactant micelle formation at concentrations well below the critical micelle concentration (cmc) of the surfactant. This lower surfactant concentration, at which polymer–surfactant complexes start to form, is referred to as the critical aggregation concentration (cac). The complexation is often presented quantitatively as binding isotherms, where the fraction or concentration of bound surfactant is plotted as a function of the free or total surfactant concentration. For slightly hydrophobic polymers, such as EHEC, the binding isotherms resemble those of surfactant micellisation on aqueous solution, with a steep increase in bound surfactant around the cac indicating a cooperative complex formation. If the polymer is made more hydrophobic, i.e. by grafting hydrophobic groups onto the backbone, the binding of surfactant to these regions becomes less cooperative. To hydrophobically modified (HM) EHEC the surfactant can bind either to the slightly hydrophobic EHEC backbone (cooperative binding) or to the hydrophobic graft domains (noncooperative binding); therefore, the binding isotherm of ionic surfactants to HM-EHEC must be described by a two-step isotherm with an initial noncooperative part followed by the cooperative part analoguous to EHEC [3].

The feature of the binding isotherm often has consequences for the macroscopic behaviour of the polymer–surfactant system. The phase behaviour, the gel swelling and the rheology of the system undergo dramatic changes when the concentration of polymer-bound surfactant changes. Therefore, we also expect that the state of adsorbed HM-EHEC polymers might change when a complexing surfactant is added to the system. Ellipsometry provides a powerful tool for the study of polymer–surfactant systems. Even though the technique cannot directly resolve several adsorbed components, the high time resolution makes it possible to distinguish between surfactant and polymer adsorption in mixed systems owing to their separate kinetics.

Experimental

Materials

Both polymers (EHEC and HM-EHEC) were supplied by Akzo Nobel Surface Chemistry. According to the manufacturer they have approximately the same molecular weight ($M_w \approx 100\ 000$) and degree of ethyl and hydroxyethyl substitution ($DS_{ethyl} = 0.6$–0.7 and $MS_{EO} = 1.8$, respectively). The hydrophobic modification of HM-EHEC consists of a low number of branched nonylphenol groups grafted onto the EHEC chain. The degree of substitution is 1.7 mol% relative to the number of glucose residues as measured by UV absorbance at 275 nm, using phenol in aqueous solution as a reference, which corresponds to 6–7 HM groups per polymer chain on average. The polymers were purified by extensive dialysis and freeze-drying.

Sodium dodecyl sulphate (SDS), obtained from BDH, was further purified by triple recrystallisation from water and stored in a freezer at −20 °C. SDS solutions were kept for no longer than a couple of hours before being used in the experiments. Octa(ethylene oxide) dodecyl ether ($C_{12}E_8$) and cetyltrimethylammonium chloride were obtained from Nikko Chemicals, Tokyo, and Merck and were used in the experiments as received.

Polished silicon wafers (boron-doped, resistivity 1–20 Ω cm) were purchased from Okmetic. They were prepared for the experiment as described elsewere [15]. Octyldimethylchlorosilane was used as a hydrophobising agent.

Ellipsometry

A detailed description of the underlying principles of ellipsometry can be found elsewhere [16]. The ellipsometer used in this investigation was a modified automated Rudolph Research thin-film null ellipsometer, type 43603-200E. The equipment allows

monitoring of Ψ and Δ with a typical accuracy of 0.001° and 0.005°, respectively, and the time resolution is 3–4 s. All measurements were performed with an incident angle of 67.2° and at the wavelength 401.5 nm in a thermostated cuvette at 25 ± 0.1 °C. Magnetic stirring was applied in all measurements at about 300 rpm.

The substrate was characterised before measurement by determining the ellipsometric angles, Ψ and Δ, in two ambient media (air followed by water). A detailed description of the method may be found in Ref. [17].

The measured Ψ and Δ in the presence of polymer and/or surfactant in solution were analysed numerically by using a four-layer model, assuming optically isotropic and planar layers. From the fit, the refractive index and the layer thickness of the adsorbed layer were obtained. The amount adsorbed was calculated from de Feijter's formula [18]. For the calculation a refractive index increment, $dn/dc = 0.15$ and 0.12 ml g^{-1} [19] for the polymers and surfactants, respectively, was used to calculate the bulk refractive index, while for the calculation of the amount adsorbed $dn/dc = 0.15$ was used throughout. The experimental errors in Ψ and Δ were used to estimate the uncertainties in n, d and Γ by fitting the adsorbed layer parameters to all combinations of the Ψ and Δ error limits.

Results and discussion

An investigation of adsorption from mixed systems requires a detailed study of the adsorption of the individual cosolutes. This was done for both polymers, EHEC and HM-EHEC as well as for SDS and $C_{12}E_8$ on both silica and hydrophobised silica. The results are not presented here, but we will shortly summarise the main features. Both EHEC and HM-EHEC adsorb to the silica–water interface even though the amounts adsorbed are rather moderate (0.5–0.6 mg m^{-2}). A pseudoplateau is established above approximately 10^{-3} wt% polymer in bulk. At the plateau, the adsorbed layers are rather thick (200–300 Å), which gives a very low apparent layer concentration (around 2 wt%). On hydrophobised silica, both polymers adsorb extensively, forming thick layers which are denser than on silica (about 1.5 mg m^{-2} and 200 Å for EHEC and about 3 mg m^{-2} and 300 Å for HM-EHEC). Here, no distinct plateau is observed for HM-EHEC, but the amount adsorbed increases up to at least 0.1 wt% in the bulk. SDS adsorbs only to hydrophobised silica, forming a monolayer with an adsorption plateau (0.8 mg m^{-2}) established at about 2 mM SDS in bulk. $C_{12}E_8$ adsorbs cooperatively to silica [20], forming a micellar structure, and noncooperatively to hydrophobised silica [21], with a monolayer as the result.

By adding surfactants to a polymer solution in contact with a surface, the interfacial layer structure may change in different ways. If the surfactant associates with the polymer it may cause synergistic adsorption with both the surfactant and the polymer at the interface or it may eliminate the polymer from the surface if the polymer–surfactant complexes are much more soluble than the polymer itself. Another scenario, which may also be

76

important for surfactants which do not associate with the polymer, is competitive adsorption. If the surfactant can effectively interact more strongly with the interface, the polymer may desorb in favour of the surfactant.

Effect of SDS addition to HM-EHEC layers preadsorbed on silica

SDS was added to a system with a HM-EHEC layer preadsorbed on silica. SDS itself does not adsorb to silica but it is clear from Figs. 1 and 2 that SDS influences the structure of a preadsorbed HM-EHEC layer. In the concentration range investigated, the amount adsorbed is only slightly affected by SDS addition, while the layer thickness increases greatly. Between 4 and 8 mM SDS (cmc ≈ 8 mM), the layer undergoes a threefold expansion.

An expansion pattern similar to what we observe was found by Rosén and Piculell [22] in an investigation of the swelling of a covalently cross-linked EHEC gel on SDS addition. They observed a twofold isotropic expansion of an 1 wt% EHEC gel from about 3 mM SDS to a maximum at 10 mM.

Changing the EHEC to HM-EHEC gave qualitatively the same result (Fig. 2a), but only if the polymer was first removed from the solution. The expansion starts at slightly higher SDS concentrations than for EHEC/SDS (above 6 mM) and the layer expands to more than 5 times the initial value. The larger expansion could possibly be attributed to two main differences between the EHEC and HM-EHEC systems. HM-EHEC has more hydrophobic character than EHEC and can therefore bind more SDS. Also, extensive binding of

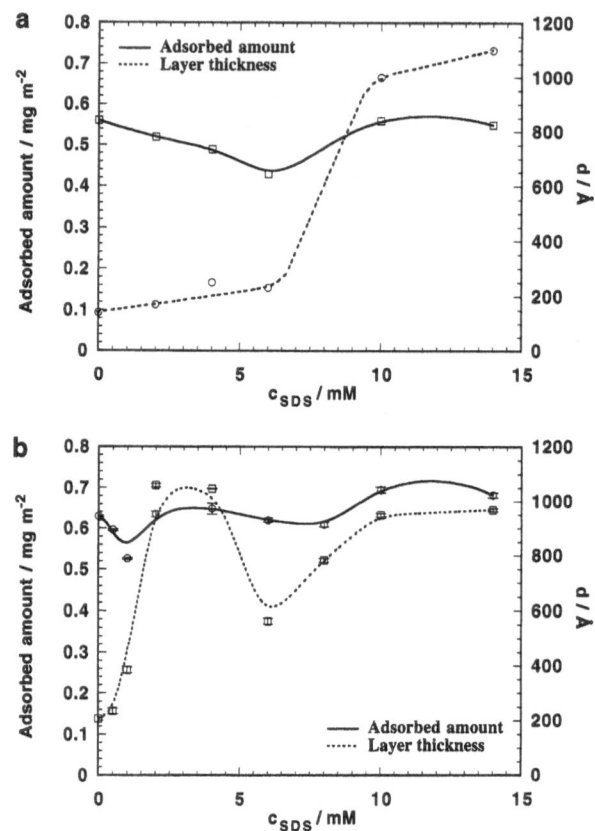

Fig. 2 a The effect of SDS addition on the apparent amount adsorbed and the thickness of a hydrophobically modified (*HM*) EHEC layer preadsorbed from a 0.01 wt% solution onto silica. The polymer was removed from the bulk solution before addition of SDS. The *lines* are drawn just to guide the eye and are not a result of data fitting. **b** The effect of SDS addition on the apparent amount adsorbed and the thickness of a HM-EHEC layer preadsorbed from a 0.01 wt% solution onto silica. The polymer was not removed from the solution before addition of SDS. The *lines* are drawn just to guide the eye and are not a result of data fitting

SDS to the hydrophobic physical cross-links of HM-EHEC eventually causes dissolution of the cross-links and thus allows a more extended layer structure. Again a parallel may be drawn between interfacial layer expansion and gel swelling. Piculell et al. [5] observed a higher degree of swelling for a HM hydroxyethylcellulose gel than the corresponding unmodified one on SDS addition.

If, on the other hand, HM-EHEC is not removed from the solution, we observe a large expansion also at SDS concentrations below 6 mM. Here, the layer swells to about 1100 Å at 2–4 mM SDS, followed by a shrinkage between 4 and 6 mM SDS. The origin of this difference is still somewhat unclear; however, surface force measurements of the HM-EHEC/SDS system indicate the existence of a "loose highly deformable layer" outside the more compact inner layer. Also, ellipsometry has shown that if HM-EHEC is removed from solution by rinsing with water (no SDS present),

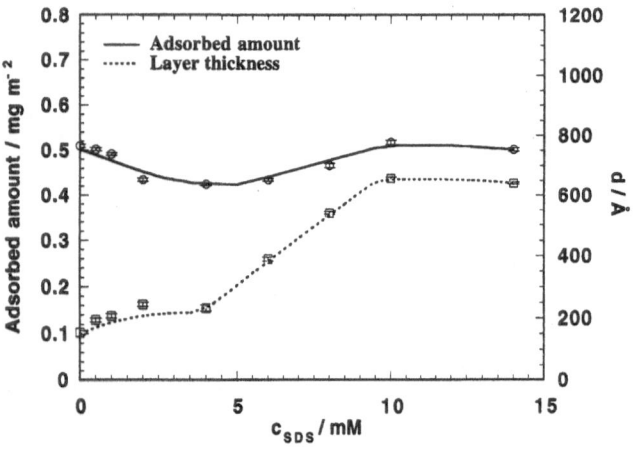

Fig. 1 The effect of sodium dodecyl sulphate (*SDS*) addition on the apparent amount adsorbed and the thickness of an ethyl(hydroxyethyl)cellulose (*EHEC*) layer preadsorbed from a 0.01 wt% solution onto silica. The polymer was not removed from the solution before addition of SDS. The *lines* are drawn just to guide the eye and are not a result of data fitting

both the amount adsorbed and the layer thickness decrease by roughly 10%. This is not seen as markedly for EHEC as for HM-EHEC, which may suggest the existence of a multilayer in the case of adsorbed HM-EHEC. If the outer HM-EHEC is anchored to the inner layer through the hydrophobic groups, SDS may solubilise these connections, causing an expansion of the layer and eventually desorption of polymer–surfactant complexes. (We note that for the EHEC–SDS system on silica, the two experimental procedures gave qualitatively the same results with differences within the experimental error.)

On SDS addition the polymer–surfactant complexes become increasingly soluble since the binding of SDS to the HM-EHEC makes the complexes negatively charged. This facilitates desorption of the complexes from the surface. Well over the cmc of SDS the adsorbed material starts to desorb, but the process is too slow (days) to be followed by ellipsometry. Since the silica is negatively charged, the local SDS concentration will be low close to the surface, which may slow the desorption process. By approaching the problem from the other side, i.e. by adding premixed polymer–surfactant solution in the concentration range 15–20 mM SDS/0.01 wt% HM-EHEC, no adsorption was observed.

The effect of adding a nonionic surfactant, $C_{12}E_8$, to a preadsorbed HM-EHEC layer is shown in Fig. 3. An increase in the amount adsorbed is observed around 0.05 mM $C_{12}E_8$, while no effect of the layer thickness is observed. The onset of adsorption corresponds well to

the cooperative adsorption of $C_{12}E_8$ on silica alone, even though the presence of HM-EHEC suppresses the amounts of $C_{12}E_8$ adsorbed. These observations suggest that the surfactants self-assemble primarily at the surface, but that the aggregate formation is somewhat sterically hindered by the presence of polymer, which lowers the amount adsorbed. For EHEC–$C_{12}E_8$ the curves look qualitatively the same as for HM-EHEC–$C_{12}E_8$.

Effect of SDS addition to HM-EHEC layers preadsorbed on hydrophobised silica

SDS was added to a system with a HM-EHEC layer preadsorbed on hydrophobised silica. The results are shown in Figs. 4 and 5. At low SDS concentrations,

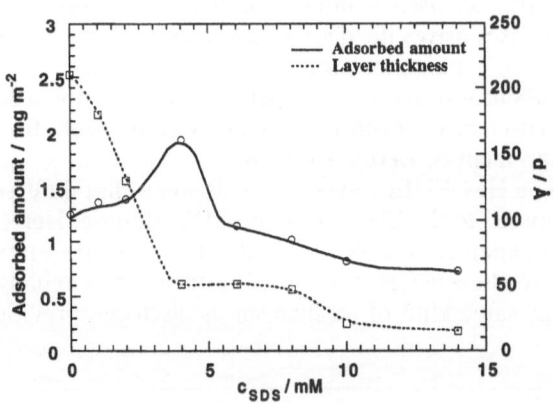

Fig. 4 The effect of SDS addition on the apparent amount adsorbed and the thickness of an EHEC layer preadsorbed from a 0.01 wt% solution onto hydrophobised silica. The polymer was not removed from the solution before addition of SDS. The *lines* are drawn just to guide the eye and are not a result of data fitting

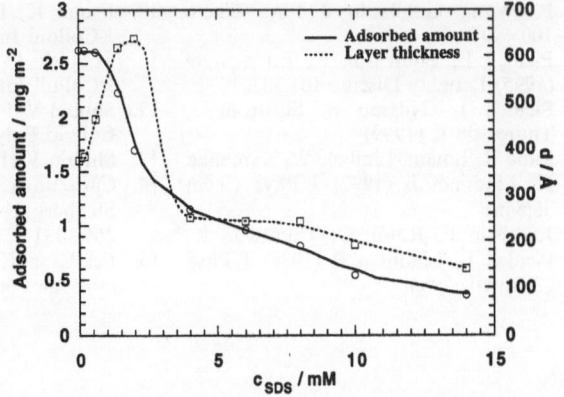

Fig. 5 The effect of SDS addition on the apparent amount adsorbed and the thickness of a HM-EHEC layer preadsorbed from a 0.01 wt% solution onto hydrophobised silica. The polymer was not removed from the solution before addition of SDS. The *lines* are drawn just to guide the eye and are not a result of data fitting

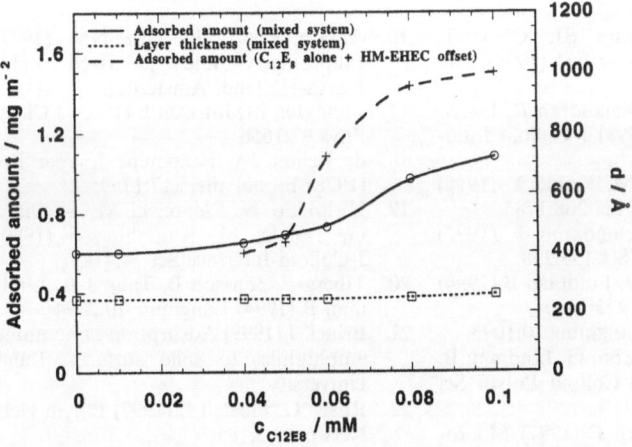

Fig. 3 The effect of octa(ethylene oxide) dodecyl ether ($C_{12}E_8$) addition on the apparent amount adsorbed and the thickness of a HM-EHEC layer preadsorbed from a 0.01 wt% solution onto silica. The polymer was not removed from the solution before addition of $C_{12}E_8$. The *lines* are drawn just to guide the eye and are not a result of data fitting. The *dashed line* shows the adsorption of $C_{12}E_8$ to silica alone (data from Ref. 20), with the modification that the EHEC plateau amount adsorbed on silica (0.6 mg m^{-2}) has been added to the original values. This has been done to facilitate comparison with the mixed system

78

0–4 mM, the effect is entirely different for preadsorbed EHEC compared to HM-EHEC. In the case of EHEC, the layer thickness decreases by a factor of 4 and the amount adsorbed still increases. A similar initial decrease was observed by Cosgrove et al. [8] for poly(ethylene oxide)/SDS on a polystyrene latex. They observed a decrease in the hydrodynamic thickness up to SDS concentrations around the cmc. The simultaneous decrease in the layer thickness and the increase in the amount adsorbed suggests that EHEC and SDS initially form mixed aggregates at the hydrophobic surface. In this regime, SDS is not able to force EHEC away from the surface, but synergistic adsorption is observed instead; however, at higher concentrations, above the cmc for SDS, essentially all EHEC is replaced by SDS (Fig. 4). Both the amount adsorbed and the layer thickness correspond to an SDS monolayer. From our results it is difficult to tell whether competitive adsorption or increased solubility of the polymer–surfactant complexes drives the desorption process. However, since there were indications for desorption above the cmc also at the silica–water interface, where SDS does not adsorb, the latter mechanism is most likely responsible for the replacement of EHEC for SDS.

The HM-EHEC layer responds somewhat differently compared to the EHEC layer on SDS addition. Here, the layer expands to twice its initial value, while the amount adsorbed decreases (Fig. 5). We attribute this expansion to the same kind of mechanism as discussed previously for the silica surface. Figure 5 also shows that SDS is not able to fully displace HM-EHEC up to 14 mM SDS, which is indicated by the still rather thick adsorbed layer.

Conclusion

It is evident from our investigation, that the structure of adsorbed HM-EHEC is significantly changed by addition of SDS. On the silica–water interface the adsorbed polymer layers show an expansion pattern, reflecting the bulk binding isotherms. The expansion was qualitatively similar for both polymers, but the HM-EHEC showed a much larger expansion than EHEC (fivefold compared to threefold expansion). The HM-EHEC expansion pattern was sensitive to whether there was polymer in the solution or not. If the polymer was removed from the solution, the expansion pattern was similar to that of EHEC, but if the polymer remained in the solution an additional expansion peak was observed at low SDS concentration. This was attributed to effects from multilayer adsorption. On hydrophobised silica, EHEC could be completely replaced by SDS, while adsorbed HM-EHEC was found at least up to 14 mM SDS.

Acknowledgements This work was financed by Akzo Nobel Suface Chemistry and the Competence Center for Amphiphilic Polymers. F.J. is grateful to Bengt Jönsson, Johanna Brinck and Krister Eskilsson for stimulating discussions and intrumental advice.

1. Piculell L, Guillemet F, Thuresson K, Shubin V, Ericsson O (1996) Adv Colloid Interface Sci 63:1
2. Sivadasan K, Somasundaran P (1990) Colloids Surf 49:229
3. Thuresson K, Söderman O, Hansson P, Wang G (1996) J Phys Chem 100:4909
4. Piculell L, Thuresson K, Ericsson O (1995) Faraday Discuss 101:307
5. Piculell L, Nilsson S, Sjöström J, Thuresson K (1999)
6. Zana R, Binana-Limbelé W, Kamenka N, Lindman B (1992) J Phys Chem 96:5461
7. Joabsson F, Rosén O, Thuresson K, Piculell L, Lindman B (1998) J Phys Chem 102:2954
8. Cosgrove T, Mears SJ, Obey T, Thompson L, Wesley RD (1999) Colloids Surf 149:329
9. Argillier JF, Ramachandran R, Harris WC, Tirrell M (1991) J Colloid Interface Sci 146:242
10. Esumi K, Iitaka M, Koide Y (1998) J Colloid Interface Sci 208:178
11. Maltesh C, Somasundaran P (1992) J Colloil Interface Sci 153:298
12. Shubin V, Petrov P, Lindman B (1994) Colloid Polym Sci 272:1590
13. Shubin V (1994) Langmuir 10:1093
14. Carlsson A, Karlström G, Lindman B, Stenberg O (1988) Colloid Polym Sci 266:1031
15. Eskilsson K, Tiberg F (1997) Macromolecules 30:6323
16. Azzam RMA, Bashara NM (1977) Ellipsometry and polarized light. North-Holland, Amsterdam
17. Landgren M, Jönsson B (1993) J Chem Phys 97:1656
18. de Feijter JA, Benjamins J, Veer FA (1978) Biopolymers 17:1759
19. Nishikido N, Shinozaki M, Sugihara G, Tanaka M, Kaneshina S (1980) J Colloid Interface Sci 74:474
20. Tiberg F, Jönsson B, Tang J-A, Lindman B (1994) Langmuir 10:2294
21. Brinck J (1999) Adsorption of nonionic amphiphiles to solid surfaces. Lund University
22. Rosén O, Piculell L (1997) Polym Gels Networks 5:185

Progr Colloid Polym Sci (2000) 116:79–83
© Springer-Verlag 2000

E. Poptoshev
A. Carambassis
M. Österberg
P. M. Claesson
M. W. Rutland

Comparison of model surfaces for cellulose interactions: elevated pH

E. Poptoshev · P. M. Claesson
M. W. Rutland (✉)
Department of Chemistry
Surface Chemistry
Royal Institute of Technology and
Institute for Surface Chemistry
PO Box 5607, 114 86
Stockholm, Sweden

A. Carambassis
School of Chemistry, University of Sydney
NSW 2006, Australia

M. Österberg
Helsinki University of Technology,
Laboratory of Forest Product Chemistry
Box 6300, 020 15 HUT, Finland

Abstract Two different substrates have been used to measure interaction forces between cellulose and between cellulose and glass at normal and high pH. Forces between microspheres of cellulose ($r = 20$–30 μm) have been measured using the colloidal probe atomic force microscopy technique. Interactions between Langmuir–Blodgett cellulose films on a hydrophobised mica substrate and a glass sphere have been determined with the noninterferometric surface force apparatus. Also, the interaction between two identical Langmuir–Blodgett cellulose films determined with the interferometric surface force apparatus is given for comparison. At low pH (5.5–6) the interaction at large separations in both systems is characterised by a double-layer repulsion with an electrosteric contribution dominating the shorter-range regime. At pH 10, the Langmuir–Blodgett cellulose film swells considerably, which generates a long-range steric repulsion. In many cases several inward steps have been observed in the force–distance curves. We attribute this to a sudden partial collapse of the swollen cellulose film. After initial compression of the steric layer (upon consecutive force runs) the long-range interaction is again dominated by a double-layer force. In contrast, measurements between two cellulose spheres have shown no excessive swelling. Only a limited increase (from about 10 nm to about 20 nm per surface) of the range of the electrosteric repulsion has been found at pH 10. The force at longer distances is in good agreement with the Poisson–Boltzmann theory, with the surface potential increasing with pH as expected.

Key words Surface forces · Cellulose · Surface interactions · Paper

Introduction

The interactions in papermaking and recycling systems have long been of interest to both scientists and industry. It is only recently, however, that developments in surface force measuring techniques have permitted direct interactions to be determined. While the practical systems are extremely complex with many additives, it is now possible to measure the interactions between model cellulose surfaces in the presence of various additives [1–5] and between cellulose and another material – for example, a model filler [6, 7] or a model ink particle.

There are three types of surfaces that have been tried as models for cellulose. The first, a spin-coating of cellulose [1] on a smooth mica substrate, broke new ground in showing the possibility for such measurement and led to a new model for the cellulose surface – the "dangling-tail model". An investigation using this substrate showed that the cellulose surface is both charged and swollen with water in an aqueous environment. Experimental difficulties with the system, however, led to two other types of second generation model surfaces being investigated, in which the approaches were very different.

In one case, multilayer films of silanated cellulose are deposited on hydrophobised mica using the Langmuir–Blodgett technique [8]. Typically ten layers are deposited and then desilylated through exposure to HCl gas. This procedure produces very homogeneous, smooth amorphous cellulose layers suitable for investigation using traditional surface force measurement [9, 10] and if interferometric techniques are used the thickness of the cellulose film and any adsorbate can be determined. The other technique utilises the capacity of the atomic force microscope (AFM) to be used as a force measuring device using the so-called colloidal probe technique [11, 12]. In this case, precipitated spheres of cellulose II are employed, with one attached to an AFM cantilever and the other to any convenient substrate. The spheres are approximately 35% crystalline [13].

Up to now most measurements have been performed at ambient pH, reflecting the novelty of the measurement systems. Of course, most paper systems operate at elevated pH and to investigate how cellulose behaves in such environments it is important that experiments to model interactions also be performed at high pH. Thus, it is necessary to evaluate the behaviour of the different model surfaces at higher pH with a view to determining whether either or both of them can be said to accurately reflect the behaviour of real cellulose and their robustness to the harsher aqueous environment.

Materials and methods

Materials

Potassium hydroxide and potassium chloride were obtained from Aldrich Chemical Company. Sodium hydroxide, sodium chloride and arachidic acid (AA) were from Sigma. All the chemicals were of analytical grade and were used without further purification. Eicosilamine (EA) was prepared according to Berg et al. [14]. Equal amounts of EA and AA were dissolved in a mixture of 95% chloroform and 5% absolute ethanol to obtain a solution with a total concentration of 2.7 mM. Trimethylsilyl cellulose (TMSC) was synthesised in our laboratory following a method described elsewhere. Cellulose spheres were purchased from Kanebo Co., Japan [13]. The particle size was between 20 and 50 μm with a crystallinity of 5–35%.

Sample preparations

For the AFM colloidal probe measurements cellulose particles were glued to an AFM cantilever (PSI, Sunnyvale, Calif.) using a commercial silicone glue (Dow Corning 734). Particle attachment was done using an Olympus optical microscope with an XYZ translation stage. One particle was glued to a tipless cantilever and another to an oxidised silicon wafer. Details of the sample preparation procedures can be found elsewhere [3].

Flat model cellulose surfaces used in a MASIF (modified version of Australian Scientific Instruments) force apparatus were prepared by the Langmuir–Blodgett technique [8]. An automatic Langmuir–Blodgett balance system from KSV Instruments, Finland, was used for the deposition. A freshly cleaved piece of muscovite mica (about 1 cm^2) was glued to a stainless-steel surface holder using Epicote 1004 epoxy resin. Hydrophobisation was achieved by depositing a mixed monolayer of AA/EA at a constant surface pressure of 30 mN/m and at a deposition rate of 5 mm/min. Ten monolayers of TMSC were then deposited on the hydrophobised substrate at a surface pressure of 15 mN/m. After the deposition was complete the TMSC film was converted to cellulose by exposing it to a humid HCl vapor for 1 min.

All the procedures regarding surface preparation were carried out inside a laminar flow cabinet in order to minimise the risk of airborne contamination.

Surface force measurements

The forces acting between a flat cellulose surface and a glass sphere were measured using the noninterferometric surface force apparatus (SFA) widely known as MASIF. A detailed description of the instrument and the technique is given in numerous earlier works [10, 15]. Here only a brief outline is given.

Both interacting surfaces are attached to piezoelectric materials. The upper surface is attached to a piezoelectric actuator and the lower (flat) surface to a bimorph force sensor. All the assembly is enclosed in a liquid cell (volume about 10 ml) and mounted on a horizontal translation stage. During a force run, the distance between the surfaces is varied by applying a ramp voltage to the piezoelectric tube actuator. Simultaneously the signal from the force sensor (charge produced upon bending) is amplified and recorded. After a hard wall contact is established, the linear displacement of the piezoelectric tube is transmitted directly to the bimorph, and the deflection can be obtained from the regime of constant compliance. The actual force is then obtained by multiplying the deflection by the bimorph spring constant.

Glass surfaces were prepared by melting one end of a borosilicate glass rod in the flame of a butane–oxygen burner until a droplet with radius of about 2 mm was formed. At the end of each experiment the radius, R, was measured with a micrometer and the force scaled by this radius was related to the free energy of interaction per unit area, G_f, according to the Derjaguin approximation [16]:

$$F/R = 2\pi G_f \ .$$

Other experimental techniques (AFM, interferometric SFA) were described previously [3, 5].

Results and discussion

To allow full comparison of the interactions at elevated pH, it is first necessary to present a brief summary of analogous measurements at ambient pH, which are published elsewhere [7]. Interaction profiles on approach between a flat cellulose plate and a glass sphere across an aqueous solution containing 0.1 and 1 mM NaCl at pH 5.5–6.0, taken from Ref. [7], are shown in Fig. 1. The resolution of this measurement is sufficient to allow the relatively small surface potential on the cellulose surface to be calculated. The potentials at both the glass and cellulose surfaces were used as fitting parameters, although typical values for the glass surface potential obtained earlier [17] were used as a guide. The results indicate that there are two regimes for the interaction. At

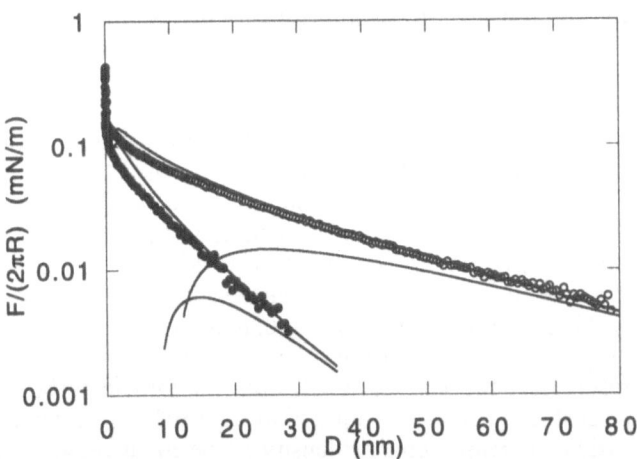

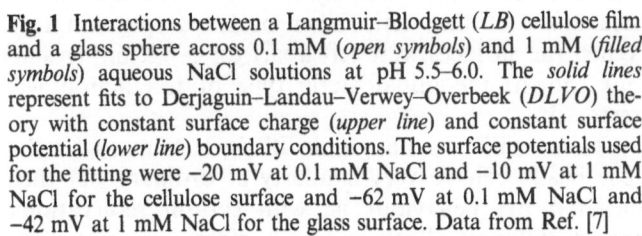

Fig. 1 Interactions between a Langmuir–Blodgett (*LB*) cellulose film and a glass sphere across 0.1 mM (*open symbols*) and 1 mM (*filled symbols*) aqueous NaCl solutions at pH 5.5–6.0. The *solid lines* represent fits to Derjaguin–Landau–Verwey–Overbeek (*DLVO*) theory with constant surface charge (*upper line*) and constant surface potential (*lower line*) boundary conditions. The surface potentials used for the fitting were −20 mV at 0.1 mM NaCl and −10 mV at 1 mM NaCl for the cellulose surface and −62 mV at 0.1 mM NaCl and −42 mV at 1 mM NaCl for the glass surface. Data from Ref. [7]

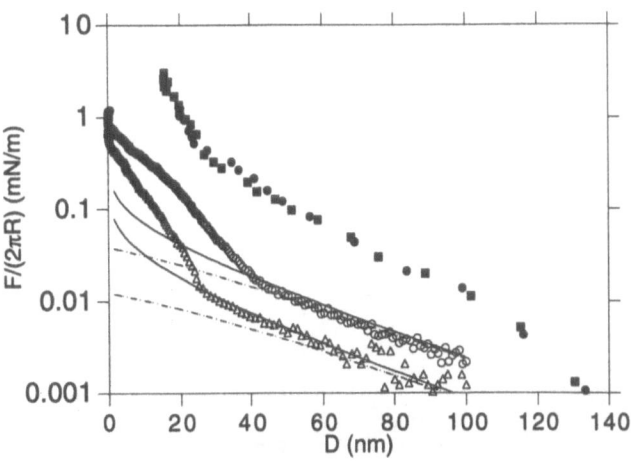

Fig. 2 Interactions between two cellulose surfaces at high pH. *Filled symbols* represent the interaction at pH 9.5 as measured between LB cellulose films with the interferometric surface force apparatus (data from Ref. [5]). *Open symbols* represent interactions between cellulose beads measured with the atomic force microscopy colloidal probe technique at pH 10.0. *Open triangles*, immediately after increasing the pH, and *open circles*, after 2 h incubation. The *lines* are fits to Poisson–Boltzmann theory assuming constant surface charge (*solid lines*) and constant surface potential (*broken lines*) The potentials used were −24 mV for the short-time case and −40 mV for the long-time case, respectively (from Ref. [3])

large separations the measured interactions appear to be adequately described by the Poisson–Boltzmann theory. At very short separations (below about 3 nm) an additional repulsive force prevents the surfaces from coming into adhesive contact. We argue that this electrosteric type of force appears as a result of surface swelling of the charged cellulose film in contact with aqueous solutions. In a swollen state, a large number of weakly charged cellulose tails extend into solution and generate steric repulsion upon compression.

The surface potential calculated for the cellulose surface at 0.1 mM (−20 mV) was of the same order as that calculated for cellulose beads [3] (data not shown) of −17 mV, although in that case the range of the electrosteric force was longer (10 nm per surface). Thus, qualitatively the two model surfaces display very similar behaviour and give a consistent picture of the charging and chain extension on the cellulose surface at ambient pH.

The results for cellulose–cellulose interactions at elevated pH from Refs. [3, 5] for the two types of substrate, measured with the AFM technique and with the interferometric SFA, are shown in Fig. 2. Once again, two regimes are observed, with the longer-ranged parts being consistent with a double-layer force and the shorter-ranged parts being an increased electrosteric force (compared to lower pH). In the case of the cellulose beads the double-layer force and the electrosteric force were found to be strongly time dependent, indicating that the charging and subsequent swelling process is quite slow in this case; however, the shape of the force profile

was not greatly different to that seen at lower pH. The surface potentials obtained (after 15 min and 2 h of −24 and −40 mV, respectively) are not true estimates of the "real" potential since the charge is spread over a thick layer; however, they do indicate semiquantitatively how the charging increases with pH and time.

In the case of the interferometric results, a similar trend was observed. The swelling and charging behaviour are both more significant at the elevated pH (in this case 9.5) compared to the cellulose data obtained from the fits in Fig. 1. However, in some cases small steps were observed in the force curves (filled circles) and the reproducibility of the measurements deteriorated. The interferometric data at first glance appear to be of greater magnitude than those for the AFM beads; however, most of this is due to differences in the definitions of zero separation. In the case of the interferometric measurements zero separation corresponds to contact of the hydrophobised mica surfaces, and in the case of the AFM beads to constant compliance. Thus, for a relevant comparison of the long-range forces the interferometric data should be offset by about 20 nm to shorter separations.

For the cellulose beads, the degree of swelling is much greater, with the electrosteric force having a range of about 20 nm per surface, defined with respect to constant compliance of the surfaces. This most probably reflects the different structures of the surfaces and the different degrees of polymerisation of the samples. (Note that it is

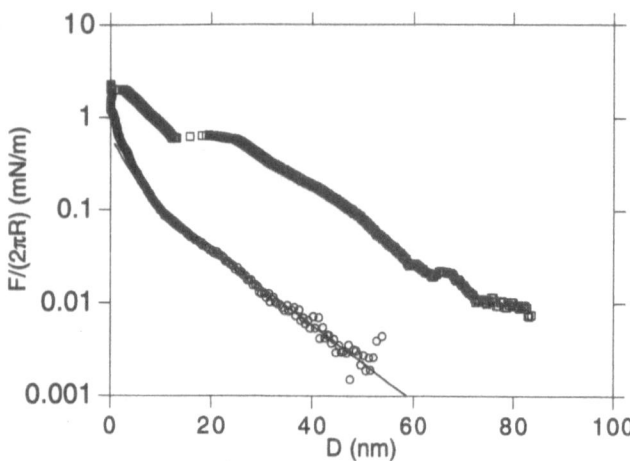

Fig. 3 Interactions Between a LB cellulose surface and a glass sphere at pH 10.0 also containing 1 mm NaCl measured with the MASIF technique. *Open squares* represent the interaction during the first approach; *open circles* represent a consecutive approach. The *solid line* is a fit to DLVO theory assuming interaction under constant surface charge boundary conditions with potentials of −35 mV at the cellulose surface and −80 mV at the glass surface

surface swelling rather than bulk swelling which is responsible for these steric forces, at least in the case of the cellulose beads, since they are 20 μm in diameter, i.e. approximately 4000 times thicker "films".)

Increasing the solution pH for the Langmuir–Blodgett film/glass system shown in Fig. 1, however, leads to a dramatic change in the force–distance profile. The interaction between cellulose and glass at pH 10 is shown in Fig. 3. When the surfaces are brought together for the first time (the so-called first approach) a long-range steric force is observed. Both the range and the magnitude of this force exceed by far those at lower pH. Up to three well-pronounced inward steps at separations of about 70, 20 and 5 nm appear in the force profile. It should be noted that such a stepwise increase of the force was repeatedly observed with different sets of surfaces and on different contact spots within one set of experiments, which excludes the possibility of artifacts; however, the number and position of the steps as well as the force magnitude at which they occurred varied substantially from experiment to experiment. We attribute this specific behaviour to a partial collapse of the swollen cellulose film induced by the compressive force.

At high pH the surface charge density of cellulose increases substantially, thus leading to increased repulsion between individual cellulose chains and between segments within a single chain, which then adopt an extended conformation. When such a layer is compressed, it appears that under a certain load the chains collapse to the next stable state. An alternative explanation would be to assume that whole cellulose monolayers are pushed out of the contact region. This is unlikely to be the case since more than ten such steps were observed

during repeated measurements, implying that all ten existing monolayers would have been pushed out, exposing the hydrophobic mica substrate; this was not observed.

After the steric layer has been compressed during the initial approaches, the force–distance profile changes (the lower curve in Fig. 3). The long-range steric repulsion is no longer present and the interaction is well described by Poisson–Boltzmann theory down to about 5 nm. Both the potentials at glass and cellulose surfaces are considerably higher than at pH 5.5–6.0, namely −82 mV on the glass and −27 mV on the cellulose. This is an expected result since the dissociation of both silanol and carboxylic surface groups is promoted at high pH. As discussed earlier, increased charge density is the main reason for the cellulose surface swelling. It should be noted that the relaxation of the steric layer occurred quite rapidly. When the surfaces were left apart for 5–10 min (results not shown) the initial steric type of interaction was restored.

Thus, while the Langmuir–Blodgett films and cellulose beads behave very similarly at ordinary pH their elevated pH behaviour is, with certain reservations, rather different. Both model surfaces increase their charge as the pH is raised and the range of the electrosteric force increases; however, the Langmuir–Blodgett films swell considerably after short equilibration times, displaying catastrophic collapse phenomena and achieving a reproducible, double-layer-like behaviour only after frequent compression cycles, after which they relax rapidly back to the extended swollen conformation. The cellulose beads display the same general behaviour as at ordinary pH, with increased range of the electric steric force. The range of this force increases slowly with time (in contrast to the rapid Langmuir–Blodgett film equilibration) and still increases even after several hours. These differences most likely reflect the different crystallinities of the two surfaces – the Langmuir–Blodgett films are thought to be quite amorphous, whereas the spheres are approximately 35% crystalline [13].

Conclusion

The behaviour of both types of model surface is entirely consistent with the observations in real paper systems that fibre "fluffing" is increased at elevated pH, enhancing subsequent paper strength. The increased charging and the extension of the cellulose chains into solution increases the repulsive interaction which would cause cellulose fibrils to separate and lead to fluffing. The two types of model cellulose surface investigated here do, however, show different equilibration and collapse behaviour associated with their differing structures. Since cellulose as it is found in the native state is highly crystalline, it might be thought that the cellulose beads constitute a "better" model. It is certainly true that they behave reproducibly from run to run; however, the mechanical treatment that cellulose fibres receive in papermaking and recycling systems means that the fibre surfaces are unlikely to preserve their native conformations, so conceivably the Langmuir–Blodgett films may be a better representation of the actual case. Further experiments, such as measuring the interaction between a real cellulose fibre and a cellulose bead, are required to adequately determine how well these substrates model the real case.

Acknowledgements E.P. acknowledges financial support from the Bo Rydins Foundation for Scientific Research. Johan Fröberg is acknowledged for providing the software for fitting Derjaguin–Landau–Verwey–Overbeek theory for uneven potentials to the experimental data.

References

1. Neuman RD, Berg JM, Claesson PM (1993) Nord Pulp Pap Res 8:96
2. Rutland MW, Carambassis A, Willing GA, Neuman RD (1997) Colloids Surf A 123–124:369
3. Carambassis A, Rutland MW (1999) Langmuir 19:5584–5590
4. Holmberg M, Berg J, Stemme S, Ödberg L, Rasmusson J, Claesson P (1997) J Colloid Interface Sci 186:369
5. Österberg M, Claesson PM (2000) J Adhes Sci Technol 14:603–618
6. Holmberg M, Wigren R, Erlandsson R, Claesson PM (1997) Colloids Surf A 129–130:175
7. Poptoshev E, Rutland MW, Claesson PM (2000) Langmuir 16:1987–1992
8. Schaub M, Wenz G, Wegner G, Stein A, Klemm D (1993) Adv Mater 5:919
9. Israelachvili JN, Adams GE (1978) J Chem Soc Faraday Trans I 74:975
10. Parker JL (1994) Prog Surf Sci 47:205
11. Ducker WA, Senden TJ (1992) Langmuir 8:1831
12. Ducker WA, Senden TJ, Pashley RM (1991) Nature 353:239
13. Okuma S, Yamagishi K, Masami H, Suzuki K, Yamamoto T (1986) Fine cellulose particles and process for production thereof. European Patent Office, vol 86/46
14. Berg JM, Eriksson LGT, Claesson PM, Börve N (1994) Langmuir 10:1225
15. Claesson PM, Ederth T, Bergeron V, Rutland MW (1996) Adv Colloid Interface Sci 67:119
16. Derjaguin B (1934) Kolloid Z 69:155
17. Poptoshev E, Rutland MW, Claesson PM (1999) Langmuir 15:7789–7794

Progr Colloid Polym Sci (2000) 116:84–94
© Springer-Verlag 2000

A. Dedinaite
P. M. Claesson
J. Nygren
I. Iliopoulos

Interactions between surfaces coated with cationic hydrophobically modified polyelectrolyte in the presence and the absence of oppositely charged surfactant

A. Dedinaite · P. M. Claesson (✉)
J. Nygren
Department of Chemistry
Surface Chemistry
Royal Institute of Technology
100 44 Stockholm and Institute for Surface
Chemistry, Box 5607, 114 86 Stockholm
Sweden

I. Iliopoulos
Laboratoire de Physico-Chimie
Macromoleculaire, UMR-7615, ESPCI
CNRS, Université Pierre et Marie Curie
ESPCI, 10 Rue Vauquelin, 75231 Paris
Cedex 5, France

Abstract The association between a cationic hydrophobically modified polyelectrolyte and an anionic surfactant was investigated in bulk solution and at a negatively charged solid surface. The bulk association was followed by measurements of turbidity and electrophoretic mobility. The maximum turbidity of the solution was found to closely coincide with the point of zero electrophoretic mobility of the aggregates. The forces acting between negatively charged mica surfaces across a dilute aqueous solution of the hydrophobically modified polyelectrolyte were monitored using surface force measurements. The presence of hydrophobic side chains on the polyelectrolyte leads to adsorption in an inner rather compact layer and an outer extended layer. After dilution only the inner layer remains adsorbed to the surface. In the next step, sodium dodecyl sulphate (SDS) was added.

It was found that the anionic surfactant is incorporated in the adsorbed layer even at very low bulk concentrations. As the surfactant concentration is increased stepwise the layer first swells and relaxes very slowly during compression. At higher SDS concentrations, desorption occurs. The interfacial properties of the hydrophobically modified polyelectrolyte alone and in mixtures with SDS are in many ways strikingly different to those of non-hydrophobically modified polyelectrolytes having a similar linear charge density. This is due to the importance of hydrophobic interactions between the hydrophobic side chains of the polyelectrolyte and between these groups and the nonpolar part of the surfactant.

Key words Surface forces · Polyelectrolyte · Surfactant · Sodium dodecyl sulphate · Polyelectrolyte–surfactant association

Introduction

The association behaviour of polyelectrolytes and oppositely charged surfactants has been extensively studied and there are many recent reviews and books covering different aspects of this fascinating topic [1–6]. The overwhelming majority of the studies on the subject are concerned with association in bulk solution. One rationale for this is the importance of polyelectrolyte–surfactant systems as rheology modifiers [5, 7], gelation agents [8], and solubilisers for sparingly soluble substances such as dyes [9, 10], perfumes, or pollutants [11, 12]. However, in many applications these mixtures, or the hydrophobically modified polyelectrolytes alone, are active at interfaces where they may control emulsion [13], dispersion [14] and foam stability [12, 15, 16], the wetting behaviour [17], and particle deposition [18]. It is fair to say that the body of knowledge about the properties of polyelectrolyte–surfactant mixtures at interfaces is too limited considering the technological importance of these systems. Among the experimental techniques that have been used in the study of interfacial properties of polyelectrolyte–surfactant mixtures and hydrophobically modified polyelectrolytes one can mention ellipsometry and reflectometry [19–23], specular neutron reflectivity [24], variable angle of incidence evanescent wave spectroscopy [25], and surface force instruments [19, 26]. Our research group has mainly used the latter technique to investigate interactions between polyelectrolytes and surfactants at solid–liquid interfaces [27–33]. One complication is the fact that the properties of the adsorbed layer in many cases depend not only on the surface and the bulk composition but also on the experimental path [20, 33, 34]. Hence, the approach towards true equilibrium is extremely slow. In particular, we have shown that the properties of the adsorbed layer depend crucially on whether it is formed by adsorption of preformed polyelectrolyte–surfactant aggregates or by allowing the surfactant to associate with preadsorbed polyelectrolyte layers [33]. In the latter case, it has been established that hardly any incorporation of surfactant in the adsorbed layer occurs until a critical surfactant concentration has been reached. This critical association concentration at the surface (cac_s) is different from the critical association concentration in bulk solution (cac_b), and it is expected to depend not only on the nature of the polyelectrolyte and the surfactant but also on the nature of the solid surface [29]. The study presented here is also concerned with interactions between preadsorbed polyelectrolyte layers and oppositely charged surfactants. It differs from our previous studies in that the polyelectrolyte used is highly hydrophobic, having 40% of the segments carrying a grafted dodecyl chain. This study is a continuation of our effort in establishing the relation between the chemical structure of polyelectrolytes and the properties of adsorbed polyelectrolyte layers in the absence and presence of oppositely charged surfactants. As will be seen in the following, the presence of the hydrophobic side chains has a significant impact on the structure of the polyelectrolyte layer before addition of

the surfactant, on the incorporation of surfactant in the adsorbed layer, and on the interactions between surfaces coated with polyelectrolyte–surfactant aggregates.

Materials and methods

The amphiphilic polyelectrolyte used in this investigation is shown in Fig. 1. It was obtained by quaternisation of a precursor, poly(vinylbenzylchloride) (PVBC), as described in Ref. [35]. In brief, the precursor PVBC was quaternised in two steps in $CHCl_3$. In the first step 40% of the VBC units were reacted with dimethyldodecylamine, while in the second one the remaining VBC units were quaternised with triethylamine. The final copolymer was purified by precipitation in dimethyl ether, followed by dissolution in water and freeze-drying. The composition of the copolymer was confirmed by NMR and elemental analysis. Owing to the synthesis route followed we expect a random distribution of the dodecyl side chains. The degree of polymerisation and the polydispersity of the precursor PVBC were estimated by size-exclusion chromatography in tetrahydrofuran. We found $DP_n \approx 100$ (molecular weight $\approx 3 \times 10^4$ g/mol) and $DP_w/DP_n \approx 2$. This polyelectrolyte will be referred to as 40 DT. Some data are also presented for another cationic polyelectrolyte, poly[2-(propionyloxy)ethyl]trimethylammonium chloride (PCMA). This polymer also carries one charge per segment but it has no grafted hydrophobic side chains. We note, however, that each cationic group is located at the end of a short side chain with partly hydrophobic character (Fig. 1). The molecular weight of the PCMA used is 1.6×10^6 g/mol (DP ≈ 8000); therefore it differs significantly from the hydrophobically modified polyelectrolyte. The surfactant, sodium dodecyl sulphate (SDS), especially pure grade for biochemical work (above 99% pure) obtained from BDH was used as received. The water was first pretreated with a Milli-RO 10 Plus system and was further purified with a Milli-Q Plus 185 system. KBr of pro analysi grade from Merck was roasted prior to use. Muscovite mica (Reliance, New York) was chosen as a substrate. It is a layered aluminosilicate mineral. Each layer is strongly negatively charged, about 2.1×10^{14} lattice charges/cm^2. The negative charge of the mica lattice stems from the fact that a quarter of the tetravalent Si atoms are substituted by trivalent Al atoms. In the crystal these charges are compensated mainly by K^+ ions located between the aluminosilicate sheets. Mica acquires a net negative charge when immersed in water or electrolyte solution. This is due to desorption of the K^+ ions located on the surface, which are only partly replaced by other positive ions present in solution.

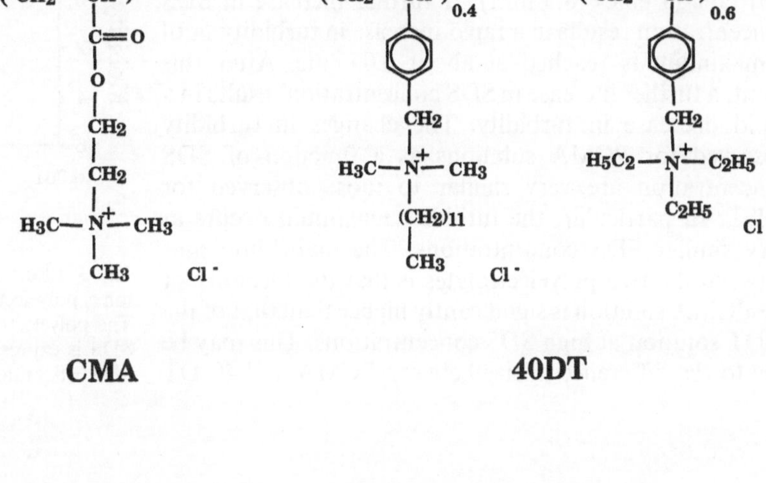

Fig. 1 The molecular structures of the monomers of the copolyelectrolyte 40 DT (see text) are shown together with that of poly[2-(propionyloxy)eth-yl]trimethylammonium chloride (PCMA). In 40 DT, 40% of the segments have a grafted hydrophobic side chain

CMA   40DT

The forces acting between muscovite mica surfaces were studied with a surface force apparatus [36] using the Mark IV model [37]. The experiment was started by measuring the surface interaction across a droplet of approximately 60 μl containing 20 ppm polyelectrolyte and 0.1 mM KBr. A beaker with the same mixture was kept in the measuring chamber to prevent evaporation of the droplet. After determining the interaction across this solution, the mixture was diluted with polyelectrolyte-free 0.1 mM KBr and during the process the measuring chamber became filled with the aqueous solution. The dilution was approximately 6000 times, i.e. the 40 DT concentration after dilution was about 0.003 ppm, which we regard as essentially polymer-free. The surface force data, obtained in a crossed cylinder configuration, are presented as the force, F, normalised by the mean undeformed geometric radius, R, as a function of surface separation, D. The distance resolution is about 2 Å. The detection limit of the normalised force in the droplet experiments is about 50 μN/m and in the filled box experiments it is about 10 μN/m.

The turbidity of the polyelectrolyte–surfactant mixtures containing 20 ppm polyelectrolyte, 10^{-4} M KBr, and various amounts of SDS was measured employing a HACH ratio turbidimeter. We emphasise that the concentration of SDS was not increased in a stepwise manner in a test tube, but a separate sample was prepared for each SDS concentration. The turbidity of each sample was measured before, T_1, and after adding the polyelectrolyte, T_2. The T_2 value was registered after different equilibration times. The results are plotted as the turbidity difference, $\Delta T = T_2 - T_1$, between the polyelectrolyte–surfactant mixture and the corresponding polyelectrolyte-free SDS solution as a function of total (bound and free) SDS concentration. The turbidity of the 20 ppm polyelectrolyte solution is similar to that of water.

The electrophoretic mobility of the polyelectrolyte–surfactant aggregates was measured using a ZetaSizer4 (Malvern Instruments, UK), employing the standard ZET5104 electrophoresis cell. The results are less accurate at very low (0.005 times the critical micelle concentration, cmc, or less) and very high (above 1 cmc) SDS concentrations owing to the low particle count rate. We chose to report the SDS concentration in units of the cmc in order to allow a ready comparison between the association concentrations of the polyelectrolyte–surfactant mixtures with the concentration at which SDS micelles are formed in bulk solution.

Results

Bulk association

The turbidity of the 20 ppm polyelectrolyte solutions as a function of SDS concentration is shown in Fig. 2. A slight increase in the turbidity of the 40 DT solution is observed when the SDS concentration has reached 0.01 cmc (1 cmc = 8.3 mM). A further increase in SDS concentration results in a rapid increase in turbidity until a maximum is reached at about 0.05 cmc. After this point, a further increase in SDS concentration results in a rapid decrease in turbidity. The changes in turbidity observed for PCMA solutions as a function of SDS concentration are very similar to those observed for 40DT. In particular, the turbidity maximum occurs at very similar SDS concentrations. The main difference between the two polyelectrolytes is that the turbidity of the PCMA solution is significantly higher than that of the 40DT solution at high SDS concentrations. This may be due to the difference in size between PCMA and 40 DT

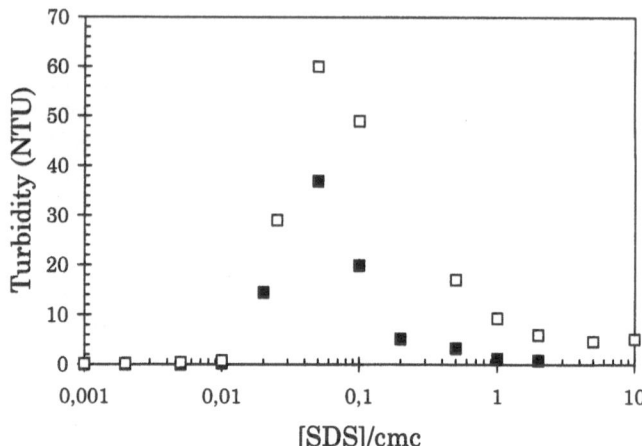

Fig. 2 Turbidity difference (nephelometric turbidity units) between solutions containing 20 ppm polyelectrolyte, 0.1 mM KBr, and different amounts of sodium dodecyl sulphate (*SDS*) and the corresponding polyelectrolyte-free solutions. The concentration of SDS is expressed in parts of the critical micelle concentration (*cmc*) of SDS in pure water (1 cmc SDS = 8.3 mM). The polyelectrolytes were 40DT (■) and PCMA (□)

since it has been noted that for PCMA the turbidity at high SDS concentrations decreases somewhat with decreasing molecular weight [32]. It may also be related to the higher hydrophobicity of 40 DT. On the other hand, the SDS concentration at the turbidity maximum for PCMA solutions seems to be independent of the molecular weight [32].

The changes in turbidity with SDS concentration can be understood by considering the electrophoretic mobility of the aggregates formed by association between the polyelectrolyte and the surfactant (Fig. 3). For both polymers the same trend is observed. The mobility of the

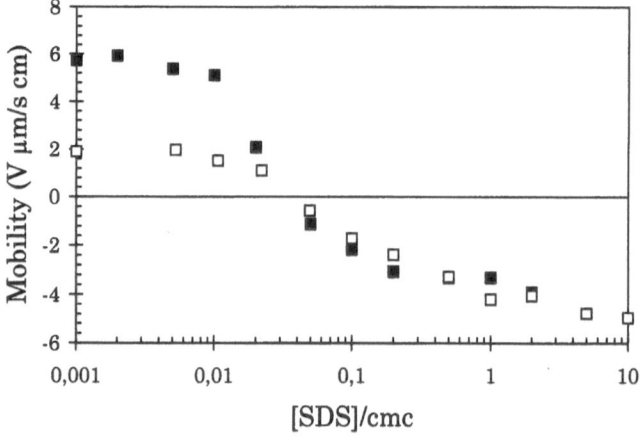

Fig. 3 Electrophoretic mobility of aggregates formed between cationic polyelectrolyte and SDS as a function of SDS concentration. The polyelectrolyte concentration was 20 ppm. The concentration of SDS is expressed in parts of the cmc of SDS in pure water (1 cmc SDS = 8.3 mM). The polyelectrolytes were 40DT (■) and PCMA (□)

aggregates is positive at low SDS concentration, zero at around 0.03–0.04 cmc, and negative at higher SDS concentrations. At low SDS concentrations we find that the mobility of PCMA–SDS aggregates is much less than that of 40 DT–SDS aggregates, which is due to the higher molecular weight and the more extended conformation of PCMA. Note that 40 DT has a compact conformation in water owing to intrachain hydrophobic aggregation. By comparing Figs. 2 and 3 we see that the turbidity starts to increase when the charge of the aggregates has decreased sufficiently, that the turbidity maximum closely coincides with the formation of uncharged aggregates, and that the decrease in turbidity at high SDS concentrations is due to a recharging of the aggregates, which now contain SDS in excess.

The data reported in Figs. 2 and 3 were obtained after approximately 30 min equilibration. We found it interesting to follow the time evolution of the turbidity since it will give some information about aggregation and sedimentation kinetics. At an SDS concentration of 0.05 cmc, close to the point of zero electrophoretic mobility, the turbidity of the 40 DT–SDS solution first increases with time and then it decreases again (Fig. 4). Just above and below this SDS concentration a slow increase in turbidity with time is observed, whereas no time dependence could be detected when the SDS concentration is either 0.01 cmc or less or 0.2 cmc or

greater. Clearly, when the aggregate charge is small enough a slow flocculation occurs. This increases the turbidity up to the point where the aggregates formed are so large that they start to sediment, which reduces the turbidity again. Similar results have previously been reported for PCMA–SDS mixtures [32].

Forces between negatively charged surfaces coated by cationic, hydrophobically modified polyelectrolytes

The forces between bare mica surfaces in aqueous 0.1 mM KBr solution are well described by Derjaguin–Landau–Verwey–Overbeek theory with a repulsive double-layer force dominating the long-range interaction, whereas a van der Waals attraction predominates at distances below 20–30 Å [31]. The forces measured between mica surfaces across a solution containing 20 ppm 40DT in addition to the 0.1 mM KBr are very different and much more complex. First, we note that no strong double-layer force is detected, showing that adsorption of the cationic polyelectrolyte results in a charge neutralisation of the mica surface. Instead a strong repulsive steric force component dominates the long-range interaction. The force measured on approach is more repulsive than that measured on separation, and the range and magnitude of the force decrease for each consecutive approach; this is illustrated in Fig. 5. Once the surfaces have been brought close together a sudden inward jump occurs from a separation of 80–110 Å to a separation of 15–20 Å. At this position a strong attractive force, with a magnitude (F/R) of 30–40 mN/m, is observed upon separation.

The forces measured when the 20 ppm 40 DT solution is diluted approximately 6000 times are shown in Fig. 6 together with a typical force curve measured prior

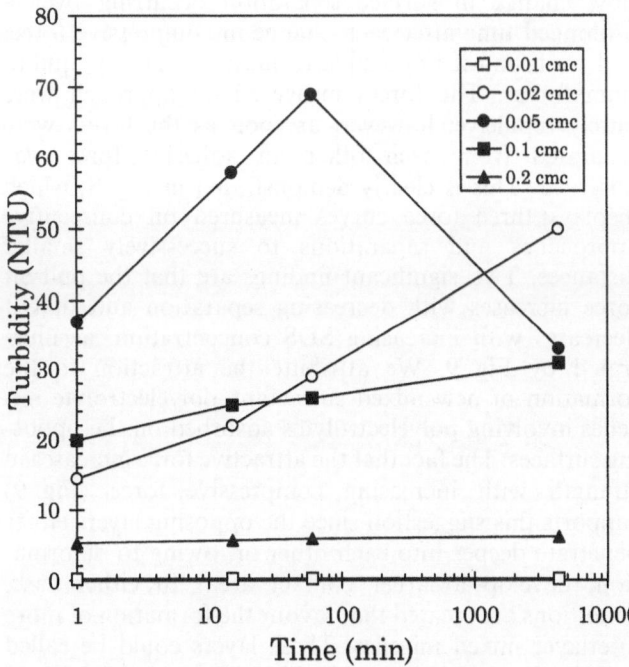

Fig. 4 Turbidity difference (nephelometric turbidity units) between solutions containing 20 ppm 40 DT, 0.1 mM KBr, and different amounts of SDS and the corresponding polyelectrolyte-free solutions. The results are plotted as a function of time after mixing the components

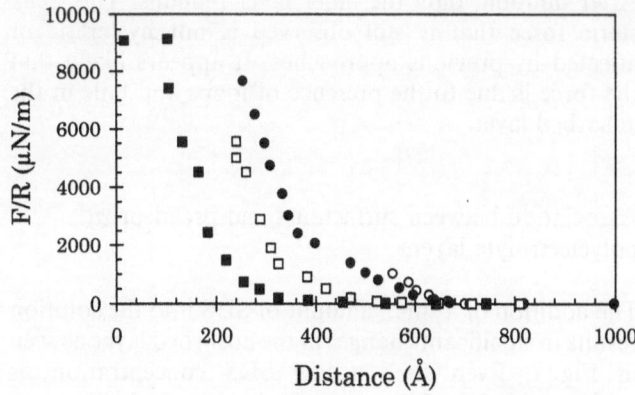

Fig. 5 Force normalised by radius as a function of surface separation. The measurements were carried out using two mica surfaces in a solution containing 20 ppm 40 DT and 0.1 mM KBr. The forces measured on four consecutive approaches to smaller and smaller separations are illustrated

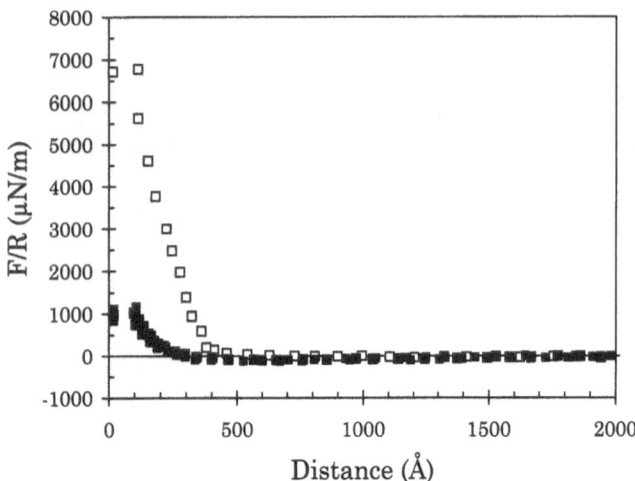

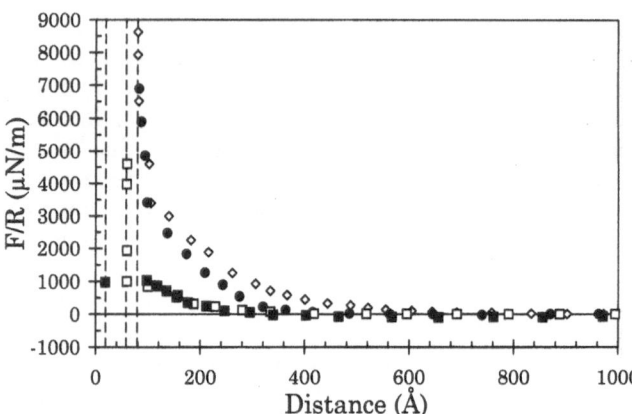

Fig. 6 Force normalised by radius as a function of surface separation. The measurements were carried out using two mica surfaces in a solution containing 20 ppm 40 DT and 0.1 mM KBr (□) and after diluting this mixture by a factor of 6000 using a polyelectrolyte-free 0.1 mM KBr solution (■)

Fig. 7 Force normalised by radius as a function of surface separation between mica surfaces precoated with an adsorbed layer of 40DT. The solution contained 0.1 mM KBr and no SDS (■), 0.005 cmc SDS (□), 0.01 cmc SDS (●), and 0.02 cmc SDS (◊). The dashed lines represent the layer thickness reached under a high force

to dilution. The most long-range part of the force curve observed after dilution is weakly attractive. This behaviour is rather uncommon. The force turns repulsive at a separation of just above 250 Å. The repulsion increases in magnitude until a separation of 100 Å is reached. From this position the surfaces jump inwards to 15–20 Å, the same as before dilution. The magnitude of the attractive normalised force measured when separating the surfaces from this position is about 60 mN/m. The results presented in Figs. 5 and 6 strongly indicate that the layer formed by adsorption from a 20 ppm 40 DT solution consists of an inner part that is strongly anchored to the surface and an outer part that is weakly attached to the inner layer. It is compression and partial removal of the outer layer that gives rise to the hysteresis and the effect of previous approaches shown in Fig. 5. After dilution, only the inner layer remains. The weak steric force that is still observed is not hysteretic or affected by previous approaches. It appears likely that the force is due to the presence of loops and tails in the adsorbed layer.

Association between surfactants and preadsorbed polyelectrolyte layers

The addition of a small amount of SDS into the solution results in significant changes in the adsorbed layer as seen in Fig. 7. Even such a low SDS concentration as 0.005 cmc (0.04 mM) causes the final layer thickness to increase by more than a factor of 2 and the adhesion force to be halved. Clearly, SDS associates with the preadsorbed polyelectrolyte layer even at this low bulk concentration. An increase in SDS concentration to

0.01 cmc results in a strong increase in the range and magnitude of the steric force (Fig. 7) and a further decrease in the adhesion force. A doubling of the SDS concentration to 0.02 cmc results in a slight further swelling.

We noted that the adsorbed layer had a pronounced viscoelastic character in the SDS concentration regime between 0.02 and 0.2 cmc, with the most characteristic feature being a slow relaxation. This was observed as a slow change in surface separation occurring over a prolonged time after each change in compressive force, and we found it impossible to measure (quasi) equilibrium forces. The forces measured on approach were purely repulsive; however, as soon as the layers were separated from each other an adhesive force was observed. This is clearly demonstrated in Fig. 8, which displays three force curves measured on consecutive approaches and separations to successively smaller distances. Two significant findings are that the pull-off force increases with decreasing separation and that it decreases with increasing SDS concentration, as illustrated by Fig. 9. We attribute the attraction to the formation of new mixed surfactant–polyelectrolyte micelles involving polyelectrolytes adsorbed on the opposing surfaces. The fact that the attractive force increases in strength with increasing compressive force (Fig. 9) supports this suggestion since the opposing layers either penetrate deeper into each other or, owing to deformation, develop a larger contact area. In either case, conditions are created that favour the formation of more interlayer mixed micelles. These layers could be called "gluelike" since they are highly viscous and sticky. It should be emphasised that the general features, as illustrated in Figs. 8 and 9, are reproducible but the details are not since they depend on the approach speed and possible previous measurements on the same spot.

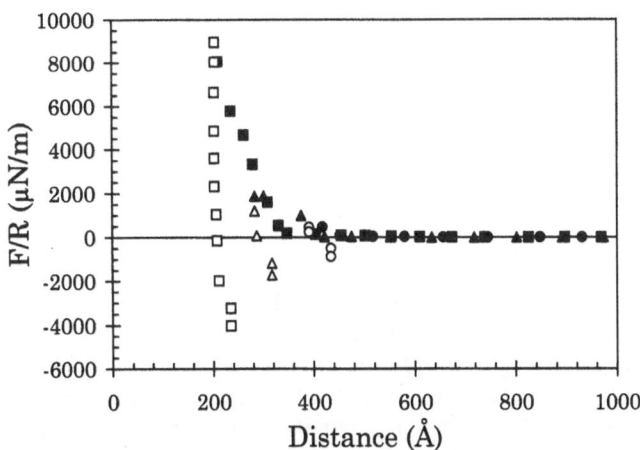

Fig. 8 Force normalised by radius as a function of surface separation between mica surfaces precoated with an adsorbed layer of 40 DT. The solution contained 0.1 mM KBr and 0.04 cmc SDS. Filled symbols represent the force measured on approach and open symbols represent forces measured on separation. Three consecutive force curves, where the surfaces have been brought closer together for each consecutive time and then separated, are illustrated

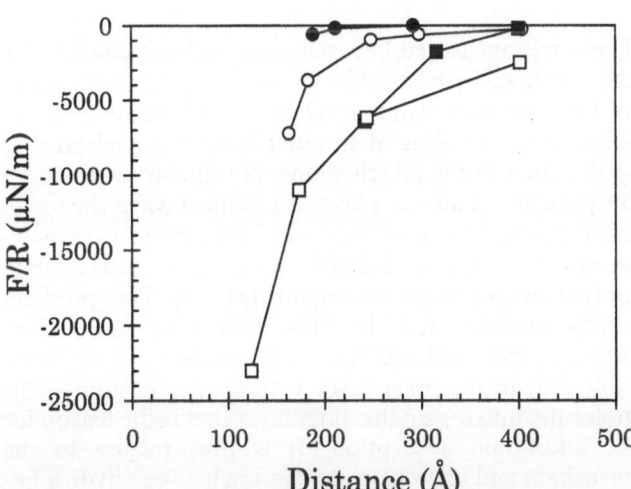

Fig. 9 The attractive force measured when separating two mica surfaces precoated by a layer of 40DT as a function of the separation distance. The solution contained 0.1 mM KBr and 0.02 cmc SDS (□), 0.04 cmc SDS (■), 0.08 cmc SDS (○), and 0.2 cmc SDS (●)

We note that similar "gluelike" layers are formed when close-to-uncharged PCMA–SDS aggregates adsorb to mica [32]. We note that the range of the forces observed is rather large considering the average size of the polyelectrolyte; however, the sample has a considerable polydispersity, which is likely to account for this effect.

The character of the layer changes again as the SDS concentration is increased to above 0.2 cmc. The measured forces are now purely repulsive both on approach and on separation. Further, the range of the steric repulsion decreases significantly with increasing SDS

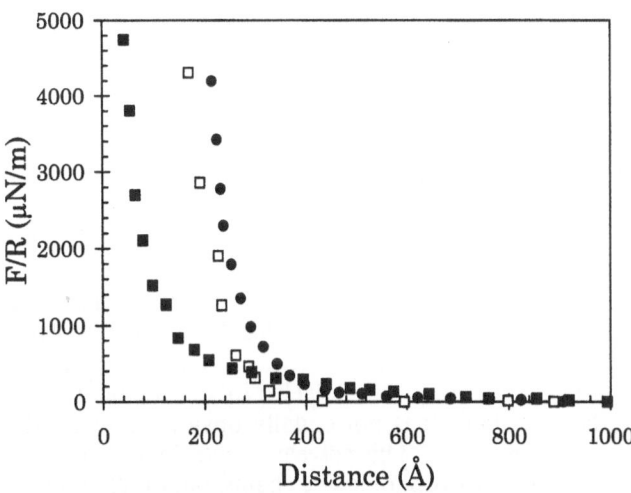

Fig. 10 Force normalised by radius as a function of surface separation between mica surfaces precoated with an adsorbed layer of 40 DT. The solution contained 0.1 mM KBr and 0.2 cmc SDS (●), 0.5 cmc SDS (□), and 2 cmc SDS (■)

concentration (Fig. 10). We attribute this to partial desorption of the polyelectrolyte. However, some polyelectrolytes also remain on the surface at an SDS concentration of 2 cmc.

Discussion

Association between polyelectrolytes and oppositely charged surfactants in bulk solution

The association of positively charged polyelectrolytes and negatively charged surfactants is partly driven by an electrostatic attraction between the opposite charges. Hydrophobic interaction between hydrophobic parts of the polyelectrolyte chain and surfactant tails is another important driving force for association. This is particularly so for the polyelectrolyte 40 DT, which contains a large number of hydrophobic dodecyl side chains. The difference in bulk association as indicated by turbidity and electrophoretic mobility measurements is small between the 40 DT–SDS and the PCMA–SDS systems, demonstrating the importance of the electrostatic driving force for association. However, when the surfactant associates with uncharged polyelectrolyte layers (i.e. the charge of the polyelectrolyte compensates the surface charge) the difference becomes apparent and SDS associates much more readily with 40 DT than with PCMA (see later).

The structures formed by association between polyelectrolytes and surfactants depend on the polyelectrolyte structure (charge, branching, hydrophobicity of the backbone, and presence of hydrophobic side chains). For polyelectrolytes with a hydrophilic backbone it is generally accepted, on the basis of, for example, fluorescence quenching measurements [38], that discrete micellar-like structures are formed along the polyelectrolyte in a "bead-and-necklace" structure. It is observed that the cooperativity of the association process decreases and the binding constant increases when hydrophobic side chains are added to a polyelectrolyte backbone [39]. The 40 DT polyelectrolyte clearly falls into this class. The hydrophobically modified polyelectrolytes self-associate to form hydrophobic microdomains in which added surfactants are easily incorporated [39]. It has been suggested that this process can be viewed as the formation of mixed micelles consisting of polyelectrolyte segments and surfactants [40]. It is not equally obvious how PCMA should be regarded. This polyelectrolyte does not contain any strongly hydrophobic side chains, but the charges are located at the end of short side chains with partial hydrophobic character. In our first article on the subject we assumed that micellar-like aggregates were formed along the PCMA chain [27]; however, our recent small-angle neutron scattering (SANS) studies indicate that this is not the case. A correlation peak, corresponding to a characteristic distance of 37–39 Å is observed, but no typical peak corresponding to assemblies of surfactant in micellar-like structures [33]. PCMA thus seems to be an intermediate case between a polyelectrolyte with a hydrophilic backbone and a hydrophobically modified polyelectrolyte or polysoap.

Considering the structural differences between PCMA and 40 DT it is surprising that the measurements of turbidity and electrophoretic mobility of 40 DT–SDS and PCMA–SDS mixtures (Figs. 2, 3) are so similar. In both cases the association starts at very low surfactant concentrations, much below the cmc in pure water, 8.3 mM. The charge reduction results in an electrophoretic mobility decrease and an increase in turbidity, which shows that the association of surfactant with the polyelectrolytes has caused a sufficient charge reduction of the complex to allow the formation of large aggregates. The polyelectrolyte–surfactant mixture thus phase-separates into a continuous water-rich, surfactant-and-polyelectrolyte-depleted solution containing dispersed polyelectrolyte–surfactant aggregates with some incorporated water. At a slightly higher SDS concentration (about 0.04 cmc) the electrophoretic mobility of the PCMA–SDS and 40 DT–SDS aggregates is close to zero. At zero mobility, the numbers of cationic and anionic groups associated with the aggregate are equal. It is plausible that the majority of these charges are due to the charged polyelectrolyte segments and the charged surfactants. Hence, the polyelectrolyte–surfactant complex is close to the charge stoichiometry. With this assumption the bound and free surfactant concentration at the point of zero mobility is easily calculated. The total SDS concentration at this point is 0.04 cmc, or 0.33 mM. The bound SDS concentration is equal to the polyelectrolyte

segment concentration, 0.1 mM for PCMA and 0.067 mM for 40 DT. Hence, the corresponding free SDS concentration is 0.23 mM (0.028 cmc) in the 20 ppm PCMA solution and 0.26 mM (0.032 cmc) in the 20 ppm 40 DT solution. The difference is small and more accurate measurements are needed before we can conclude if it is significant. Electrostatic forces oppose incorporation of more surfactant into the uncharged aggregate. Nevertheless, it is clear from the increasingly negative electrophoretic mobility curve that further surfactant association with the aggregates occurs. This is caused by hydrophobic attraction between surfactant hydrocarbon tails and hydrophobic regions of the aggregates (i.e. there is a partition of the extra surfactant between the aqueous phase and the hydrophobic regions of the aggregates). We note that the electrostatic free energy cost of creating large aggregates increases with the charge density of the aggregate. One thus expects that the average size of the aggregates decreases, which would explain the decrease in turbidity.

Forces between polyelectrolyte-coated surfaces

The forces measured between mica surfaces coated with the cationic, hydrophobically modified polyelectrolyte, 40 DT, are rather complex (Fig. 5), and we suggest that this is due to binding of an outer layer of polyelectrolyte to the inner firmly attached one. No similar evidence for the presence of an outer layer is obtained when the forces are measured between mica surfaces coated with polyelectrolytes such as PCMA, which does not contain grafted hydrophobic side chains [41, 42]. This provides strong evidence for the view that it is interactions between the hydrophobic side chains of adsorbed molecules in the inner layer with similar groups in the molecules making up the outer layer that is the reason for the additional adsorption. It is thus related to the intrachain and interchain association between hydrophobically modified polyelectrolytes that occurs in bulk solution [5, 39]. The formation of an outer layer of 40 DT is thus due to hydrophobic interactions, the same driving force that results in the formation of "bilayers" of cationic surfactants on highly negatively charged surfaces at high enough surfactant concentrations [43].

The adsorption of hydrophobically modified polyacrylates to hydrophobic surfaces has been studied with ellipsometry [22] and reflectometry [23] and no evidence for "bilayer" adsorption has been reported. This is not in conflict with our results. On hydrophobic surfaces the polymer binds with the hydrophobic groups towards the surface, and the surface acquires a net charge that prevents further adsorption. In our case it is the cationic groups that adsorb to the surface and the surface charge is essentially neutralised. The hydrophobic units have hardly any affinity for the polar surface and they can easily associate with similar groups in polyelectrolytes that are present in solution. This leads to the buildup of the outer layer. We also note that ellipsometric and reflectometric measurements are rather insensitive to the presence of a dilute outer layer, whereas the presence of such layers will have a significant impact on the long-range forces as determined from surface force measurements. A multilayer model, which is consistent with our results, has been proposed for the adsorption of hydrophobically modified polyacrylamide onto clays [44].

After dilution, when only the inner layer remains on the surfaces, a long-range attraction is observed. No similar attraction is present between layers of, for example, PCMA after dilution [27], and it is very tempting to explain this attraction as being due to the presence of the hydrophobic units; however, in our opinion the range of the force is too long to be a "hydrophobic" interaction. Instead, it seems likely that it has an electrostatic origin, i.e. it is an attractive double-layer force or a force due to correlations between negative and positive patches on the two surfaces. Measurements at different salt concentrations, which have not yet been done, can easily be used to test this hypothesis. It is also worth commenting on the presence of a steric force in the distance regime 300–100 Å. Such a steric force is not commonly observed when non-hydrophobically modified polyelectrolytes with a high charge density are used [27]. Hence, the conformations of the adsorbed hydrophobically modified polyelectrolytes are more extended normal to the surface compared to non-hydrophobically modified polyelectrolytes. We suggest that this is related to the more compact conformation of 40 DT in bulk solution, where hydrophobic interactions between the side chains promote intrachain self-association [35]. A similar self-association at the surface will result in a more extended layer. We note that when a small amount of oppositely charged surfactant is added to a non-hydrophobically modified polyelectrolyte and the complexes are allowed to adsorb to mica, a polymer conformation that is more extended normal to the surface is obtained, compared to the situation without any surfactant [32]. Again, we relate this to the change in conformation in bulk solution owing to the presence of hydrophobic side chains: in this case the hydrophobic groups come from the surfactant associated with the polyelectrolyte.

Association between anionic surfactants and cationic polyelectrolytes adsorbed to negatively charged surfaces

When comparing the cmc with the cac between PCMA and SDS in bulk (cac_b) and at a negatively charged mica surface (cac_s) we found the trend cmc > cac_s > cac_b [29]. The reasons for this difference are, first, the polyelectrolyte adsorbed to the surface is less flexible than the polyelectrolyte in bulk solution. Second, most of the charged polyelectrolyte segments are in close contact with the negatively charged surface and SDS has to compete with the surface sites when associating with polyelectrolyte segments. Third, there is a difference in counterion entropy for the processes. The counterion entropy increases when polyelectrolytes associate with surfactants in bulk solution. On the other hand, association at an uncharged polyelectrolyte-coated surface results in a decrease in counterion entropy. This is due to the recharging of the surface that confines the counterions in the diffuse electrical double layer. The importance of the counterion entropy for polyelectrolyte–surfactant association is discussed in more detail by Wallin and Linse [45–47]. We note that association between SDS and adsorbed layers of the hydrophobically modified polyelectrolyte 40 DT on mica occurs much more readily than between SDS and adsorbed PCMA, as will become clear from the discussion later. The reason for this is suggested to be that SDS can bind to hydrophobic microdomains in the adsorbed 40 DT layer, whereas no such regions are present in the PCMA layer.

Forces measured at low SDS concentrations (0.01 cmc or lower)

The results shown in Fig. 7 demonstrate that some surfactants are incorporated in the adsorbed layer even at the lowest SDS concentration used (0.005 cmc). More SDS is incorporated in the layer, which swells significantly as the surfactant concentration is increased further. This is very different compared to when SDS is added to a preadsorbed layer of PCMA, where no effect is observed until the SDS concentration has reached 0.1 cmc [27]. Clearly, the presence of the hydrophobic side chains in 40 DT promotes the uptake of surfactant within the preadsorbed polyelectrolyte layer. For this system it is not meaningful to talk about a cac between SDS and the polyelectrolyte adsorbed on the surface, whereas such a concept has a meaning for non-hydrophobically modified polyelectrolytes [29]. The same is true for bulk association. A cac between oppositely charged surfactant and non-hydrophobically modified polyelectrolytes is generally observed owing to a polyelectrolyte-induced self-association of the surfactant [48], whereas when highly hydrophobically modified polyelectrolytes are used a more gradual uptake of surfactant occurs owing to incorporation of surfactants in already existing hydrophobic microdomains [39, 49].

Forces measured at intermediate SDS concentrations (between 0.02 and 0.2 cmc)

At intermediate SDS concentrations the most pronounced features of the preadsorbed 40 DT layers are

92

their slow equilibration in response to a change in surface separation and the development of adhesive force as soon as the two opposing layers come into contact.

The slow relaxation time indicates a corresponding slow change in the internal structure of the adsorbed layer. We note that a slow relaxation, i.e. long relaxation time, in bulk polymer solutions results in a high shear viscosity at low frequencies [50], and in analogy with this one may say that the adsorbed layer is highly viscous. As discussed previously, hydrophobically modified polyelectrolytes and oppositely charged surfactants form mixed micelles that may include segments from more than one polyelectrolyte. This association strongly affects the viscosity of concentrated polyelectrolyte solutions. Normally, the viscosity first increases dramatically owing to the formation of mixed micelles involving surfactant and more than one polyelectrolyte [7, 51]. At higher surfactant concentrations the viscosity decreases again because each polyelectrolyte chain becomes saturated with surfactants. The adsorbed layer is concentrated in polyelectrolyte, and mixed micelles connecting different polyelectrolytes together will be formed when SDS is present in the right amount. Upon compression this network is disturbed. We propose that the slow relaxation is due to the difficulty in establishing the new free-energy minimum of the network confined between the two surfaces. No similar effects are observed when SDS is added to preadsorbed polyelectrolytes that are not hydrophobically modified [27, 30]. One reason for this may be that below cac_s hardly any SDS is incorporated in the adsorbed layer, whereas above cac_s the amount of SDS in the layer is high. However, similar force curves and adhesive behaviour as reported in Figs. 8 and 9 have been observed between surfaces coated with close-to-uncharged aggregates formed by non-hydrophobically modified polyelectrolytes and surfactants [32]. These aggregates were first formed in bulk solution and were then allowed to adsorb to the surface.

Forces measured at high SDS concentration (above 0.2 cmc)

The adhesion between the adsorbed layers has nearly vanished when the SDS concentration has reached 0.2 cmc. This indicates that each hydrophobic side chain in the adsorbed layer is solubilised in aggregates consisting of several SDS molecules and a few hydrophobic side chains. At even higher SDS concentrations the adhesion is even lower or is completely absent. The adsorbed layer is close to saturated with SDS. The adsorbed 40 DT–SDS complexes now have a high negative charge and desorption starts. As a result, the range of the forces decreases with SDS concentration. Again, the results presented in this study are very different to those observed for preadsorbed layers of

PCMA associated with SDS [27]. The concentration regime with an SDS concentration of 0.2 cmc or greater is above the cac_s for PCMA–SDS complexes on negatively charged mica surfaces and the adsorbed layer is strongly swelled. The force curve determined for the PCMA–SDS system displays pronounced oscillations with separation, indicating that there is a characteristic distance (about 40 Å) within the layer that reflects its structure. In the earlier article we interpreted this as the dimension of the SDS micelles associated with the polyelectrolyte [27]. Recent SANS measurements have confirmed the presence of a characteristic distance of 37–39 Å in PCMA–SDS aggregates formed in bulk solution, but the SANS data do not give any evidence for the presence of SDS micelles in these aggregates [33]. At present, it is not clear how to best describe the internal structure of the PCMA aggregates formed in bulk solution or the internal structure of the PCMA–SDS layers present on mica surfaces; however, it seems that it is similar to the highly ordered mesomorphous polyelectrolyte–surfactant phases characterised by Antonietti and coworkers [52–54]. Desorption of PCMA–SDS complexes from mica surfaces is, in contrast to the 40 DT–SDS complexes, limited even when the SDS concentration is 2 cmc [27]. This may be due to the higher molecular weight of PCMA compared to 40 DT or it may reflect a difference in the amount of SDS incorporated in the adsorbed layer.

One way of summarising the results for the 40 DT–SDS system is to plot the measured adhesion force and the compressed layer thickness as a function of the SDS concentration (Figs. 11, 12). First, we note that in the absence of SDS the adhesion force between mica coated with PCMA (65–250 mN/m) [27, 33] is higher than between mica coated with 40 DT (60 mN/m). Hence, the

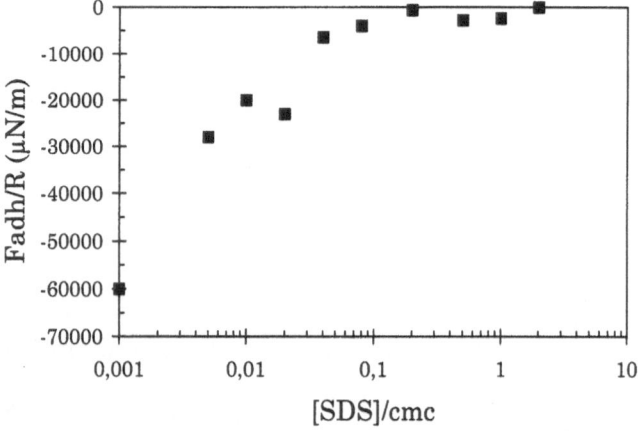

Fig. 11 Maximum adhesion force normalised by radius as a function of SDS concentration. The concentration of SDS is expressed in parts of the cmc of SDS in pure water (1 cmc SDS = 8.3 mM). The point plotted at an SDS concentration of 0.001 cmc really corresponds to zero SDS concentration

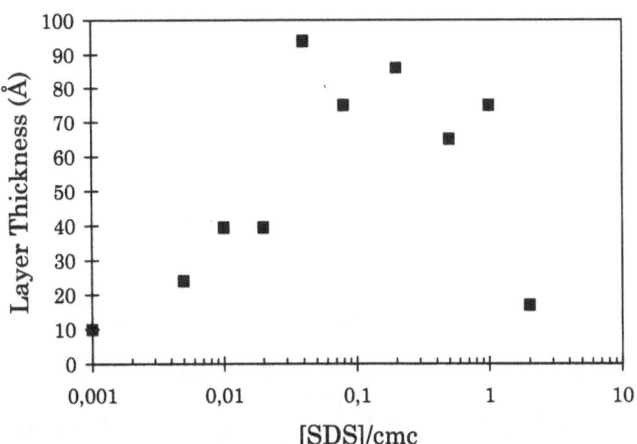

Fig. 12 Layer thickness, i.e. half the surface separation, reached under a high compressive load ($F/R = 10.000$ μN/m) as a function of SDS concentration. The concentration of SDS is expressed in parts of the cmc of SDS in pure water (1 cmc SDS = 8.3 mM)

less hydrophobic polyelectrolyte gives rise to the highest adhesion force. The likely explanation is that the dominating contribution to the adhesion comes from a bridging attraction due to polyelectrolytes bound to both surfaces. This force contribution is thus stronger in the case of PCMA, which forms a thinner adsorbed layer than 40 DT. The contribution from hydrophobic interactions between adsorbed polyelectrolytes, present in case of 40 DT, is thus not strong enough to compensate for the reduction in polymer bridging that result from the increased bulkiness of the polyelectrolyte.

When the SDS concentration is increased we find that the adhesion force decreases (Fig. 11) and that the compressed layer thickness increases (Fig. 12). The adhesion that is observed is also present between the layers when the surface separation is rather large (Fig. 9). This observation indicates that bridging due to polymers binding to both surfaces is less important and that the dominating mechanism now is association between the polymer layers themselves caused by the formation of mixed micelles between SDS and hydro-

phobic side chains from the polyelectrolyte. The analogous situation in bulk gives rise to a strong increase in the viscosity of solutions containing both hydrophobically modified polyelectrolytes and surfactants. In bulk solution, the viscosity decreases again when more SDS is added because the individual hydrophobic side chains are solubilised in individual SDS micelles. An analogous event takes place in the adsorbed layers, which explains the low adhesion force observed at and above 0.1 cmc.

Conclusion

The interactions between negatively charged surfaces coated with a cationic, hydrophobically modified polyelectrolyte have been determined in the absence and the presence of SDS. Several differences, compared to the analogous experiment using non-hydrophobically modified polyelectrolytes, have been observed. The presence of hydrophobic side chains makes it possible for the polyelectrolyte to adsorb in a "multilayer" on each surface. The inner layer is rather compact, whereas the outer one has a substantial extension normal to the surface. When SDS is added, it is incorporated in the adsorbed layer even at the lowest concentration studied (0.005 cmc). As the SDS concentration is increased, more SDS is incorporated, which first leads to a swelling, and a layer displaying a slow relaxation is obtained. Further increases in SDS concentration to above 0.2 cmc results in desorption. This is very different compared to non-hydrophobically modified polyelectrolytes where no, or very limited, incorporation of SDS occurs below a critical SDS concentration, the cac$_s$. Hence, for non-hydrophobically modified polyelectrolytes it is useful to talk about a cac at the surface, whereas this concept is not applicable for the hydrophobically modified polyelectrolyte studied here.

Acknowledgements The authors would like to thank the Human Capital and Mobility Program of the European Union for financial support in the framework of the Network "Water-soluble polymers", contract CHRX-CT94-0655.

References

1. Goddard ED, Ananthapadmanabhan KP (1993) Interactions of surfactants with polymers and proteins. CRC, Boca Raton
2. Kwak JCT (1998) Polymer–surfactant systems. Surfactant science series, vol 77. Dekker, New York
3. Piculell L, Lindman B (1992) Adv Colloid Interface Sci 41:149
4. Piculell L, Guillemet F, Thuresson K, Shubin V, Ericsson O (1996) Adv Colloid Interface Sci 63:1
5. Iliopoulos I (1998) Current Opin Colloid Interface Sci 3:493
6. Hansson P, Lindman B (1996) Curr Opin Colloid Interface Sci 1:604
7. Magny B, Iliopoulos I, Zana R, Audebert R (1994) Langmuir 10:3180
8. Leung PS, Goddard ED (1991) Langmuir 7:608
9. Goddard ED, Hannan RB, Matteson GH (1977) J Colloid Interface Sci 60:214
10. Leung PS, Goddard ED (1985) Colloids Surf 13:47
11. Lee BH, Christian SD, Tucker EE, Scamehorn JF (1991) Langmuir 7:1332
12. Goddard ED, Ananthapadmanabhan KP (1998) In: Kwak JCT (ed) Polymer–surfactant systems. Surfactant science series, vol 77. Dekker, New York, pp 21–64
13. Perrin P, Monfreux N, Lafuma F (1999) Colloid Polym Sci 277:89
14. Magdassi S, Rodel BZ (1996) Colloids Surf 119:51
15. Bergeron V, Langevin D, Asnacios A (1996) Langmuir 12:1550
16. Asnacios A, Espert A, Colin A, Langevin D (1997) Phys Rev Lett 78:4974
17. Somasundaran P, Cleverdon J (1985) Colloids Surf 13:73

18. Berthiaume MD, Jachowicz J (1991) J Colloid Interface Sci 141:299
19. Shubin V, Petrov P, Lindman B (1994) Colloid Polym Sci 272:1590
20. Furst EM, Pagac ES, Tilton RD (1996) Ind Eng Chem Res 35:1566
21. Pagac ES, Prieve DC, Tilton RD (1998) Langmuir 14:2333
22. Poncet C, Tiberg F, Audebert R (1998) Langmuir 14:1697
23. Göbel JG, Besseling NAM, Cohen Stuart MA, Poncet C (1999) J Colloid Interface Sci 209:129
24. Creeth A, Staples E, Thompson L, Tucker I, Penfold J (1996) J Chem Soc Faraday Trans 92:589
25. Neivandt DJ (1998) PhD thesis. University of Melbourne, Melbourne
26. Ananthapadmanabhan KP, Mao GZ, Goddard ED (1991) Colloids Surf 61:167
27. Claesson PM, Dedinaite A, Blomberg E, Sergeyev VG (1996) Ber Bunsenges Phys Chem 100:1008
28. Kjellin URM, Claesson PM, Audebert R (1997) J Colloid Interface Sci 190:476
29. Claesson PM, Dedinaite A, Fielden M, Kjellin URM, Audebert R (1997) Prog Colloid Polymer Sci 106:24
30. Claesson PM, Fielden M, Dedinaite A, Brown W, Fundin J (1998) J Phys Chem B 102:1270
31. Fielden ML, Claesson PM, Schillén K (1998) Langmuir 14:5366
32. Dedinaite A, Claesson PM (2000) Langmuir 16:1951
33. Dedinaite A, Claesson PM, Bergström M (2000) Langmuir 16:5257
34. Neivandt DJ, Gee ML, Tripp CP, Hair ML (1997) Langmuir 13:2519
35. Fundin J, Iliopoulos I, Lampin M Colloid Polym Sci (in press)
36. Israelachvili JN, Adams GE (1978) J Chem Soc Faraday Trans 1 74:975
37. Parker JL, Christenson HK, Ninham BW (1989) Rev Sci Instrum 60:3135
38. Zana R (1998) In: Kwak JCT (ed) Polymer–surfactant systems. Surfactant science series, vol 77. Dekker, New York, pp 409–454
39. Anthony O, Zana R (1996) Langmuir 12:3590
40. Linse P, Piculell L, Hansson P (1998) In: Kwak JCT (ed) Polymer-surfactant systems. Surfactant science series, vol 77. Dekker, New York, pp 193–237
41. Dahlgren MAG, Waltermo Å, Blomberg E, Claesson PM, Sjöström L, Åkesson T, Jönsson B (1993) J Phys Chem 97:11769
42. Dahlgren MAG, Claesson PM, Audebert R (1994) J Colloid Interface Sci 166:343
43. Claesson PM, Kjellin URM (1999) In: Binks BP (ed) Modern characterization methods of surfactant systems. Surfactant science series, vol 83. Dekker, New York, p 255
44. Volpert E, Selb J, Candau F, Green N, Argillier JF, Audibert A (1998) Langmuir 14:1870
45. Wallin T, Linse P (1996) Langmuir 12:305
46. Wallin T, Linse P (1996) J Phys Chem 100:17873
47. Wallin T, Linse P (1997) J Phys Chem 101:5506
48. Anthony O, Zana R (1996) Langmuir 12:1967
49. Borisov OV, Halperin A (1996) Macromolecules 29:2612
50. Larson RG (1999) The structure and rheology of complex fluids. Oxford University Press, New York
51. Guillemet F, Piculell L (1995) J Phys Chem 99:9201
52. Antonietti M, Conrad J, Thünemann A (1994) Macromolecules 27:6007
53. Antonietti M, Wenzel A, Thünemann A (1996) Langmuir 12:2111
54. Antonietti M, Maskos M (1996) Macromolecules 29:4199

Progr Colloid Polym Sci (2000) 116:95–99
© Springer-Verlag 2000

INTERFACIAL PROCESSES

K. Lunkenheimer
J. C. Earnshaw
W. Barzyk
V. Dudnik

Transitional behaviour of amphiphiles at the air/water interface under equilibrium and dynamic conditions

K. Lunkenheimer (✉) · V. Dudnik
Max-Planck-Institut für Kolloid- und
Grenzflächenforschung, Am Mühlenberg 2
14476 Golm/Potsdam, Germany

J. C. Earnshaw
The Queen's University of Belfast
Belfast BT7 1NN, UK

W. Barzyk
Institute of Catalysis and Surface
Chemistry, ul. Niezapominajek No. 1
30-239 Cracow, Poland

Dedicated to the memory of Prof. J.C. Earnshaw, who died on the occasion of a tragic accident

Abstract According to an earlier hypothesis of one of us on the surface equation of state, extended-chain surfactants usually adsorb in (at least) two different surface configurations depending on the surface area available per adsorbate molecule. Evidence for the continuous transition between the different surface configurations in the equilibrium state is provided by thermodynamic (surface tension), rheological (surface elasticity as measured by surface laser light scattering) and electric (surface potential) surface properties of surface chemically pure amphiphiles. The transition between the two alternative surface configurations was also observed in surface potential kinetics of n-dodecanoic acid in 0.005 M HCl.

Key words Adsorption isotherm · Surface configuration · Decanoic acid · Surface tension · Surface dilational elasticity

Introduction

To describe the equilibrium, dynamic, kinetic, and rheological properties of surfactant adsorption layers one first needs a convenient surface equation of state to enable the bulk concentration to be related to the surface excess. To do so the equations of Langmuir–Szyskowski and of Frumkin [1–3] have always been used for describing either "ideal" or "real surface behaviour", obviously satisfactorily.

However, only since the advent of "surface-chemical purity", enabling one to provide and to guarantee sufficient surfactant purity not only for the bulk but also for the interfacial phase, have doubts arisen.

When we tried to describe the equilibrium surface tension (σ_e) versus concentration (c) isotherms of various homologous series of amphiphiles we realized that you cannot match these isotherms satisfactorily using either of these equations when you cover the entire adsorption interval.

Therefore the idea was put forward to take into consideration different surface configurations of the adsorbed amphiphile depending on the grade of its surface coverage [4]. To do so we assumed that the adsorbates' surface configuration can occur in two alternative kinds: a flat arrangement at comparatively low surface coverage and a (more or less) upright configuration at high surface densities. According to this, the transition from the low to the high surface coverage configuration does not occur as an all-or-nothing change but occurs successively within a definite concentration interval the width of which depends on the amphiphile's geometric conditions. This interval is called the "transitional region".

In fact, this hypothesis led to a much better matching of the experimental σ_e versus c isotherms. By using it, any experimentally determined σ_e versus c isotherm can be matched with high precision. The whole σ_e versus c isotherm is described mathematically by the sum of two contributions according to Eq. (1)

$$\Delta\sigma_e(c) = \alpha\Delta\sigma_e^I + (1-\alpha)\Delta\sigma_e^{II} \ . \tag{1}$$

$\Delta\sigma_e^I$ represents the Traube–Henry equation,

$$\Delta\sigma = kc = RT\Gamma \ , \tag{2}$$

and $\Delta\sigma_e^{II}$ stands for the Langmuir–Szyszkowski equation,

$$\Delta\sigma = RT\Gamma_\infty \ln(1 + c/a_L) \qquad (3)$$

and/or the Frumkin equation

$$\Delta\sigma = -RT\Gamma_\infty \ln\left(1 - \frac{\Gamma}{\Gamma_\infty}\right) + a'\left(\frac{\Gamma}{\Gamma_\infty}\right)^2 , \qquad (4)$$

with

$$a' = \Gamma_\infty H^s \qquad (5)$$

and

$$c = a_L\Gamma/(\Gamma_\infty - \Gamma)\exp(-2H^s\Gamma/RT\Gamma_\infty) . \qquad (6)$$

Γ_∞, a_L, and H^s denote the saturation adsorption, surface activity, and surface interaction parameters, respectively.

The transitional region is described by the function $\alpha(c)$, which is equal to 1 at the onset and equal to 0 at the end of the transitional interval. The transition function is defined by

$$\alpha = 0.5[1 - p(X_t)] \qquad (7)$$

with

$$X_t = (\ln c - \ln c_t)/\beta , \qquad (8)$$

where c_t and β stand for the transition concentration (50% value) and the width of the transition range, respectively. Various transition functions, such as $\tan(c,\beta)$, $\tanh(c,\beta)$, or polynomials have been applied. We usually used the latter. The different transition functions do not change the transition in principle; however, the width of the transitional interval is somewhat dependent on the kind of the transition function (Fig. 1). The function α is determined by matching the experimental σ_e versus c isotherm using Eq. (1). This represents a nonlinear problem of minimization which is solved by an iteration procedure.

Meanwhile a lot of fundamental features of surfactant adsorption have been discovered by this approach, such as

1. A dependence of the homologous surfactants' cross-sectional areas on chain length [5, 7, 8].
2. Even/odd effects of the adsorption parameters within homologous series of surfactants [5–8].
3. Discrete counterion effects [9].

Although this idea turned out to be so successful we were eager to look for more support from surface properties other than surface tension to learn about the nature of the transitional region.

The surface potential and laser light scattering seemed promising for observing changes in the adsorbate's surface configuration.

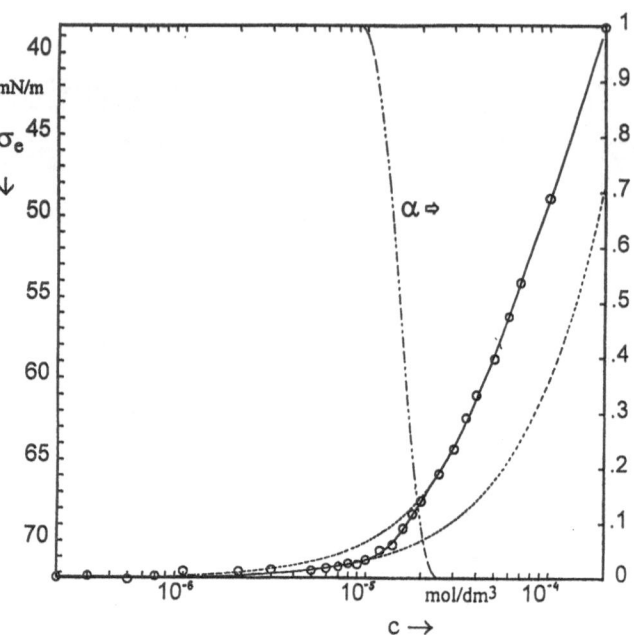

Fig. 1 Equilibrium surface tension (σ_e) and transitional function (α) with respect to the concentration of n-decanoic acid in 0.005 M HCl. The *dashed lines* show extrapolations of the single Henry (*right*: $\alpha = 1$) and/or Frumkin (*left*: $\alpha = 0$) equation of the two-state approach to the surface equation of state [4] (cf. Eq. (1)) The *dash–dot line* represents the transition function, $\alpha(c)$

Experimental

The surface tension was determined using an automatic Lauda TE-1M tensiometer, taking into consideration modifications necessary in applying it to surfactant solutions [10].

Stock solutions were purified by an automatically operating high-performance purification apparatus described in Ref. [11] to get "surface chemically pure" surfactant solutions. The grade of purity was monitored by applying the criterion derived in Ref. [12].

Surface potential (ΔV) measurements were performed using the vibrating plate method (KSV 1000 SPD). The vibrating electrode (disk of 2-cm diameter and 0.5-mm thickness) operated at a frequency of 90 Hz.

The scattering of light by thermally excited capillary waves has become an accepted technique to probe the dynamic behaviour of liquid interfaces [13]. The laser light scattering experiments refer to the damping of thermally excited ripples occurring in the surface of every liquid with amplitudes of several angstroms. The surface of a liquid supporting an adsorption layer sustains various modes of fluctuation, which are excited by thermal excitation. Two are of present concern: capillary waves and dilatational waves. The formers are primarily governed by the surface tension, σ, while the dilatational waves are subject to the surface dilatational elastic modulus, E. In principle, E is equivalent to the Gibbs modulus; however, the surface waves are sensitive to surface properties appropriate to the frequencies of the capillary fluctuation, which may differ from their equilibrium values owing to relaxation processes, such as diffusional exchange and molecular reorientation in the adsorption layer.

The thermally excited capillary waves on a liquid surface can be studied by quasielastic light scattering. They are also coupled to the dilatational waves and, therefore, the surface light scattering is sensitive to dilatational surface properties.

Using a sophisticated technique (heterodyne technique) it is possible to determine the rheological surface parameters of shear and dilation in addition to the surface tension from the wave number and the damping coefficient.

Light from an Ar^+ ion laser (488 nm) is incident on the liquid surface [14]. The light is scattered by the capillary waves of different wave numbers at different angles. To enable the small frequency shifts caused by the capillary waves to be determined the so-called heterodyne technique was used. By using photon correlation, the temporal evolution of the scattered light was measured. The surface properties are obtained from the analysis of the correlation functions (wave frequency, damping, frequency shift) [15, 16].

Light scattering observations were made over the range $403 \leq q \leq 2192$ cm^{-1}, where q denotes wave number.

The correlation functions are Fourier transforms of the spectra of the thermally excited capillary waves of the experimental selects q. They can be analysed to yield unbiased estimates of the propagation frequency, ω_0 and damping constant, δ, of the waves. There are various options of analysis to get the values for the surface properties. It is not possible to determine four properties (surface elasticity and viscosity of dilation and shear) from the wave frequency and damping without a priori assumptions. For the decanoic acid adsorption layer the analysis was facilitated by the fact that the observed values of ω_0 and δ could be reproduced by assuming that the surface tension equals the classically measured surface tension (ring method) and that the surface shear viscosity is zero. In addition, it was found that the dilatational viscosity is zero and/or negligibly small. Hence it is evident that the deviations of the measured surface elasticity values, E, from the theoretical values, E_{th}, are not an artifact of the data analysis, but are a real feature of the light-scattering results [20].

Results

The experimental values of the σ_e versus c isotherm of aqueous solutions of n-decanoic acid in 0.005 M HCl together with the course of the transition function $\alpha(c)$ are shown in Fig. 1.

The equilibrium surface potential ΔV_e versus c isotherm is given together with the measured and the calculated surface dilational elasticity E versus c isotherms in Fig. 2. The latter quantity is defined as the quotient of the surface tension change, dσ, and the imposed, relative surface area change, dA/A (and/or the surface concentration dΓ/Γ), according to

$$E = d\sigma/d\ln A = -d\sigma/d\ln\Gamma . \qquad (9)$$

Using this relation the theoretical value of the surface dilational elasticity, E_{th}, called the Gibbs modulus, can be calculated using Frumkin's equation and the adsorption parameters which are obtained from the evaluation of the experimentally determined σ_e versus c isotherm (see Introduction). The $E_{th}(c)$ isotherm is calculated from these adsorption parameters using the Frumkin equation as

$$E_{th} = RT\Gamma_\infty\Gamma/(\Gamma_\infty - \Gamma) - 2H^s\Gamma_\infty(\Gamma/\Gamma_\infty)^2 . \qquad (10)$$

On comparing Figs. 1 and 2 one clearly notices that the strongest change in the surface potential occurs within the transitional region. Comparatively small slopes of the surface potential are observed where – according to our

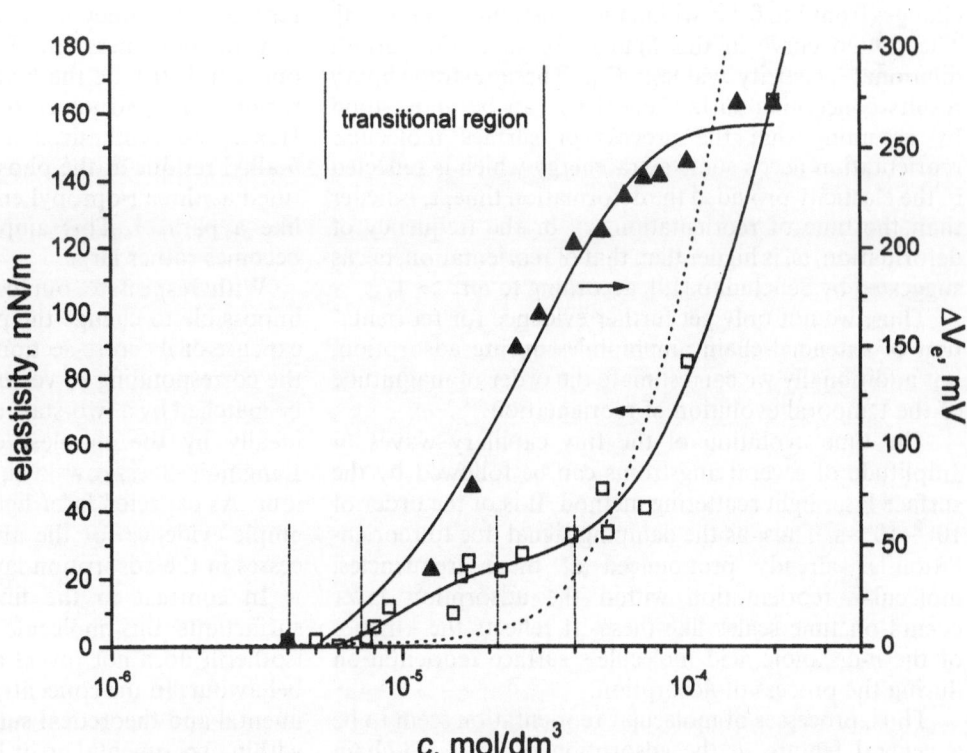

Fig. 2 Equilibrium surface potential (ΔV_e) and surface dilational elasticity (E) of aqueous solutions of n-decanoic acid in 0.005 M HCl with respect to concentration. *Dashed line*: equilibrium surface elasticity calculated using Eq. (7) and the adsorption parameters of n-decanoic acid. *Dash–dot lines* denote the region of the onset and/or of the end of the transitional interval when various mathematical functions are used [23]: *small region*: best fit by polynomial, *broad region*: best fit by tanh

hypothesis – the adsorbate molecules maintain their surface configuration, i.e. at very low and very high coverages. The decisive contribution to the overall ΔV_e value occurs just within that region where the adsorbed decanoic acid molecules depart from their flat surface arrangement. This reorientation is accompanied by a corresponding change in the molecular dipole, which causes most of the resulting surface potential change [17].

At the time when we investigated this problem Earnshaw and coworkers reported rather strange results in the surface dilational elasticity, E, of hexadecyltrimethylammonium bromide and sodium dodecyl sulphate solutions measured by laser light scattering experiments. Their observed elasticity values were systematically too high at medium concentrations compared with those calculated from the corresponding adsorption parameters by using the surface equation of state. Surprisingly, the elasticity values agreed well (within measuring error) at low and high concentrations [18].

Discussion between our groups led to the conclusion that these discrepancies might be caused by some relaxation connected with reorientation processes in the adsorption layers. This possibility had already been taken into consideration by Scheludko [19]. In conclusion, we started a common investigation with the already "classical" nonionic surfactant n-decanoic acid to avoid complications arising from ionic charge and possible artifacts due to surface-active trace impurities. The results are also plotted in Fig. 2. Within experimental error the surface potential and the surface elasticity reveal their peculiar behaviour exactly in that concentration region where α changes from 1 to 0, i.e. within the transitional region [23]. The dotted curve in this figure represents the surface dilational elasticity values, E_{th}. These extraordinary results concerning surface elasticity can be understood by assuming that the process of surface molecular reorientation needs some extra energy which is reflected in the elasticity provided the deformation time, τ, is faster than the time of reorientation, τ_r, or the frequency of deformation, ω, is higher than that of reorientation, ω_r, as suggested by Scheludko [19], according to $\omega\tau_r \gg 1$.

Thus, we not only get further evidence for reorientation of extended-chain amphiphiles during adsorption, but additionally we can estimate the order of magnitude of the temporal evolution of reorientation.

The time evolution of the tiny capillary waves of amplitude of several angstroms can be followed by the surface laser light scattering method. It is of the order of 10^{-6}–10^{-5} s. Thus, as the damping signal due to reorientation is already pronounced at these frequencies, molecular reorientation within the adsorption layer occurs on time scales like these. It reflects the kinetics of the n-decanoic acid molecule's surface reorientation during the process of adsorption.

Thus, processes of molecular reorientation seem to be a general feature in the adsorption of extended-chain amphiphiles. If this was true, the process of reorientation ought to be reflected in the kinetics of the surfactant's adsorption. Measurement of the surface potential is favoured for this purpose because it gives the most sensitive signal with respect to molecular reorientation.

However, the drawback of the vibrating plate method applied is that we can start running the experiment only after half a minute or so after having prepared the experimental setup with the solution. To overcome this disadvantage we need a surfactant of very strong surface activity which adsorbs at very low bulk concentrations. This, in turn, means that there will be adsorption times long enough to be investigated by surface potential.

For this experiment n-dodecanoic acid in 0.005 M HCl was used. The dependence of the dynamic surface potential on time is illustrated in Fig. 3. For the lowest concentrations an inflexion point occurs in the $\Delta V(t)$ dependence. This is assumed to be due to the change of the surfactant's dipole moment at the interface owing to the molecule's reorientation when the surface coverage increases. At higher concentrations the relaxation times concerned are faster than the time available for following them using the measuring device.

These, however, are the first results we have obtained. Further investigations are in progress.

Finally, we tried to get further evidence using some molecular geometric considerations.

Usually, in terms of molecular structure, the term "surfactant" denotes a polar molecule consisting of a hydrophobic part and some hydrophilic, polar group. The former usually consists of an extended chain representing either n-alkyl- or n-perfluoroalkyl chains, siloxane oligomers, etc. However, it is imaginable that one can distribute the hydrophobic residue of the same formula in a structure other than an extended chain. Hence, we synthesized a nonionic molecule with the n-alkyl residue at the phosphorous oxide residue substituted as three isopropyl entities. It possesses a geometry like a parasol. This amphiphile's cross-sectional area becomes rather large.

With respect to our concern the point is that it is impossible to change the parasol molecule's girth at the expense of its cross-sectional area [21]. As a coincidence the corresponding σ_e versus c isotherm does not need to be matched by a two-state approach but can be described ideally by the simplest case imaginable, i.e. by the Langmuir–Szyszkowski equation of ideal surface behaviour. As expected laser light scattering experiments give ample evidence for the absence of reorientational processes in the adsorption layer.

In contrast to the findings obtained with n-alkyl surfactants this molecule's surface elasticity versus c isotherm does not reveal any indication of transitional behaviour. In the concentration region concerned experimental and theoretical surface elasticity values coincide within experimental error [22].

Fig. 3 Surface potential (ΔV) versus time of various solutions of n-dodecanoic acid in 0.005 M HCl: 8.0×10^{-6} mol/dm^3 ($\triangledown$), 6.0×10^{-6} mol/dm^3 ($\triangle$), 4.0×10^{-6} mol/dm^3($\bigcirc$), ($\square$) 2.0×10^{-6} mol/dm^3

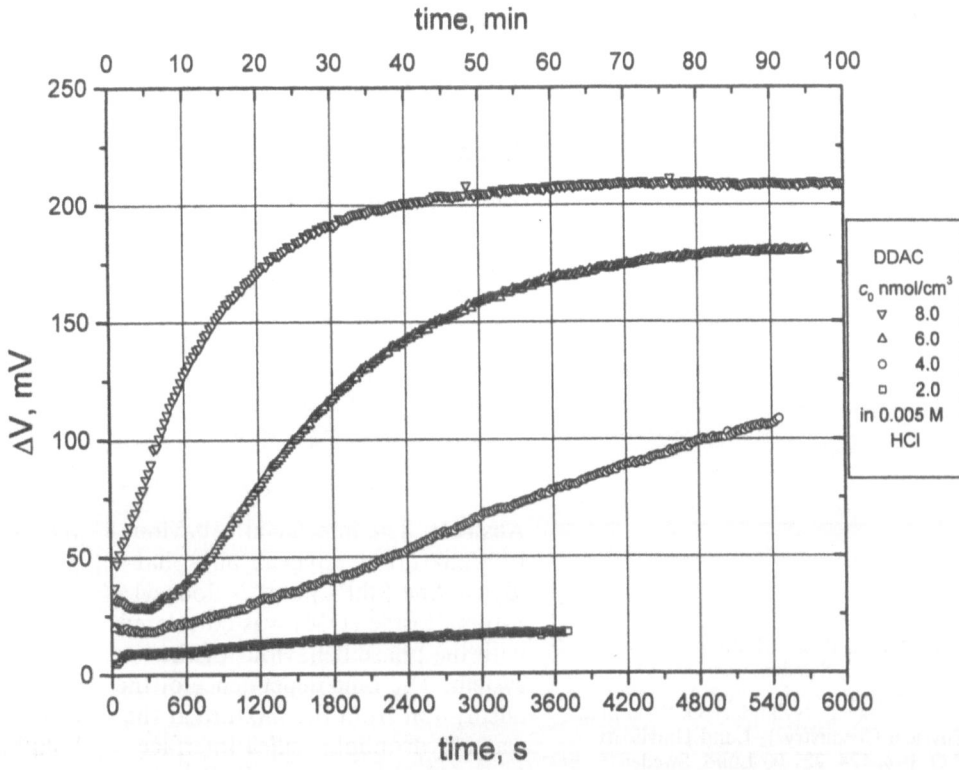

Conclusions

The investigations on the adsorption properties of soluble surfactants using surface tension, surface potential, and surface light scattering confirmed our thermodynamic hypothesis about processes of the n-alkyl chains' conformational changes in the adsorption layer. Additionally, new insight about the time scale of these processes could be gained.

Thus, it can be generalized that adsorption layers of extended-chain amphiphiles experience a reorientation at medium surface coverages owing to their special chemical structure. The molecular reorientation occurs within time scales of 10^{-6}–10^{-5} s. This orientation is caused by the n-alkyl chains' propensity to adopt different conformations in the adsorption layer with respect to the available surface area per molecule adsorbed.

The conformational changes can also be revealed in the surfactants' adsorption kinetics if their surface activity is strong enough to have a time window available for measuring the kinetics.

References

1. Langmuir I (1918) J Am Soc 40:1361
2. von Szyszkowski B (1908) Z Phys Chem 64:385
3. Lucassen-Reynders EH (1976) Prog Surf Membr Sci 10:253
4. Lunkenheimer K, Hirte R (1992) J Phys Chem 96:8683
5. Lunkenheimer K, Burczyk B, Piasecki A, Hirte R (1991) Langmuir 7:1765
6. Lunkenheimer K, Laschewsky A (1992) Prog Colloid Polym Sci 9:239
7. Lunkenheimer K, Czichocki G, Hirte R, Barzyk W (1995) Colloids Surf A 101:187
8. Lunkenheimer K, Haage K, Hirte R (1999) Langmuir 15:1052
9. Lunkenheimer K (1998) Novel results on the adsorption properties of soluble amphiphiles at fluid interfaces. Invited lecture presented at the Gordon Research Conference "Chemistry at Interfaces", Kimball Union Academy, Meriden, July 1998
10. Lunkenheimer K, Wantke K-D (1981) Colloid Polymer Sci 259:354
11. Lunkenheimer K, Pergande H-J, Krüger H (1987) Rev Sci Instrum 58:2313
12. Lunkenheimer K, Miller RJ (1987) Colloid Interface Sci 120:176
13. Langevin D (1992) Light scattering by liquid surfaces and complementary techniques. Dekker, New York
14. Earnshaw JC, McGivern RC (1987) J Phys D 20:82
15. Winch PJ, Earnshaw JC (1988) J Phys E 21:287
16. Earnshaw JC, McGivern RC (1988) J Colloid Interface Sci 123:36
17. Barzyk W, Pomianowski A, Lunkenheimer K (1997) Bull Pol Acad Sci Chem 45:189
18. Earnshaw JC, McCoo E (1995) Langmuir 11:1087
19. Scheludko A (1966) Abh Dtsch Akad Wiss Berlin Kl Chem Geol Biol 6b:531–541
20. Earnshaw JC, Nugent CP, Lunkenheimer K (1996) J Phys Chem 100:5004
21. Vold MJJ (1984) Colloid Interface Sci 100:224
22. Earnshaw JC, Grattan M, Lunkenheimer K, Rosenthal U (2000) J Phys Chem B 104:2709
23. Hirte K, Lunkenheimer K (1996) J Phys Chem 32:13786

Progr Colloid Polym Sci (2000) 116:100–106
© Springer-Verlag 2000

J. Fagefors
K. Wannerberger
T. Nylander
O. Söderman

Adsorption from soybean phosphatidylcholine/*N*-dodecyl *β*-D-maltoside dispersions at liquid/solid and liquid/air interfaces

K. Wannerberger[1] (✉)
AstraZeneca R & D Lund
221 87 Lund, Sweden

J. Fagefors · T. Nylander · O. Söderman
Physical Chemistry 1, Lund University
P.O. Box 124, 221 00 Lund, Sweden

Present address:
[1]Ferring AB, P.O. Box 30047
200 61 Limhamn, Sweden
e-mail: kristin.wannerberger@ferring.se
Tel.: +46-36-1000
Fax: +46-40-154795

Abstract The interfacial behaviour of a mixture of soybean phosphatidylcholine (SbPC) and *N*-dodecyl *β*-D-maltoside (DM) was correlated with the phase behaviour of the system. The time dependence of the adsorption from the mixture at the water/hydrophobic silica interface was studied by means of ellipsometry. The adsorption of DM was reversible, as the adsorbed layer was removed by rinsing with water. The desorbable fraction from adsorbed DM–SbPC mixtures decreased as the DM–SbPC composition approached the solubility limit and close to the two-phase region no desorption occurred. This indicates that the adsorbed layer mainly consists of SbPC. Hence this approach can be used to create a pure phospholipid layer on a solid surface.

Key words Ellipsometry · *N*-Dodceyl *β*-D-maltoside · Adsorption · Soybean phosphatidylcholine · Dispersion

Introduction

In the pharmaceutical industry, water-insoluble drugs are often delivered as emulsions or, in the case of solid particles, as suspensions. In order to stabilise a colloidal dispersion, we need to add stabilising agents that reduce the interfacial tension between the dispersed phase and the medium. Suspension stabilisers act by increasing the wettability of the particles. In addition, they can contribute to the reduction of particle aggregation as well as to a lowering of the sedimentation rate [1]. Formulation development is necessary when new active substances are to be introduced, when a change in the dose is requested or when there is a demand for alternative stabilising agents.

Phospholipids are examples of natural stabilisers, where, in particular, phosphatidylcholine (PC) (Fig. 1a) is used as an emulsifier and is then often added to the oil phase [2, 3]. However, the difficulty in using these lipids to stabilise suspensions is to adsorb them on the particle surfaces, as PC has very low solubility in water. One way is to solubilise the PC using a surfactant. Surfactant solutions above their critical micelle concentration (cmc) can solubilise water-insoluble compounds, for example, fats and oil, to form mixed micellar aggregates with the water-insoluble compounds [4]. In this way the solution concentration of PC can be increased in order to promote the adsorption to the particle surfaces. Here we used an approach for depositing well-defined and pure lipid surface layers by coadsorption with maltoside surfactants near the solubility limit of the lipid, a method originating from recent work by Tiberg et al. [5], who developed this technique for depositing bilayers on silica. In the nonionic surfactant *N*-dodecyl *β*-D-maltoside (DM) the polar head group consists of two glucose rings coupled to an acyl chain with 12 carbons (Fig. 1b). The reason for choosing DM was that sugar surfactants, especially alkyl glycosides, have frequently been used to solubilise membrane phospholipids [6–10]. In addition, the large hydrophilic head group of DM is expected to be more efficient in stabilising particle suspensions compared to the monoglucose equivalent.

The aim of this work was to study the adsorption from DM and PC–DM water mixtures to hydrophobic surfaces. The adsorption was followed by ellipsometry.

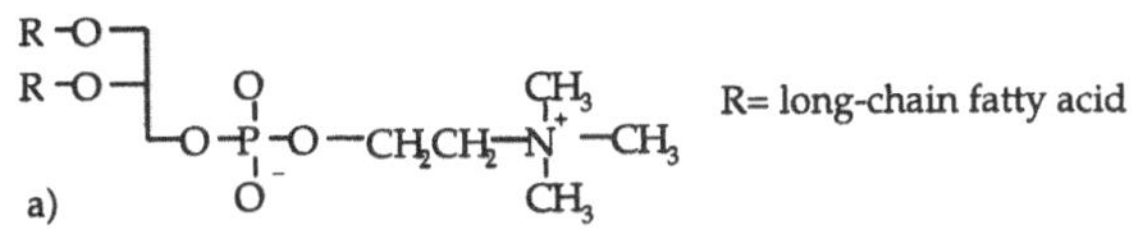

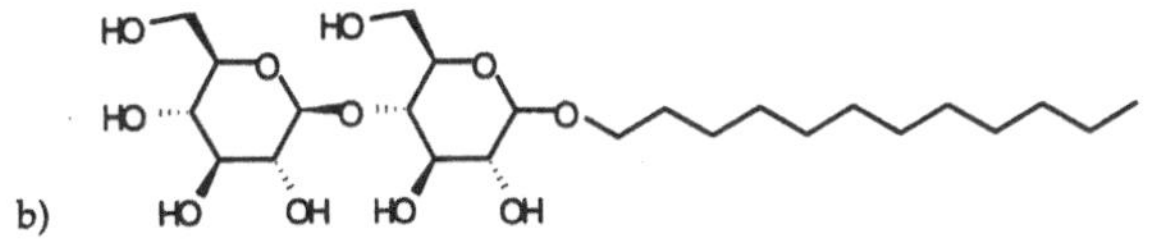

Fig. 1 Molecular structure of **a** soybean phosphatidylcholine (*SbPC*) and **b** *N*-dodecyl β-D-maltoside (*DM*)

The interfacial behaviour of the mixture was correlated with the phase behaviour of the system; therefore, the relevant part of the phase diagram for the PC–DM–water system was determined. In this way we could determine how much PC DM could solubilise. The effect of adding PC on the cmc of DM was studied by surface tension measurements.

Materials and methods

DM ($M_w = 511$ g/mol) was purchased from Sigma and soybean PC (SbPC) (Epicuron 200) (batch no. 1-8-9021) ($M_w \approx 785$ g/mol) was obtained from Lucas Meyer Co., Germany. The fatty acid composition for Epicuron 200 is C16:0 = 13.3%, C18:0 = 3%, C18:1 = 10.2%, C18:2 = 66.9% and C18:3 = 6.6% [11]. SbPC stock dispersions were prepared by dispersing 1.5 g SbPC in 50 ml water by using a Heidolph Diax 900 homogeniser (Heidolph, Germany) for 2 min. The SbPC stock dispersion was then added to 20-ml test tubes containing 0.2–1.5 g DM/l water solution. The samples were mixed by a vibramix for 30 s and equilibrated overnight at 25 °C.

The water used in the surface tension measurement was deionised and passed through a PURELAB Plus UV/UF UV/ultrafiltration water purification system, (USF, Ransbach-Baumbach, Germany). The water used in the ellipsometry experiments was deionised and passed through a Milli-Q water purification system (Millipore). All vessels and cuvettes etc. were cleaned using Decon 90 (Decon Laboratories, Sussex, UK) and thereafter were thoroughly rinsed with water. The silica plates were obtained from Stefan Klintström, University of Linköping, Linköping, Sweden. The wafers were polished and thermally oxidized to a SiO$_2$ layer thickness of about 300 Å and were then cut into slides with a width of 12.5 mm.

Phase behaviour

The phase boundary was established by visual inspection of different samples. When the concentration of DM exceeds the cmc, DM starts to solubilise the SbPC (transparent sample). If more SbPC is added the phase boundary is reached and a two-phase region appears (turbid sample).

Surface tension measurements

The surface tension was measured continuously as function of time or concentration using a tensiometer (Sigma 70, KSV instruments, Helsinki, Finland) fitted with a de Noüy platinum ring [4]. The ring

was thoroughly cleaned in ethanol and burned in a flame before it was used. All experiments were performed at 25 °C.

Determination of the cmc for DM

The cmc for DM was determined by using the plateau values from measurements of the surface tension versus time at different concentrations.

Surface tension of aqueous solution dispersions of DM and SbPC

In order to determine the effect of the solution composition on the adsorption at the air/water interface, the surface tension was measured versus time for a solution of 0.2 g DM/l and three SbPC–DM–water mixtures. The concentrations of SbPC in the three SbPC–DM–water mixtures were 0.0075, 0.015 and 0.05 g/l and the concentration of DM was 0.2 g/l for all three samples. The samples were prepared as described earlier. The results were compared with the surface tension observed when spreading SbPC at the air/water interface. The technique to determine the equilibrium spreading pressure for phospholipids has been described by Phillips and Chapman [12]. A flake of SbPC was applied to the water surface in 20 ml water and the surface tension was measured until the SbPC had spread to the equilibrium layer on the air/water interface.

The cmc of SbPC–DM water mixtures

The cmc for a mixture of SbPC and DM (molar ratio 1:4.4) was determined. For this purpose a dispenser with a 10-ml syringe was used to add 1-ml portions of a concentrate (0.105 g SbPC/l and 0.3 g DM/l) to water during stirring. Between each addition, the surface tension was measured until a stable value was found. Owing to limitation of the syringe volume, the measurements were done in nine steps with a pause to refill the syringe with water. The pause between the two first steps was 2 h, while the pause between subsequent steps was 5 min. The lengths of these intermissions were not found to affect the results.

Ellipsometry

Ellipsometry was used to follow the adsorption of DM water solutions and SbPC–DM–water mixtures to hydrophobic silica plates. The ellipsometer used in this study was an Optrel Multiskop (Optrel, Berlin, Germany) with a Nd:Yd laser ($\lambda = 532$ nm) [13]. The instrument was fitted with a 5-ml cuvette equipped with a magnetic stirrer (300 rpm) and thermostated at 25 °C as described in Ref. [14]. The ellipsometer measures the change of polarised light in terms of the phase shift, Δ, and the amplitude ratio, Ψ, when the light is reflected against a smooth surface [15]. When a film is present on the surface, the thickness and the refractive index of the adsorbed layer can be measured. From these data the amount adsorbed can be obtained [16]. The calculations of the amount adsorbed were based on a dn/dc value of 0.169 cm^3/g [17].

Surface preparation

Hydrophobised silica plates were used as the surface in this work. The silicon wafers were cleaned and hydrophobised in the following way [17, 18]. First the plates were placed in a mixture of 180 ml water, 40 ml H$_2$O$_2$ (30%) and 40 ml NH$_3$ (25%) at 80 °C for about 5 min and then in a mixture of 180 ml water, 40 ml H$_2$O$_2$ and 40 ml HCl (32%) at 80 °C for about 10 min. The wafers were rinsed 10 times in water and 2 times in ethanol and then cleaned in a plasma cleaner (Harrick Scientific Corporation, model PDC-3XG) for a period of 5 min at a pressure of 0.03 mbar and a power of 30 W. After cleaning, the wafers were placed in a closed reaction vessel, together with a small vessel containing 2 ml octyldimethyl-chlorosilane overnight. Excess reagents were removed from the wafers by rinsing them with tetrahydrofuran in an ultrasonic bath for 15 min, followed by rinsing with water 10 times. Finally the wafers were treated with Decon 90 in an ultrasonic bath for 15 min, rinsed with water 10 times and then sonicated in ethanol for 15 min. The water contact angle of these surfaces was, in an earlier work, found to be 104° and 96° for the advancing and receding angles, respectively [19].

Adsorption isotherm of DM

The adsorption isotherm was recorded with stepwise addition of DM. This procedure is justified if the adsorption is reversible. To test if the adsorption process was indeed reversible 1 ml 1.0 g DM/l was added to a cuvette with 4 ml water (0.2 g DM/l in the cuvette) and the changes in Δ and Ψ were recorded. After 30 min the cuvette was rinsed by flushing with water and it was found that DM was completely desorbed from the surface. The adsorption isotherm was then determined as follows. The initial volume of water in the cuvette was 4 ml, to which 0.1 ml DM stock solution was added every 10 min. To cover the whole concentration range, four different DM stock solutions of 0.2, 0.6, 1.2 and 2.3 g/l were prepared. The final concentration was 0.2 g/l. After the adsorption isotherm was complete extensive rinsing with water was performed. The adsorption isotherm was constructed on the basis of the data obtained.

Adsorption of DM–SbPC–water mixtures

The adsorption from mixtures of DM–SbPC–water to hydrophobic surfaces was followed in a similar way. A mixture of 1 ml of aqueous solution of SbPC and DM (range 0.1–0.3 g/l and 1 g/l, respectively) was added to the cuvette with 4 ml water while the changes in Δ and Ψ were recorded. The kinetics of adsorption was followed for at least 1 h before the rinsing with water was performed.

Results and discussion

Phase behaviour

A schematic phase diagram was determined for the SbPC–DM–water system in the water-rich region (99.95–100.00% water) (Fig. 2).

A concentration of 0.2 g DM/l could solubilise 0.07 g SbPC/l, which corresponds to a DM/SbPC molar ratio of 4.4. To our knowledge, no results concerning solubilisation of SbPC by DM have been reported. Our results can be compared with a similar study using DM and a solution of egg PC/phosphatidic acid (PA) (9:1 molar ratio) by Ribosa et al [10]. Ribosa et al. found that DM could solubilise egg PC/PA at a DM/(PC/PA) molar ratio of 2.2. The reason why they could solubilise more egg PC/PA with DM is probably the 6 times higher absolute concentration of DM used in their study. Further, as they used a PC/PA mixture, the presence of the charged lipid (PA) may facilitate the solubilisation.

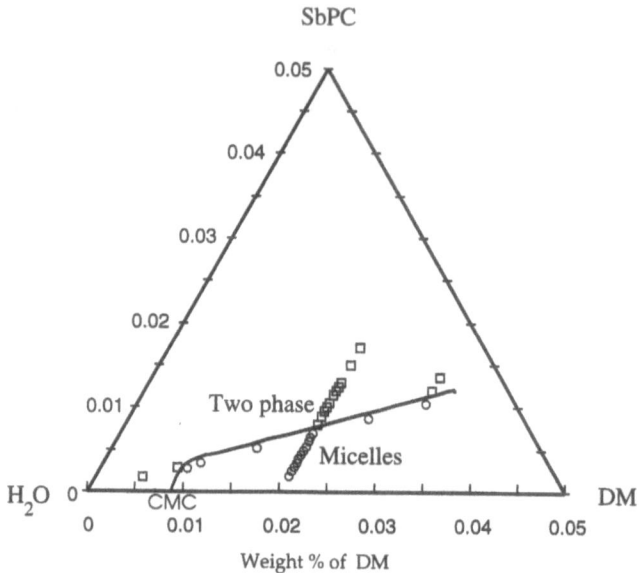

Fig. 2 Partial phase diagram for the DM–SbPC–water system. One-phase samples and two-phase samples are represented by □ and ○, respectively

Surface tension measurements

Determination of the cmc for DM

The determination of the cmc for DM is based on surface tension measurements. One measurement was done for each concentration and the surface tension values are presented in Fig. 3 as a function of concentration. The cmc was determined to be 0.093 g/l (1.82×10^{-4} M) at 25 °C. In previous work by Drum-

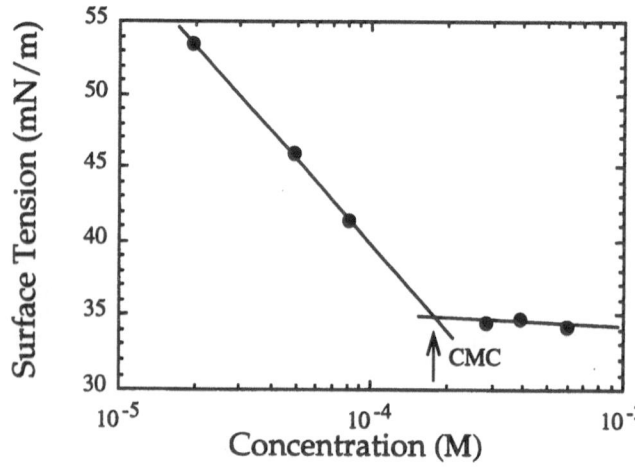

Fig. 3 Surface tension versus logarithm of the concentration of DM. The *arrow* indicates the critical micelle concentration (*cmc*)

mond et al. [20], the cmc was found to be 0.077 g/l at 25 °C. The absence of minima in Fig. 3 indicates that the sample is free from surface-active impurities, as discussed in a number of studies [21–23]. The presence of impurities, which are more surface active than the surfactant, causes a decrease in the surface tension at concentrations below the cmc of the pure surfactant. As the cmc is reached the impurities are solubilised, which leads to an increase in surface tension. In such a case a minimum in the surface tension versus concentration is observed just below the cmc.

Surface tension of aqueous solution dispersions of DM and SbPC

As shown in Table 1 the surface tension increases to 35.5 mN/m when a sufficient amount of SbPC (DM/SbPC molar ratio of 6.2) is added to an aqueous solution of 0.2 g DM/l. This is similar to the equilibrium surface tension value of 35.6 mN/m when SbPC is spread at the air/water interface. It is slightly higher than the plateau value for pure DM, which is about 34.4 mN/m. Although the values are close, the data indicate that the adsorbed surfactant/lipid layer at the air/water interface contains a large amount of SbPC when the SbPC–DM–water mixture has a composition close to the phase boundary. Phillips and Chapman [12] observed equilibrium spreading tension for the dioleoyl PC of 25 mN/m, which is substantially lower than in the present study. Here it should be noted that our sample contains a mixture of lipids (see Materials and methods). In addition, Phillips and Chapman [12] made their spreading experiment with 0.1 M NaCl.

The cmc of SbPC–DM–water mixtures

The cmc in mixtures of SbPC–DM–water (1:3 weight ratio and 1:4.4 molar ratio) in the one-phase region was determined to about be 0.095 g DM/l. This is comparable to the cmc for pure DM. Thus, the presence of SbPC does not seem to have a large influence on the DM micelle formation. At present we cannot explain this somewhat unexpected behaviour. Additional experiments, using different DM/SbPC ratios are therefore needed.

Ellipsometry

Adsorption of DM to a hydrophobic surface

The amount of DM adsorbed to a hydrophobic surface versus time from an aqueous solution containing 0.2 g DM/l is shown in Fig. 4. Flushing the cuvette with water leads to complete desorption of the surfactant, which shows that the adsorption process is reversible. The adsorption isotherm is shown in Fig. 5 and reached a plateau value at 0.09 g/l, which is comparable with the cmc for DM (Fig. 5). The data for the adsorption isotherm could be fitted to a Langmuir isotherm:

$$\Gamma = \frac{\Gamma_{max} K c}{Kc + 1} \, ,$$

where K is the binding constant, c is the bulk concentration and Γ_{max} is the plateau value of the surface excess (amount adsorbed). This is justified as the adsorption of DM was found to be reversible and the interaction between the adsorbed DM can be assumed to be small. The third condition for using the Langmuir isotherm is that all the binding sites on the surface are equal. The homogeneity of trialkylsilane monolayers chemically grafted to a silicon surface has been discussed thoroughly in the literature and has been found to be quite dependent on the mode of surface preparation. In a recent study, Fadeev and McCarthy [24] investigated

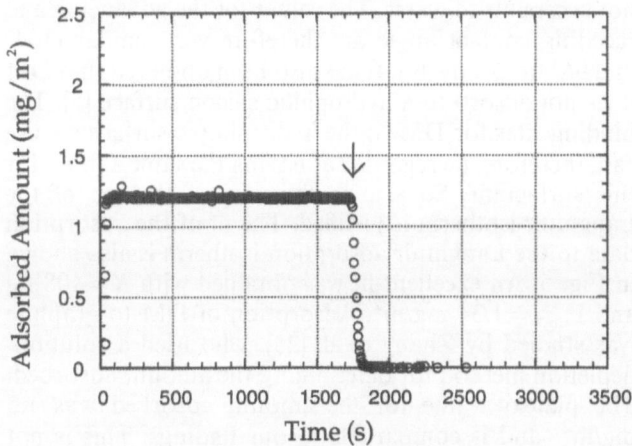

Fig. 4 Adsorption from 0.2 g DM/l solution to a hydrophobic surface. The *arrow* indicates the start of rinsing with water

Table 1 Plateau values of the surface tension of *N*-dodecyl β-D-maltoside (*DM*)–soybean phosphatidylcholine (*SbPC*) mixtures of different compositions

DM concentration (g/l)	SbPC concentration (g/l)	Molar ratio (DM/SbPC)	γ (mN/m)
0.2	0.0	0.0	34.4
0.2	7.5×10^{-3}	41.0	34.9
0.2	1.5×10^{-2}	20.5	35.0
0.2	5.0×10^{-2}	6.2	35.5
0.0	SbPC (spread from solid)	0.0	35.6

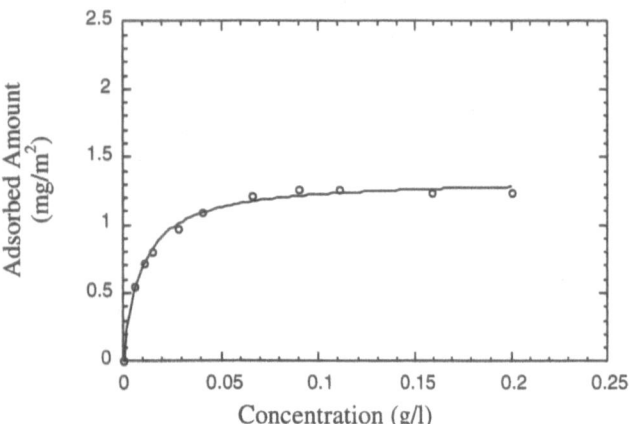

Fig. 5 Adsorption isotherm for DM on a hydrophobic surface

wetting properties of a number of different alkylsilanes monolayers formed from gas-phase and solution-phase reactions with a silicon surface. They found that the modifications with octyldimethylchlorosilane in the gas phase, that is the same procedure used in our study, gave surfaces with the lowest contact angle hysteresis. In fact their contact angle data, with very low hysteresis, was almost exactly the same as for our surfaces. When comparing alkyldimethylchlorosilanes of different chain lengths, no significant differences in the contact angle were found. They therefore suggested that, under these optimal conditions, the residual free silanol groups are not accessible to water. The values for the advancing and receding contact angle are therefore very similar (104° and 96°, respectively). It has also been observed that DM does not adsorb to a hydrophilic silicon surface [5]. The binding sites for DM to the hydrophobic surface we use can, therefore, be regarded as having the same affinity for this surfactant. So also in this respect the use of the Langmuir isotherm is justified. The fit of the adsorption data to the Langmuir adsorption isotherm is also shown in Fig. 5. An excellent fit was obtained with $K = 108$ l/g and $\Gamma_{max} = 1.34$ mg/m^2. Adsorption of DM to graphite was studied by Zhang et al. [25], who used a solution-depletion method for determining the amount adsorbed. The plateau value for the amount adsorbed was 1.5 mg/m^2 and is comparable to our findings. This is not unexpected as the hydrophobic graphite and methylated silica have similar values of water contact angles ($\theta = 86°$ for graphite [26]). Adsorption to methylated silica gives an estimated value for the area per molecule of 68 Å^2. The corresponding value of Zhang et al. [25] was 55 Å^2.

Adsorption of DM–SbPC mixtures to hydrophobic surfaces

The goal of this work was to find a transparent formulation in the mixed micellar region but which is

close enough to the phase separation so that a pure and stable SbPC layer is formed on the hydrophobic surface even after extensive dilution with water. The kinetics of adsorption from DM–SbPC mixtures is shown in Fig. 6. With increasing SbPC content the plateau value for the amount adsorbed before rinsing as well as after rinsing increases (Table 2). This is probably due to the increased amount of SbPC adsorbed in the layer. The maximum amount adsorbed (2.3 mg/m^2) is found to be at a DM/SbPC weight ratio of 0.2/0.06 (Fig. 6). If we assume a surface area per molecule for SbPC of 60 Å^2, which is the head group area of SbPC observed in liquid-crystalline lamellar phases [27], and a molecular weight of SbPC of 785 g/mol a monolayer of SbPC will correspond to an adsorbed amount of 2.2 mg/m^2. The close agreement indicates that we have a more or less close-packed monolayer of SbPC on the hydrophobised silica surface. The phase-separation limit occurs at the DM/SbPC weight ratio of 0.2/0.07 (Fig. 2). The effects of increasing the total surfactant concentration were investigated and are presented in Figs. 7 and 8 for weight ratios of DM/SbPC between 5 and 10. The amounts adsorbed before and after rinsing were higher for 0.3/0.03 DM/SbPC compared to those with less surfactant solution. This indicates that more SbPC is adsorbed to the surface (Table 2). The composition of the adsorbed layer, shown in Table 2, is calculated in the following way: It was assumed that only DM desorbs completely from the adsorbed layer, while the desorption of SbPC was considered to be negligble. Under these conditions the amount desorbed corresponds to the amount of DM in the layer at the plateau before rinsing. Figures 7 and 8 also show that a higher content of SbPC leads to slower adsorption kinetics. This is even more pronounced at a SbPC/DM weight ratio of 5 (Fig. 8). It is clear that when

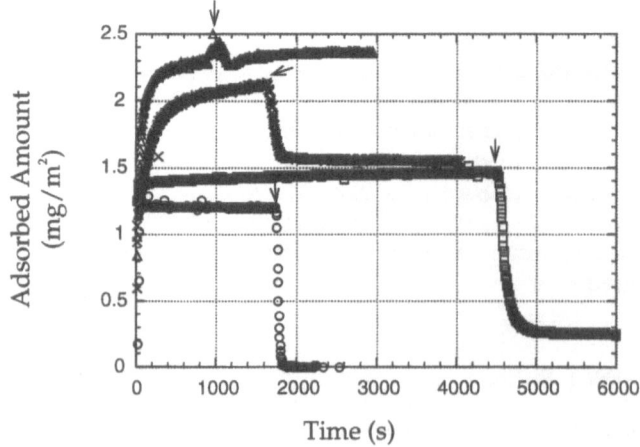

Fig. 6 Adsorption from four mixtures of 0.2 g DM/l and SbPC to a hydrophobic surface versus time. The SbPC concentrations were 0, 0.02, 0.04 and 0.06 g/l ($\bigcirc$, $\square$, $\times$ and $\triangle$). The *arrows* indicate the start of rinsing with water

Table 2 Amounts adsorbed to hydrophobic surfaces from mixtures of different concentrations of DM and SbPC before and after rinsing

	DM concentration (g/l)	SbPC concentration (g/l)	Amount adsorbed (mg/m^2) before rinsing	Molar ratio (DM/SbPC)		Amount adsorbed (mg/m^2) after rinsing
				Solution	Adsorbed layer	
1	0.2	None	1.25	0	DM	0
2	0.2	0.001	1.35	308	0	0
3	0.2	0.02	1.45	15.4	7.4	0.25
4	0.2	0.04	2.15	7.7	1.5	1.55
5	0.2	0.06	2.3	5.1	0	2.35
6	0.3	0.03	1.65	15.4	1.6	0.8
7	0.3	0.06	1.3 (not plateau value)	7.7	0.13	1.2

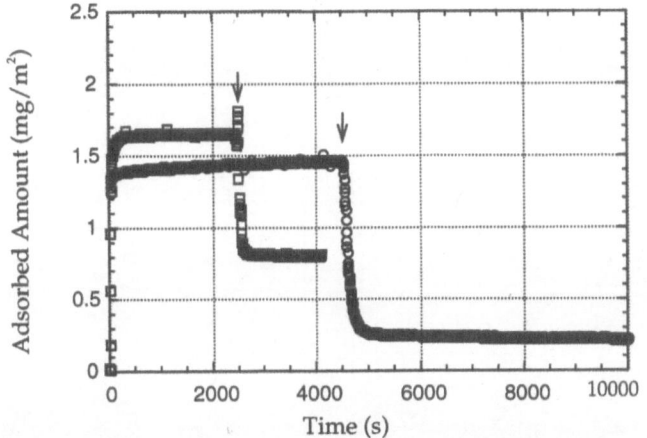

Fig. 7 Adsorption from a mixture of 0.3 g DM/l and 0.03 g SbPC/l (□) and from a mixture of 0.2 g DM/l and 0.02 g SbPC/l (○) to a hydrophobic surface. The *arrows* indicate the start of rinsing with water

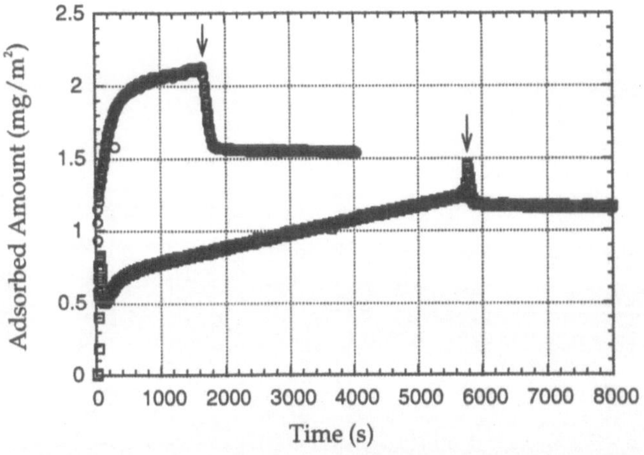

Fig. 8 Adsorption from a mixture of 0.3 g DM/l and 0.06 g SbPC/l (□) and from a mixture of 0.2 g DM/l and 0.04 g SbPC/l (○) to a hydrophobic surface. The *arrows* indicate the start of rinsing with water

we approach the phase boundary the kinetics of adsorption becomes slower. This indicates that "precipitation" or phase separation of SbPC on the solid surface occurs. Consequently the desorbable fraction decreases.

Conclusion

We have demonstrated a technique to cover a solid (hydrophobic) surface with a monolayer of SbPC. The method is based on the dispersion of the phospholipid by a sugar surfactant, i.e. DM. As the precipitation limit of the lipid in the surfactant solution is approached, a layer enriched in SbPC is formed. In contrast to DM, which is completely reversibly adsorbed, it is very difficult to remove this layer by rinsing.

Acknowledgements Fredrik Tiberg is acknowledged for stimulating discussions. This work was supported by the Swedish Research Council for Engineering Sciences (TFR).

References

1. Aulton ME (1988) Pharmaceutics – the science of dosage form design. Churchill Livingstone, New York
2. Larsson K, Friberg SE (1997) Food emulsions. Dekker, New York
3. Larsson K (1994) Lipids. Oily Press, Dundee, UK
4. Shaw DJ (1992) Introduction to colloid and surface chemistry. Butterworth-Heinemann, Oxford
5. Tiberg F, Harwigsson I, Malmsten M (2000) Eur J Biophys 29:196
6. Jackson ML, Schmidt CF, Lichtenberg D, Litman BJ, Albert AD (1982) Biochemistry 21:4576
7. Keller M, Kerth A, Blume A (1997) Biochim Biophys Acta 1326:178
8. Carion-Taravella B, Chopineau J, Ollivon M, Lesieur S (1998) Langmuir 14:3767
9. Paternostre M, Meyer O, Garabielle-Madelmont C, Lesieur S, Ghanam

M, Ollivon M (1995) Biophys J 69:2476

10. Ribosa I, Sanchez-Leal J, Comelles F, Garcia MT (1997) J Colloid Interface Sci 187:443

11. Shinoda K, Araki M, Sadaghiani A, Khan A, Lindman B (1991) J Phys Chem 95:989

12. Phillips MC, Chapman D (1968) Biochim Biophys Acta 163:301

13. Harke M, Teppner R, Schulz O, Motschmann H, Orendi H (1997) Rev Sci Instrum 68:3130

14. Arnebrant T, Barton K, Nylander T (1987) J Colloid Interface Sci 119: 383

15. Azzam RMA, Bashara NM (1989) Ellipsometry and polarized light. North Holland, Amsterdam

16. De Feijter JA, Benjamins J, Veer FA (1978) Biopolymers 17:1759

17. Tiberg F, Landgren M (1993) Langmuir 9:927

18. Eskilsson K, Tiberg F (1997) Macromolecules 30:6323

19. Nylander T, Tiberg F (1999) Colloids Surf B 15:253

20. Drummond CJ, Warr GG, Grieser F (1985) J Phys Chem 89:2103

21. Vijayendran BR (1977) J Colloid Interface Sci 60:418

22. Lunkenheimer K, Czichocki G (1993) J Colloid Interface Sci 160:509

23. Wannerberger K, Wahlgren M, Arnebrant T (1996) Colloids Surf B 6:27

24. Fadeev AY, McCarthy TJ (1999) Langmuir 15:3759

25. Zhang L, Somasundaran P, Maltesh C (1997) J Colloid Interface Sci 191:202

26. Adamson A, Gast A (eds) (1997) Physical chemistry of surfaces. Wiley, New York

27. Rand RP (1971) Biochim Biophys Acta 241:823

Progr Colloid Polym Sci (2000) 116:107–112
© Springer-Verlag 2000

INTERFACIAL PROCESSES

W. Barzyk
K. Lunkenheimer
P. Warszyński
A. Pomianowski
U. Rosenthal

Ideal (Langmuir-type) surface and electric properties of *tert-* isopropylphosphine oxide at the air/water interface

P. Warszyński (✉) · W. Barzyk
A. Pomianowski
Institute of Catalysis
and Surface Chemistry
Polish Academy of Sciences
Kraków, Poland
e-mail: ncwarszy@cyf-kr.edu.pl
Tel.: +48-12-4252841
Fax: +48-12-4251923

K. Lunkenheimer
Max-Planck-Institut für Kolloid-
und Grenzflächenforschung
Potsdam, Germany

U. Rosenthal
University of Rostock
Institute of Organic Chemistry
Rostock, Germany

Abstract The surface pressure (π) and electric surface potential (ΔV) of *tert*-isopropylphosphine oxide (i-Pr_3PO) were measured at the air/water interface. Solutions matching the criterion of "surface-chemical purity" were investigated. Measurable effects on π and ΔV are revealed at 1×10^{-4} mol/dm^3 i-Pr_3PO and extend throughout three concentration decades. The initial rise of the surface potential with the concentration is several times steeper compared to that of the surface tension. The dependence of the surface tension on the concentration of i-Pr_3PO is well described by a Langmuir-type isotherm. Furthermore, the relation between ΔV and the surface concentration (Γ_L) calculated using the Langmuir isotherm is linear. Such consistency with the Langmuir-type adsorption throughout the entire surface activity range is an uncommon feature among soluble surfactants. The ideal surface properties of i-Pr_3PO are mainly due to the ellipsoidal hard-sphere-like structure of the molecule delimited by the iso- propyl chain, which determines the closest packing density of the adsorbate. This is justified by the minimum cross-sectional area of about 80 Å^2/ molecule adsorbed, which corresponds to a sphere delimited by the length of the fully extended isopropyl chain (5 Å). The other reason for the ideal surface behavior is the compatibility of the dimensions of an i-Pr_3PO molecule with an average water cluster (of 13 H$_2$O). The slope (ε_0 dΔV/dΓ_L) determines the effective dipole moment (per molecule) $\mu_\perp/ \varepsilon_s = 1.2 \pm 0.1$ D. This relatively high value of $\mu_\perp/\varepsilon_s$ is a result of the dipole moments of the highly polar $\equiv$P$=$O head group and of six –CH$_3$ terminal groups. Owing to the radial distribution of the terminal groups around the head group, the effective dipole moment is practically inde- pendent of the molecule's conforma- tion at the surface.

Key words Alkylphosphine oxide surfactant · Surface pressure · Electric surface potential · Effective dipole moment

Introduction

The common purpose of the theoretical description of adsorption of surface-active substances is to find reliable values of the standard functions of adsorption: the Gibbs free energy, ΔG^0, the entropy, ΔS^0 and the Helmholtz free energy, ΔH^0, for comparing surface properties of different surfactants. The most convenient way to do so, is to describe the experimental results in terms of one of the basic equations of surface state, reviewed in Refs. [1, 2]; however, for the majority of soluble surfactants, none of these theoretical surface pressure isotherms can be fitted to experimental results without notable deviation. In such a case, an isotherm fit may be improved by applying different approaches, also reviewed in Refs. [1, 2].

The possible methods are as follows:

1. A series expansion of an adsorption isotherm (by introducing virial coefficients).
2. Fitting the basic isotherm with some parameters which are a function of the surfactant concentration in the bulk and at the interface.
3. Superimposing two (or more) adsorption isotherms.

The latter approach was explored in earlier work of one of the authors [3, 4], who superposed two of these basic equations of surface state – Henry–Langmuir and/or Henry–Frumkin isotherms. Recently, another approach was proposed by one of the authors [5], who introduced the conformational term to the Frumkin isotherm (as to the universal equation for liquid interfaces). This term is attributed to the changes in the conformation of the adsorbate with increasing adsorption density. An attempt has also been made to correlate the changes in the conformations with the experimentally observed variations of the electric surface potential [6, 7].

The surfactants containing the phosphoryl head group ($\equiv$P$=$O), like alkylphosphine oxides, were the subject of recent investigations by one of the authors [8, 9]. They are of particular interest with respect to the influence of the branched hydrophobic chain (characterizing a "bulky" molecule) on the surface and electric properties of a soluble monolayer at the air/water interface. As shown previously [8, 9], the surfactants of the phosphine oxide type show a medium surface activity, extending throughout a relatively wide region, usually of about three concentration decades.

In this article we report the results of surface tension and electric surface potential measurements that we obtained for *tert*-isopropylphosphine oxide (*i*-Pr$_3$PO). Its peculiarity, as compared to other "bulky" surfactants, is the hydrophobic part composed of three short isopropyl chains; the substituents filling three valences of the phosphoryl head group, with the phosphorous atom in a tetrahedral arrangement. As illustrated in Fig. 1, the *i*-Pr$_3$PO structure corresponds to a hard sphere like an ellipsoid flattened on the side of the head group (the pivot of the molecule). This structure should not undergo any conformational changes at the interface. The length of the isopropyl chain, which is about 5 Å in the fully extended state, determines the distance between the centers of the *i*-Pr$_3$PO molecules at the highest packing density. As this limiting distance almost exceeds the range of van der Waals interactions (around 10 Å [10]), one can expect that van der Waals interactions between the adsorbate molecules are negligible. Moreover, the radius of an *i*-Pr$_3$PO molecule is comparable to that of a water cluster (composed of 13 water molecules, on average [11]); therefore, the adsorbate may easily replace the cluster from the air/water interface. Such an exchange

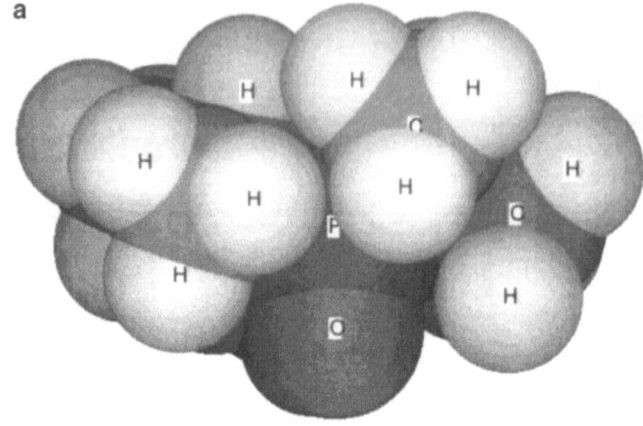

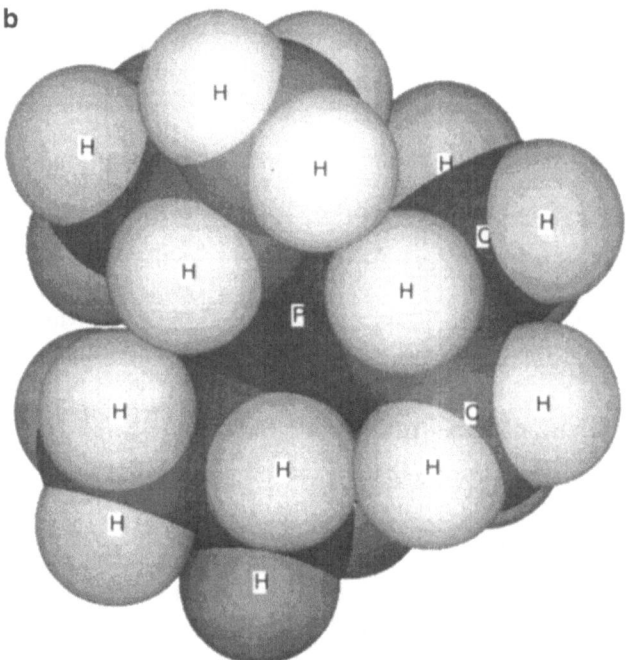

Fig. 1 Structural model of *tert*-isopropylphosphine oxide (*i*-Pr$_3$PO)

does not induce a noticeable structural change of water, so it is characterized by a very low standard entropy of adsorption. We note here that the *i*-Pr$_3$PO molecule is in a state of "hydrophobic hydration" [11], which means a "clathrate hydrate", i.e., a state of embedding in a water cluster under dynamic equilibrium.

The molecular properties of *i*-Pr$_3$PO suggest it should be a suitable surfactant to exhibit ideal surface behavior within the entire range of the surface activity region; however, the ideal properties have to be verified not only by the good suitability of a Langmuir-type isotherm to the experimental surface tension results (which is not the ultimate criterion), but also by the analysis of the adsorption data as a function of the temperature; The latter has to fulfill the corresponding dependence of the entropy of ideal surface mixing on temperature, given in Refs. [12, 13].

Experimental

i-Pr$_3$PO was synthesized according to the method reported elsewhere [14]. Prior to the measurements, the stock solution (2×10^{-1} mol/dm^3 *i*-Pr$_3$PO in water) was purified in the automatic purification apparatus described in detail in Ref. [15].

The state of the "surface-chemical purity" of the solution was achieved after having performed 500 purification cycles. (For more details on the purification procedure see Ref. [15]). Triple distilled water in the quartz distiller was used.

Surface tension measurements were performed with the de Noüy ring technique by using an automatic Lauda TE-1M tensiometer. All the requirements for the reliable measurements of the surface tension of soluble surfactants were taken into account [16].

The surface potential measurements were performed using a surface potential meter (SPD-1000) provided by KSV (Helsinki, Finland). The principle of the vibrating plate technique is reviewed by us elsewhere [17], where a special procedure is described for the application of the method to free solution surfaces to avoid numerous interferences. The "solutions replacement" procedure of measurements developed by us is described elsewhere [18].

All measurements were performed at room temperature of 22 ± 0.1 °C. The surface potential was measured at 22 ± 3 °C.

Results

Side and bird's-eye views of a free molecule of *i*-Pr$_3$PO shown as a hard-sphere model (Chem3D software) are presented in Fig. 1a and b. The molecule is in the state of minimized heat of formation (59.9 kcal/mol) and of a minimum steric energy (1:19.2 kcal/mol).

Fig. 2 shows the dependence of equilibrium values of the surface pressure (π) and the electric surface potential (ΔV) on the concentration of *i*-Pr$_3$PO in solution. The experimental data are compared with the results of calculations. Curve 1 represents the Langmuir-type surface equation of state,

$$\pi = -RT\Gamma_\infty \ln\left(1 - \frac{\Gamma}{\Gamma_\infty}\right) , \qquad (1)$$

and/or the corresponding isotherm,

$$\beta^* c = \frac{\Gamma}{1 - \frac{\Gamma}{\Gamma_\infty}} , \qquad (2)$$

where Γ and c are the surface and bulk surfactant concentration, respectively, with the best-fitted parameters $\Gamma_\infty = (1.95 \pm 0.05) \times 10^{-6}$ mol/m^2 and $\beta = 0.63 \pm 0.05$ m (Table 1);

Curve 2 represents the corresponding Γ versus c dependence. It also shows the dependence of the surface coverage (defined as $\Theta = \frac{\Gamma}{\Gamma_\infty}$) as a function of the surfactant bulk concentration (c) and, simultaneously, the dependence of the surface potential, ΔV versus c. This convergence strongly suggests that $\Delta V \sim \Gamma$. It should be stressed that all the dependencies shown in Figs. 2 and 3 are scaled to cover exactly the maximum effect measured, i.e., they are related to π_{max}, ΔV_{max} and Γ_∞ (Table 1), respectively. The error bars on the open symbols denote the average of the ΔV values taken after measurement times of 30 and 60 min, after the formation of the free solution surface. The initial regions of the isotherms are displayed in the inset of Fig. 2 as a linear function of concentration.

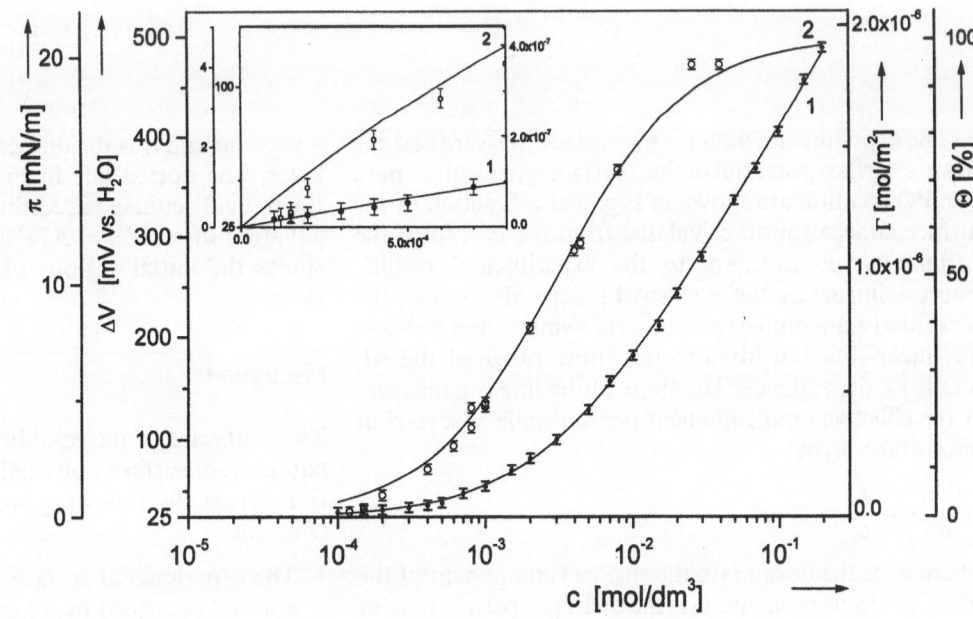

Fig. 2 Equilibrium isotherms of surface pressure, π, (filled symbols) versus concentration, c, and of electric surface potential, ΔV, (open symbols), versus c, of *i*-Pr$_3$PO solution. The *solid curves* show the results of the calculation: *curve 1* – the Langmuir-type π–c isotherm with the best-fitted parameters: $\Gamma_\infty = 1.95 \times 10^{-6}$ mol/m^2 and $\beta = 0.63$ m; *curve 2* – the surface excess, Γ_L, versus c and the corresponding Θ versus c dependencies. (The dependencies are shown normalized to the maximal effects measured; c.f. Table 1). The *inset* shows the initial region of the isotherms as a linear function of concentration

Table 1 Adsorption parameters of *tert*-isopropylphosphine oxide at the air/water interface

Parameter	Value	Unit/measurement precision
Maximum surface pressure (equilibrium value)	22.2	± 0.2 mN/m
Maximum electric surface potential change, ΔV_{max} (1 h)	473	± 5 mV
Maximum surface excess, Γ_∞ (parameter of Langmuir isotherm)	$(1.95 \pm 0.02) \times 10^{-6}$	mol/m^2 $\pm$ 1%
$RT\Gamma_\infty(25\,^\circ C)$	4.84	mN/m $\pm$ 1%
The adsorption equilibrium constant, β (parameter of Langmuir isotherm)	$(6.3 \pm 0.3) \times 10^{-1}$	m $\pm$ 5%
The standard enthalpy of adsorption, ΔG^0	-0.28 ± 0.01 (-1.2 ± 0.06)	kcal/mol (kJ/mol) $\pm$ 5%
The effective dipole moment, $\mu_\perp/\varepsilon_s = \varepsilon_0\, d\Delta V/d\Gamma_F$ [a]	1.2 ± 0.1	D $\pm$ 10%
Concentration at $1/2\Gamma_\infty$, $c_{\Theta/2}$	$(3.1 \pm 0.1) \times 10^{-3}$	mol/dm^3 $\pm$ 5%

[a] With dielectric permittivity of vacuum, $\varepsilon_0 = 8.85 \times 10^{-12}$ F/m

Fig. 3 π (filled symbols) and ΔV (open symbols) of *i*-Pr$_3$PO solution as a function of Γ_L derived from the Langmuir-type isotherm. The *solid lines* denote the results of the calculation: *curve 1* – the corresponding Langmuir-type dependence, π versus Γ_L; *curve 2* – the linear function fit to the ΔV versus Γ_L dependence. (All parameters are given as quantities normalized to the maximal effects measured). The *dotted line* (*curve 3*) shows the theoretical course of the Henry surface equation of state at 25 °C. The *inset* displays the initial region of the isotherm

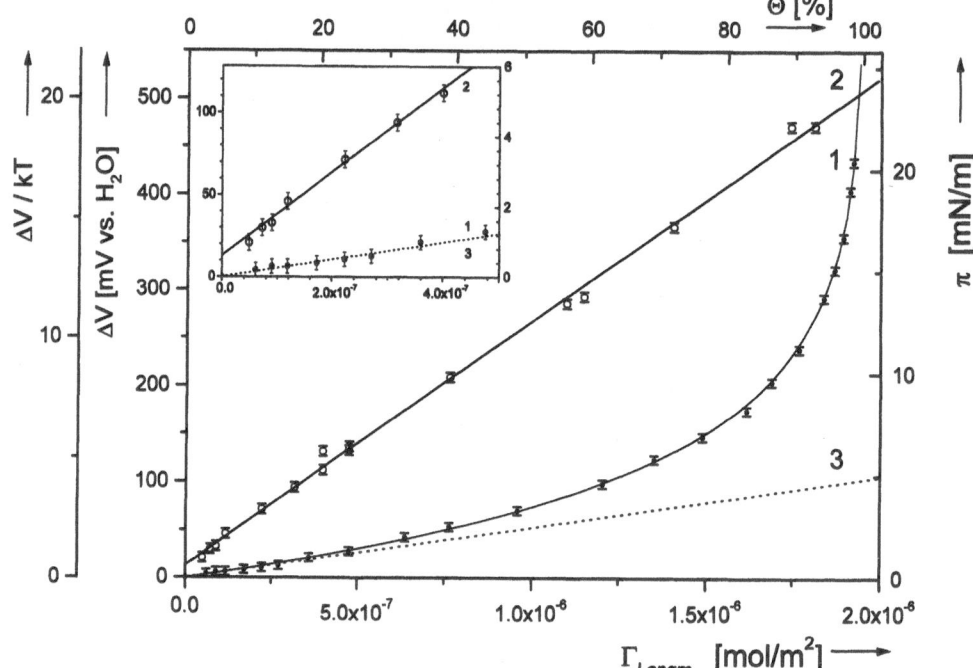

The experimental data for the surface pressure and the electric surface potential of the "surface chemically" pure *i*-Pr$_3$PO solution are shown in Fig. 3 as a function of the surface concentration calculated from the best fit of the Langmuir-type isotherm to the experimental results. Curve 1 illustrates the π versus Γ_L dependence, i.e., the best-fitted Langmuir-type isotherm, while curve 2 shows the linear function fitted to the entire range of the ΔV versus Γ_L dependence. The slope of the line is a measure of the effective dipole moment per molecule adsorbed at the surface, $\mu_\perp/\varepsilon_s$:

$$\varepsilon_0 \frac{d\Delta V}{d\Gamma} = \frac{\mu_\perp}{\varepsilon_s} = \text{const} \quad , \tag{3}$$

where $\mu_\perp$ is the normal (to the surface) component of the effective dipole moment, ε_0 is the dielectric permittivity of

a vacuum and ε_s is the dielectric constant of the adsorbed layer. The dotted line in Fig. 3 (curve 3) illustrates the theoretical course according to the Henry surface equation of state, $\pi = RT\Gamma$ at 25 °C. The inset of Fig. 3 shows the initial regions of the isotherms in the same ratio.

Discussion

The isotherms of the equilibrium surface tension and the equilibrium surface potential as a function of concentration reveal the following surface properties of *i*-Pr$_3$PO (Fig. 2):

1. The experimental surface pressure results, π versus c, are well described by a Langmuir-type isotherm (solid

curve 1). It indicates that the conditions of ideal mixing of water and surfactant molecules at the surface are fulfilled.

2. The Langmuir-type sigmoidal shape of the surface excess versus concentration dependence, Γ_L versus c, is also well reflected in the electric surface potential isotherm, ΔV versus c (between solid curve 2 and the markers in Fig. 2). The corresponding curves, ΔV versus c and Γ_L versus c, practically coincide provided the scales of the dependencies are related exactly to the maximum effects measured (the parameters of Γ_∞ and ΔV_{max} mentioned previously). The two dependencies superpose if the onset of the ΔV coordinate is fixed at 15 mV and the onset of the Γ coordinate is fixed at zero (inset of Fig. 2). The extent of the correction in ΔV (15 mV) is suggested by the results of Fig. 3 extrapolated to zero (inset of Fig. 3). Note that by fixing the onset of the ΔV coordinate at 15 mV, we formally exclude the initial rise in ΔV from the effect which is a linear function of the surface excess of i-Pr$_3$PO. A reason for a slight initial contribution to ΔV could be a reorientation of the surface water dipoles around the bulky molecules of i-Pr$_3$PO; however, to elucidate it, a separate investigation is required.

3. By comparing the initial regions of the π versus c and the ΔV versus c isotherms (inset of Fig. 2), one can evaluate the range of linear π versus c dependence. It is maintained up to a concentration of about 8×10^{-4} mol/dm^3 i-Pr$_3$PO. This is followed by a slightly narrower range in which the ΔV versus c and the Γ versus c dependencies can be approximated by a linear function up to 5×10^{-4} mol/dm^3 i-Pr$_3$PO.

4. The previously mentioned properties of i-Pr$_3$PO are well substantiated by the π and ΔV dependencies as a function of Γ_L (Fig. 3). Here, the correspondence between ΔV and the calculated Γ_L values is shown explicitly as a linear ΔV versus Γ_L dependence covering the entire adsorption range (solid curve 2). The gradient, $d\Delta V/d\Gamma$, shows the constant effective dipole moment of the i-Pr$_3$PO adsorbate, $\mu_\perp/\varepsilon_s = 1.2 \pm 0.1$ D. Notice that this value is by about 5 times greater than that reported for an insoluble monolayer of a straight-chain molecule (in the range 0.25 ± 0.05 D/molecule [19]). For example, we found that the effective dipole moment of n-decanoic acid in 5×10^{-3} mol/dm^3 HCl at the air/solution interface was $\mu_\perp/\varepsilon_s = 0.20 \pm 0.05$ D [20]. The relatively high value of $\mu_\perp/\varepsilon_s$ of i-Pr$_3$PO is a result of the sum of the dipole moments of the highly polar head group $\equiv$P$=$O and of the six $-$CH$_3$ terminals. Owing to the radial distribution of terminal groups around the head group, the effective dipole moment is practically independent of the conformation of the molecules at the surface.

5. By comparing the initial regions of the isotherms (the concentration range below 1×10^{-3} mol/dm^3 i-Pr$_3$PO), one can notice that the relative increase in ΔV with concentration is several times steeper compared to the increase in π with c (Fig. 2). If we describe the results as a function of the surface excesses, Γ_L (Fig. 3), the greater effect of adsorption on ΔV compared to π is encountered in the surface coverage range up to about 80%. For greater coverage the relationship is reversed since the $d\pi/d\Gamma$ gradient increases steeply to the maximum value, Γ_∞, and the $d\Delta V/\Gamma$ gradient remains constant throughout the entire adsorption range of i-Pr$_3$PO.

One can notice in Fig. 3 that the Henry surface equation of state well describes the adsorption coverage up to about 25% of that available at the maximum. It is worth mentioning here that for other soluble surfactants investigated, such as decanoic acid [20], n-octyl-β-D-glucopyranoside [21] and/or sodium n-dodecyl sulphate [7], considerably narrower regions of consistence with the perfect 2D gaseous state were found; these never exceeded 3% of the maximum coverage. This exceptionally wide range of the perfect 2D gaseous state is a clear hint for the ideal mixing of surfactant and water within the interfacial layer.

Finally, it is worth discussing here the initial region of the electric surface potential isotherm – the ΔV versus c and/or the ΔV versus Γ dependencies (Figs. 2, 3). The latter suggests that at a low surface concentration, slight increase in ΔV, is superposed on the effect of the adsorption. By extrapolating the ΔV versus Γ dependence to zero surface excess, one obtains $\Delta V_{\Gamma=0} = 15$ mV at the maximum (inset of Fig. 3). The initial rise in ΔV in the range of a fraction of a kT unit ($kT = 25$ mV) can be ascribed to the reorientation of surface water dipoles by i-Pr$_3$PO molecules approaching the interface before the adsorption becomes detectable. The appearance of the initial effect in the ΔV versus Γ_L isotherm seems to be favored for i-Pr$_3$PO in particular. This is because the surface activity region sets in at a relatively high concentration, about 1×10^{-4} mol/dm^3 i-Pr$_3$PO. As a consequence, a relatively great number of i-Pr$_3$PO molecules occupy the sublayer adjacent to the interface (equivalent to the bulk concentration accounting for 2D); thus, they constitute a background accompanying the adsorption surface excess. At a very low surface excess (the Henry region), the background may noticeably contribute to the ΔV measured. Probably, the initial effect on ΔV is exceptionally well measurable for i-Pr$_3$PO owing to the bulky structure of the surfactant. As mentioned earlier, the extensive "hydrophobic hydration" favors strong disorientation of surface water dipoles occurring below the onset of the adsorption region.

Conclusions

The π versus c isotherm of i-Pr$_3$PO at the air/solution interface is well described by a Langmuir-type equation. This implies ideal surface properties of i-Pr$_3$PO within the entire adsorption range. This uncommon case among soluble surfactants is explained in terms of the structure of the i-Pr$_3$PO molecule, which enables it to be replaced by a water cluster at the interface – the exchange involves a minimal entropy change. Additionally, owing to the ellipsoidal hard sphere structure of the molecule there are no conformational changes in the adsorption layer as the surface concentration of i-Pr$_3$PO increases. This observation is also supported by recent measurements of the dynamic surface properties of i-Pr$_3$PO solutions [5, 9], which indicate that no transition occurs in the adsorption layer. The Henry equation of surface state describes a relatively wide range of the coverage of i-Pr$_3$PO, up to about 25%. This behavior, uncommon for soluble surfactants, strongly supports the idea of perfect mixing.

The highest packing density of i-Pr$_3$PO ($\Gamma_\infty = 1.95 \times 10^{-6}$ mol/m^2, corresponding to a spherical cross-sectional area of about 85 Å^2/molecule) corresponds to a radius of about 5 Å – well consistent with the length of the fully extended isopropyl chain.

The electric surface potential of i-Pr$_3$PO is a linear function of the surface excess, Γ_L. The linear ΔV versus Γ_L dependence shows a constant effective dipole moment, $\mu_\perp/\varepsilon_s = 1.2 \pm 0.1$ D, within the entire adsorption region, which further indicates the lack of any conformational changes with increasing surface concentration. The value of the effective dipole moment is 5 times higher than that encountered for straight-chain surfactants (of about 0.25 D). This can be explained as a result of the summing of the dipole moments of several bonds, i.e., of the six –CH$_3$ terminal groups and of the $\equiv$P$=$O head group.

References

1. Damaskin BB, Petrii OA, Batrakov WW (1968) Adsorption of organic compounds on electrodes. Nauka, Moscow
2. Frumkin AN, Petrii OA, Damaskin BB (1980) In: Bockris JO'M, Conway BE, Yeager E (ed) Comprehensive treatise of electrochemistry, vol 1. Plenum, New York, p 5
3. Lunkenheimer K, Hirte RJ (1992) J Phys Chem 96:8683
4. Hirte RJ, Lunkenheimer K (1996) Phys Chem 100:13786
5. Warszyński P, Lunkenheimer K (1999) J Phys Chem 103:4404
6. Warszyński P, Barzyk W, Lunkenheimer K, Pomianowski A (1999) Conformational state of surfactant molecules at the air/water interface studied by surface tension and electric surface potential. Lecture at the First Nordic–Baltic Meeting on Surface and Colloid Science, August 21–25, Vilnius, Lithuania
7. Barzyk W, Warszyński P, Pomianowski A, Lunkenheimer K (1998) A transition in soluble monolayers (SDS) at the air–water interface. Abstract of poster. Proceedings of the symposium Electrochemistry at the turn of the twentieth century: Industry bio- and surface science, Kraków, May 24–27. Polish Academy of Sciences
8. Lunkenheimer K, Haage K, Hirte R, (1999) Langmuir 15:1052
9. Earnshaw JC, Grattan M, Lunkenheimer K, Rosenthal U (2000) J Phys Chem B 104:2709
10. Israelachvili JN (1985) Intermolecular and surface forces. Academic, London
11. Franks F (1984) Water. University of Cambridge
12. Defay R, Prigogine I, Bellemans A, Everett DH (1966) Surface tension and adsorption. Longman, London
13. Lucassen-Reynders E-H (1976) Prog Membr Sci 10:253
14. Rosenthal U (1995) Synthesis of tert-isopropylphosphine oxide. University of Rostock
15. Lunkenheimer K, Pergande H-J, Krüger H (1987) Rev Sci Instrum 58:2313
16. Lunkenheimer K, Wantke KD (1981) Colloid Polym Sci 259:354
17. Barzyk W, Lunkenheimer K (2000) Application of the vibrating plate (VP) for measuring the electric surface potential of solutions; I Part. The principle of measurement and interferences, prepared for publication in Rev Sci Instrum
18. Barzyk W, Lunkenheimer K (2000) Application of the vibrating plate (VP) to soluble monolayers; II Part. The procedure of the electric surface potential measurement preventing interferences by a surface-active impurities", prepared for publication in Rev Sci Instrum
19. Taylor DM, Oliveira ON, Morgan H (1990) J Colloid Interface Sci 139:176
20. Lunkenheimer K, Pomianowski A, Barzyk W (2000) Surface and electric properties of a model soluble surfactant, n-decanoic acid in 5 * 10^{-3} M HCL, prepared to publication in Bull Pol Ac: Chem
21. Barzyk W, Lunkenheimer K, Warszyński P, Pomianowski A (2000) Bull Pol Ac: Chem 48:153

Progr Colloid Polym Sci (2000) 116:113–119
© Springer-Verlag 2000

A. Hamraoui
K. Thuresson
T. Nylander
K. Eskilsson
V. Yaminsky

Dynamic wetting and dewetting by aqueous solutions containing amphiphilic compounds

A. Hamraoui (✉) · K. Thuresson
T. Nylander · K. Eskilsson
Physical Chemistry 1, Center for Chemistry
and Chemical Engineering
Lund University, P.O. Box 124
221 00 Lund, Sweden
e-mail: ahmed.hamraoui@fkem1.lu.se
Fax: +46-46-2224413

V. Yaminsky
Department of Applied Mathematics
Research School of Physical Sciences and
Engineering, Institute of Advanced Studies
Australian National University, Canberra
ACT 0200, Australia

Abstract The kinetics of wetting and dewetting of a glass surface by an aqueous solution containing an amphiphilic compound, $C_{18}OE_{84}$, has been investigated by following the capillary rise of the liquid. Typically, an overshoot of the liquid entering the capillary was observed, before relaxation towards the equilibrium height occurred. At concentrations much below and above the critical micellar concentration (cmc) the magnitude of the overshoot is small, while at intermediate concentrations (close to the cmc) the overshoot is more pronounced. The kinetics of the relaxation towards the equilibrium height depends on the concentration, with an increase in the relaxation rate at concentrations above the cmc. The observations are explained by a reduced (nonequilibrium) surface excess of the surfactant at the liquid/vapor interface. The transport of surfactant molecules to this surface depends on parameters such as concentration and diffusion constants (molecular weights). Since the concentration of the surfactant used is low, owing to the low cmc, the actual concentration in the capillary can be significantly lower as the surfactant is depleted from solution owing to adsorption on the walls of the capillary. To account for this effect, the adsorption isotherm for the solid/liquid interfaces was determined by ellipsometry. At high concentrations the change in the height of the capillary rise with time has a smooth profile, while for low and intermediate concentrations (below and close to the cmc) the behavior is more complex. An abrupt stop at a height corresponding to a certain value of the surface tension is observed. This behavior is discussed in terms of dynamic contact angle and surface tension effects due to the kinetics of surfactant adsorption that occurs close to the three-phase line at the solid/vapor and solid/liquid interfaces. The wetting kinetics varies significantly, depending on the preparation and treatment of the capillaries before the experiment.

Key words Kinetics of wetting · Capillary rise · Dynamic contact angle · Surfactant adsorption · Ellipsometry

Introduction

Wetting phenomena involve the interaction of a liquid with a solid. Common applications where these phenomena occur are when spreading droplets on a surface, penetration of a liquid into a porous medium, foaming and emulsification in the presence of particles, in painting, in coatings, or when displacing one liquid with another. Here the adhesive forces acting between the liquid and the solid surface favor spreading, while the cohesive forces within the liquid counteract spreading, and the balance between these forces determines the contact angle, θ. This balance is described by the Young equation that relates the contact angle to the surface free energies, γ, $\gamma_{SV} - \gamma_{SL} = \gamma_{LV} \cos \theta$, in a system containing solid (S), liquid (L), and vapor (V). For a system in mechanical equilibrium the wetting tension is $\tau = \gamma_{LV} \cos \theta$. In general adsorption of surface-active compounds can proceed at all three interfaces around the three-phase line. The Gibbs–Reynders form of the equation, $\tau^0 - \tau = \Pi_{SV} - \Pi_{SL}$, relates a change in the wetting tension to surface pressures, $\Pi = \gamma^0 - \gamma$, arising from adsorption at solid/liquid and solid/vapor interfaces. Thus, in the case of a surfactant solution, a complete description of the wetting properties of the solution requires knowledge of the adsorption of the surfactant molecules at all three interfaces. These equations, originally derived for systems in thermodynamic equilibrium, may be used to account for more complicated dynamic wetting phenomena, where the interfacial tensions change with time [1].

It is well documented that advancing and receding contact angles usually do not coincide. For (perfectly) pure liquids the hysteresis is referred to heterogeneity, roughness, or deformation of the surface at the three-phase line. If the speed by which a liquid moves over a substrate is important the contact angle becomes dynamic in character. Impurities located at the water/vapor interface change the surface tension of the liquid, and even trace amounts may have a significant influence on wetting. These are not the only reasons by which the advancing and receding contact angles differ. By mechanisms similar to that of Langmuir–Blodgett deposition the solute may be transferred, for instance, onto the Wilhelmy plate. Then already on the second immersion the advancing contact angle can change significantly from that of the first immersion. Such a situation is frequently observed for surfactant solutions. Here, surfactant molecules may have a tendency to assemble at the three-phase line (solid/liquid/vapor). In particular, if the headgroup of the surfactant molecule interacts favorably with the substrate and the tail reacts unfavorably with the liquid, the molecules adsorb at the solid/vapor interface. In situations where the pure liquid has zero contact angle with the substrate, the surfactant adsorption at the solid/vapor interface can cause a dewetting transition [2]. Furthermore, under conditions when the surfactant solution is forced over the substrate (e.g. immersing a Wilhelmy plate) stick–slip behavior may appear. Non-equilibrium wetting effects develop when the liquid front moves too fast for the surfactant molecules to desorb and readsorb to maintain adsorption equilibrium at the three-phase line. For example, the liquid jumps to parts of the plate as yet untouched by this adsorption [2].

If the area of a liquid/vapor interface expands or contracts faster than the rate by which adsorption equilibrium is restored the surface excess, or adsorption, Γ, of the surfactant may deviate substantially from the equilibrium value. In the present investigation we followed the rise of a solution containing a surfactant with a polymeric-sized poly(oxyethylene) headgroup, $C_{18}OE_{84}$, in a glass capillary. To rationalize the dynamic wetting and dewetting results we have to take into account that the surfactant molecules adsorb both at the solid/vapor and at the solid/liquid interfaces, and that the amount adsorbed at the liquid/vapor interface varies. Furthermore, the capillaries used in this study were prepared in two different ways, and it was found that the surface preparation had a strong influence on the wetting kinetics.

Experimental

Materials and sample preparation

The synthesis of the poly(oxyethylene) monoalkyl ether surfactant has been described elsewhere [3]. The hydrophobic part of this surfactant is an unsaturated C_{16}–C_{18} chain, and the headgroup has, on average, 84 repeating ethylene oxide units. The polydispersity index was found to be $M_w/M_n \approx 1.08$, by using size-exclusion chromatography. The surfactant was purified before use by dialyzing a solution with about 5 wt% surfactant against Millipore water for several days to remove low-molecular-weight impurities (such as salt and oligo(oxyethylene)). The solution was then freeze-dried. The recovered compound ($C_{18}OE_{84}$, $M_w \approx 3950$ g/mol), which has a diblock structure, may be viewed as a surfactant molecule with a polymeric-sized headgroup. In aqueous solutions it has a critical micellar concentration (cmc) of 1.0×10^{-3} wt% [2]. The water to prepare all solutions was deionized, passed through a Millipore water purification unit, and equilibrated with a pinch of charcoal. The desired surfactant concentration was obtained by diluting a stock solution and equilibrating the samples at least overnight before measurement. Carefully cleaned glass tubes (see the cleaning procedure used for the capillaries as described later), sealed with Teflon-tightened screw caps, were used for the solution preparation.

The material for the glass capillaries was Duran capillary tubing with a radius of 0.295 mm. Two different preparation methods were used. First, to clean the capillaries they were washed in a hot (70 °C) solution of sodium dichromate in sulfuric acid and rinsed with charcoal-treated Millipore water. Capillaries with a "pre-wetted" surface were obtained by strongly shaking the cleaned capillaries to remove rinsing water, while capillaries with a dry surface were obtained by removing the remaining water by heating with a burner. The capillaries were used only once, and all measurements were performed with freshly prepared capillaries.

Methods

Capillary rise experiments

The experimental setup used to follow the capillary rise has been described in detail elsewhere [4]. In short, a stroboscope lamp was used to obtain multiple exposures on each image when the capillary rise was recorded by a video camera. In this way we were able to capture the initial fast rise of the liquid, with a time resolution of the order of 1/100 s. At the highest observed speed (about 200 mm/s), this corresponds to a 2-mm displacement of the meniscus. Each recorded video frame was analyzed using the NIH image analyzing software (Nation Institute of Health, USA) on a Macintosh PowerPC connected to the videotape recorder. The height of the meniscus in the capillary was measured in the center of the tube. All experiments were carried out in a temperature-controlled room at 25 °C.

Ellipsometry

Polished silicon test slides (p-type, boron-doped, resistivity 10–20 Ω cm^{-1}) were purchased from Okmetic. The wafers were oxidized thermally in a saturated oxygen atmosphere at 920 °C for 1 h, followed by annealing and cooling in a flow of argon. This procedure renders an approximately 300-Å-thick surface SiO$_2$ layer on the silicon wafer. The oxidized wafers were then cut into slides with a width of 12.5 mm and cleaned according to the procedure described in Ref. [5]. Just before use the substrates were treated in a plasma cleaner (Harrick Scientific Corporation, model PDC-3XG) at a power of 30 W for 5 min.

The used ellipsometer was an automated Rudolph Research thin-film ellipsometer, type 43603-200E, equipped with high-precision step motors from Berger–Lahr, type VRDM 566, and controlled by a personal computer. The experimental setup as well as a description of the procedure for in situ characterization of thin films adsorbed on layered substrates are given in Ref. [5].

Adsorption isotherm

In order to make accurate measurements of the properties of the adsorbed layer, knowledge of the substrate properties is first required. Since four unknown parameters (n_2, k_2, n_1, and d_1) must be determined, at least four measured parameters are needed. The method is described in detail in Ref [5]. The data resulting from these measurements were then used to calculate the complex refractive index, $N_2 = n_2 - jk_2$, of the bulk silicon as well as the thickness, d_1, and the refractive index, $N_1 = n_1$, of the top silica layer. The adsorption isotherm was recorded by sequentially adding surfactant from stock solutions. When no further changes in the measured parameters (Ψ and Δ) were observed, the next aliquot of the stock solution was added. Stirring was performed with a magnetic stirrer at about 300 rpm and the cuvette temperature was kept constant during the measurement (25 ± 0.1 °C). The average refractive index, n, and the mean thickness, d, of the adsorbed layer were calculated numerically from the measured Δ and Ψ and from substrate properties in different ambient media [5]. The calculated values of n and d were also used to obtain the amount adsorbed or the surface excess (Γ) according to the formula derived by de Feijter et al. [6]

$$\Gamma = \frac{(n_1 - n_0)d}{\left(\frac{dn}{dc}\right)} ,$$

where $\left(\frac{dn}{dc}\right) = 0.16$ ml g^{-1} is the refractive index increment of the surfactant.

The adsorption isotherm was constructed from the plateau values of Γ (versus time) corresponding to different surfactant concentrations.

Results

The results from prewetted capillaries are shown in Fig. 1. The height of the liquid is plotted against time, for

the capillary rise of solutions with various concentrations of C$_{18}$OE$_{84}$. The expected equilibrium height, h_e, of a liquid in a capillary can be obtained from the balance between the Laplace pressure,

$$\Delta p_L = \frac{2\gamma}{R} = \frac{2\gamma \cos \theta}{r} ,$$

and the hydrostatic pressure,

$$\Delta p = \Delta \rho g h ,$$

as

$$h_e = \frac{2\gamma \cos \theta}{\Delta \rho g r} . \tag{1}$$

R is the radius of curvature of the meniscus, r is the inner radius of the capillary tube, γ is the surface tension of the liquid, θ is the contact angle, g is the acceleration due to gravity, and $\Delta \rho$ is the density difference between the liquid and the vapor phases. Thus, as expected, solutions that contain the surfactant do not reach to the same height as for pure water, but stop at a lower level. Typically, a nonequilibrium height, h_{ne}, is reached, which is higher than h_e predicted from the equilibrium surface tension (Eq. 1). Depending on the surfactant concentration the meniscus can stay at this level for several seconds before it begins to relax to the equilibrium height, h_e. This delay time increases with C$_{18}$OE$_{84}$ concentration below the cmc and decreases with increasing surfactant concentration above the cmc. It seems that at low and intermediate C$_{18}$OE$_{84}$ concentrations (below and near the cmc) the rise of the liquid is abruptly stopped at approximately the same value of h_{ne}. This height corresponds to a certain wetting tension ($\tau = \gamma \cos \theta$). At higher concentrations the profiles are much smoother and no "pinning" at the maximum height, h_{ne}, is observed (Fig. 1b). The wetting tension calculated from h_{ne} and h_e is shown in Fig. 2. This figure also shows the equilibrium surface tension isotherm (data reproduced from Ref [2]). The behavior at short times is displayed in Fig. 1c. Note that at low concentrations of C$_{18}$OE$_{84}$ the initial rate of the capillary rise is approximately the same as for pure water, while at higher concentrations of C$_{18}$OE$_{84}$ the initial rate is smaller.

The corresponding data for the dry capillaries is presented in Fig 3a. It is found that the speed is always smaller than in the prewetted capillaries. Above the cmc of the surfactant the liquid reaches h_e without overshooting to a nonequilibrium height. This is apparent from Fig. 3b, where data for both the prewetted and the dry capillaries are shown for the C$_{18}$OE$_{84}$ solution with a concentration of 500 cmc.

To estimate the adsorption of C$_{18}$OE$_{84}$ at the solid/liquid interface, the adsorption isotherm at the silica/water interface was recorded by in situ ellipsometry. The

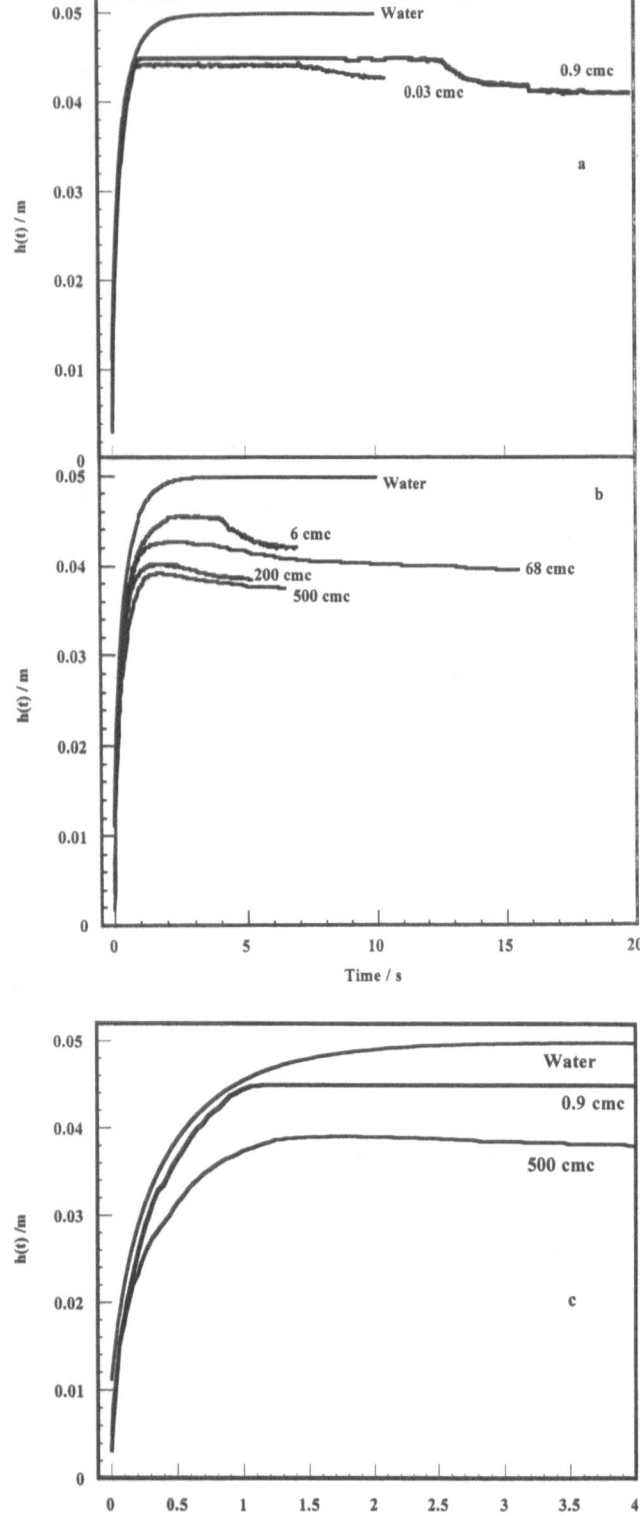

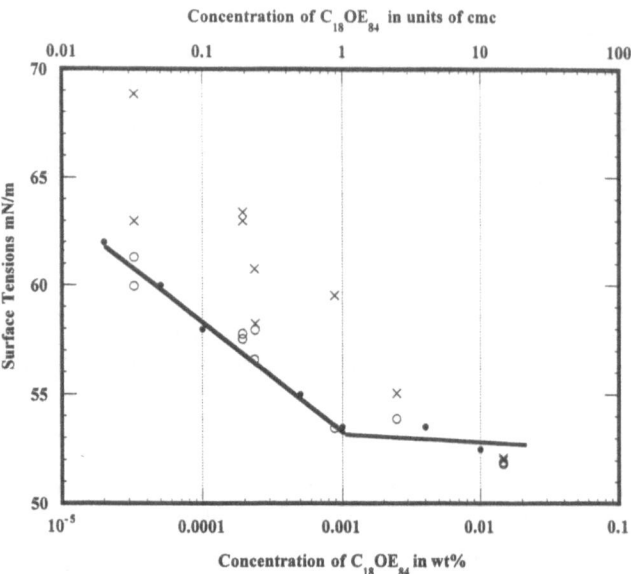

Fig. 2 Surface tension isotherm for the $C_{18}OE_{84}$ surfactant (data reproduced from Ref. [2] The wetting tensions calculated from h_{ne} (*crosses*) and from the longest observation times (*circles*) are included

Fig. 1a–c The height of the liquid in the capillary versus time for aqueous solutions of $C_{18}OE_{84}$ at different concentrations. **a** $c \leq$ critical micellar concentration (*cmc*); **b** $c > $ cmc; **c** the initial part of the capillary rise for pure water, and for one low and one high concentration of $C_{18}OE_{84}$. The curve for pure water is included

results are shown in Fig. 4. The amount adsorbed reaches a plateau value at 1.5×10^{-6} M (about 6×10^{-4} wt% or 0.6 cmc).

Discussion

The Washburn–Lucas (W-L) equation [7, 8] is frequently used to describe the height of the liquid versus time:

$$2\pi r \gamma \cos \theta = \pi r^2 \rho g h(t) + 8\pi \eta h(t) \frac{\partial h(t)}{\partial t} \; . \qquad (2)$$

The contribution from inertia is neglected, which is a good approximation with the small diameters of the capillaries used in this study [4]; Here, $h(t)$ is the meniscus height of the liquid in the capillary at time t and ρ and η are the density and the viscosity of the liquid, respectively. In recent work we found that the W-L equation fails to give an adequate description of experimental data [4] for the rise of pure water in a glass capillary, where $\gamma = 72$ mN/m and $\cos \theta$ is expected to be constant and close to unity. In particular significant deviations were found at the initial rise of the capillary, i.e., when the rate is high. Under these conditions, the capillary rises more slowly than predicted by the W-L equation (Eq. 2). If we use Eq. (2) to calculate an apparent dynamic contact angle, it results in high values (as much as 80° with a dry capillary, which corresponds to dynamic wetting tension of about 12.5 mN/m) at the initial stage of the capillary rise (high speed). In a previous publication this dynamic contact angle was discussed in terms of a friction coefficient, β, of water flowing over the substrate [4].

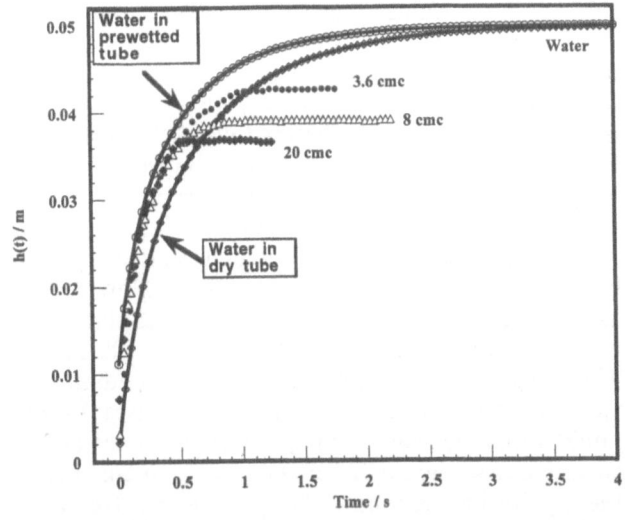

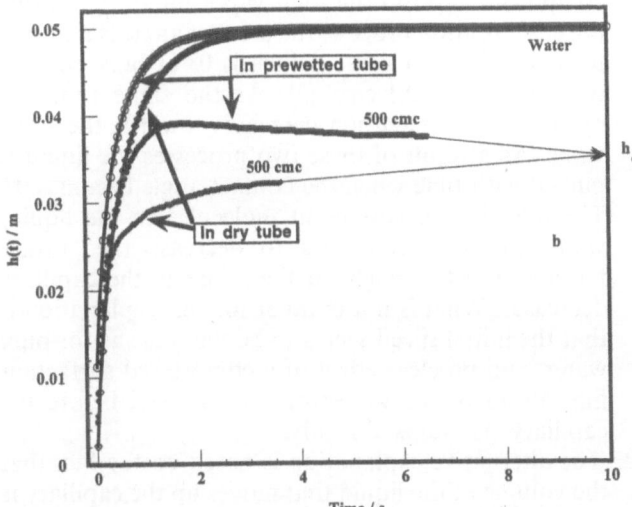

Fig. 3 Capillary rise in a dry capillary for various concentrations of $C_{18}OE_{84}$. The curve for pure water is included and the *full lines* represent fits to Eq. (3) of the data for pure water. The rise of pure water and one high concentration of $C_{18}OE_{84}$ (500 cmc) in prewetted and dry capillaries. The *arrows* indicate the equilibrium height of the surfactant solution

The relation between the dynamic contact angle and the friction coefficient is discussed in detail elsewhere [4]. β was introduced in a modified W-L equation as

$$2\pi r\left(\gamma - \beta\frac{\partial h(t)}{\partial t}\right) = \pi r^2\rho g h(t) + 8\pi\eta h(t)\frac{\partial h(t)}{\partial t} \quad . \quad (3)$$

By using Eq. (3) (the modified W-L equation) the rise of pure water could be described adequately. Molecular kinetic theories may be used to rationalize the friction in terms of the formation of a wetting film or a precursor film. This liquid layer may be compared with the stagnant layer that normally occurs close to a surface over which a liquid is flowing. The formation of such a

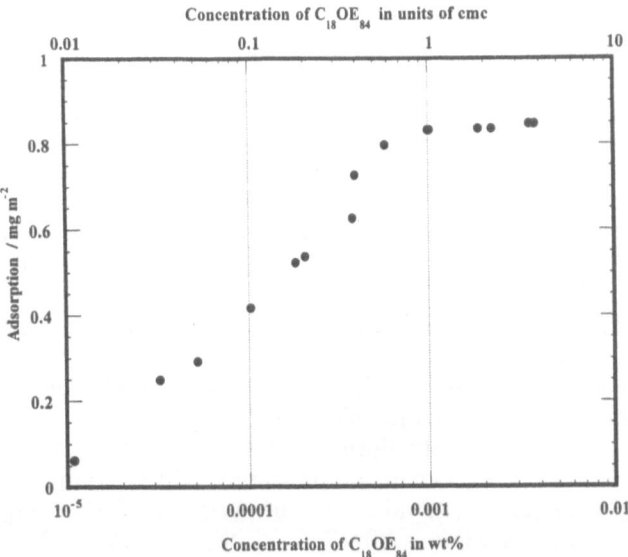

Fig. 4 Adsorption isotherm for the $C_{18}OE_{84}$ surfactant on hydrophilic silica obtained from ellipsometry measurements

film depends on the molecular interaction between the liquid and the surface and therefore the resistance to flow depends on the properties of the substrate. Consequently, this resistance was found to increase when the glass tube was dried. This observation also confirms the presence of a thin aqueous film ahead of the rising meniscus in the prewetted tubes.

In the present investigation pure water was used as a reference liquid and the influence of the amphiphilic compound $C_{18}OE_{84}$ on the capillary rise was followed. When a surface-active compound is added to the solution the surface tension is expected to decrease, and therefore the force acting on the liquid rising up the capillary is expected to be lower. This should decrease the equilibrium height of the liquid in the capillary as well as the initial speed of capillary rise. The surface tension and the contact angle are, however, expected to change with time during the capillary rise, as discussed further later. Since, we do not have any data for either the surface tension or the contact angle versus time on the relevant time scale, it is not useful to apply the W-L equation in any of its forms for the surfactant system. We start our discussion with the results for the prewetted capillaries shown in Fig. 1. While the equilibrium height indeed is decreased, the initial speed seems to be the same as that for pure water, and (at intermediate times) the liquid passes a height, h_{ne}, that is significantly higher than the equilibrium value, h_e. One way to rationalize such a profile is the following. h_{ne} corresponds to a wetting with a (vanishing) small contact angle (θ), and the meniscus height decreases to h_e because θ increases (and the wetting tension decreases). This results from an increasing surface pressure owing to the adsorption of surfactant molecules at the solid/vapor side of the three-phase line. An explanation for the high initial speed could be that the amphiphilic compound decreases the friction coefficient between the liquid and the substrate in such a way that the dynamic contact angle during capillary rise becomes small. This suggestion is, however, not compatible with results presented in a previous publication [2]. It was found that at concentrations above 0.1 cmc the contact angle on a glass surface was 0°. Furthermore, the wetting tension calculated from h_{ne} gives a much higher value, while a calculation based on h_e gives a value of the surface tension that corresponds to that of the surface tension isotherm (Fig. 2). An alternative explanation for the observation is the following. Under dynamic conditions, which apply during capillary rise, the surface excess of the surfactant is significantly reduced compared to the equilibrium value, i.e., the surface of the moving meniscus is almost "free" from the surfactant. This explains the high initial speed, which, within experimental error, was the same as for pure water. A lower surface excess, and a higher surface tension, also explains why the liquid reaches h_{ne}, which is much higher than that suggested by the equilibrium surface tension. A reduction in the surface excess during dynamic conditions may be enhanced by adsorption of $C_{18}OE_{84}$ molecules on the capillary walls. In this way the surfactant molecules are removed from the liquid/vapor interface by lateral diffusion towards and adsorption at the capillary surface. Indeed, the adsorption isotherm shown in Fig. 4 supports this explanation. Significant adsorption of $C_{18}OE_{84}$ to a silica surface at the solid/liquid interface was observed above concentrations corresponding to about 0.1 cmc. We note that this is a typical behavior of nonionic surfactants with a poly(oxyethylene) headgroup. A common explanation for this behavior is a favorable interaction of the headgroup with the solid surface.

At low to intermediate concentrations the rise of the liquid stops rather abruptly, and in this concentration range the height h_{ne} is rather independent of the $C_{18}OE_{84}$ concentration.

Two mechanisms might explain this observation:

1. Surfactant molecules that diffuse laterally towards the solid surface either must be left behind the moving liquid front and form an adsorbed layer at the solid/liquid interface or may stick to the three-phase line and be pushed ahead of the rising liquid. In the latter case the surfactant molecules can orient in such a way that the poly(oxyethylene) headgroup interacts with both water and the solid surface, while the hydrophobic part is exposed to air. The surfactant molecules that are pushed ahead of the front of the liquid form a compressed layer. A surface pressure is built up from the solid/vapor side of the three-phase line. At a certain point the advancing contact angle attains

such a high value that the capillary rise stops (abruptly). This observation parallels the stick–slip behavior that was observed when a Wilhelmy plate is dipped for the first time into a $C_{18}OE_{84}$ solution [2]. A finite contact angle was observed when the surfactant layer was compressed by the advancing wetting front. However, when the contact angle became too big, as an effect of the forced immersion, the liquid jumped to an untouched part of the Wilhelmy plate where the contact angle again reduces to zero. Now let us assume that the liquid that moves up the capillary has essentially the same composition as the bulk liquid. The surface tension relaxation to the equilibrium value proceeds when the $C_{18}OE_{84}$ diffuses to the liquid/vapor interface. This process continues as the surface excess (liquid/vapor interface) remains below the equilibrium amount. The process involves both diffusion from the solution and diffusion from the compressed layer at the solid/vapor interface. During the equilibration process, the contact angle is adjusted to its equilibrium value, which is 0° at bulk concentrations above 0.1 cmc [2]. At the same time the dynamic surface tension decreases towards the static value. As a result of these two processes the liquid is pinned with time when the contact angle becomes 0°. The diffusion of surfactant molecules to the liquid/vapor interface continues to decrease the surface tension, and the height of the liquid in the capillary decreases. What is not in favor for this explanation is that the initial speed seems to be the same as for pure water, and no clear effect of a compressed surfactant film ahead of the waterfront is observed before capillary rise stops abruptly.

2. The alternative explanation is based on the fact that the volume of the liquid that moves up the capillary is rather small and that the inner surface of the capillary where the $C_{18}OE_{84}$ surfactant is likely to adsorb is rather large compared to that volume. From the adsorption plateau as determined by ellipsometry on a silica surface ($\Gamma \approx 0.85$ mg/m^2) and the radius of the capillary (r = 0.295 mm) we can estimate the amount which can be depleted from the bulk solution by adsorption on the capillary walls. It seems that in a concentration range up to 1 cmc, the average concentration of $C_{18}OE_{84}$ in the rising liquid should not be much above 0.1 cmc. This is approximately the concentration where the amount adsorbed at the solid/liquid interface reaches half the plateau value (Fig. 4). Furthermore, the rising liquid continuously meets an untouched part of the capillary, and therefore a concentration gradient of the surfactant is expected. Close to the bottom of the capillary the liquid is expected to have essentially the bulk composition and the solid/liquid interface can be closer to the adsorption equilibrium, while close to the liquid/vapor region a small concentration close to

0.1 cmc may be expected. Indeed, if we take the h_{ne} value observed for a $C_{18}OE_{84}$ bulk concentration of 0.9 cmc and assume that $\theta \approx 0°$ we obtain an effective surfactant concentration of only 0.07 cmc from the surface tension isotherm in Fig. 2. This depletion scenario also explains why the time lag under which the liquid remains pinned at h_{ne} before it relaxes to h_e is longest at concentrations close to the cmc. (Fig. 1). Here it can be expected that the adsorbed fraction of $C_{18}OE_{84}$ molecules is large and that the concentration gradient extends over a rather long distance. Thus, the equilibrium time becomes long. On the other hand, the shorter equilibrium times at low and high $C_{18}OE_{84}$ concentrations follows from the consideration that at low concentrations the adsorption at the solid/liquid interface becomes less significant (cf. isotherm in Fig. 4). At high concentrations the depletion of the $C_{18}OE_{84}$ molecules from the bulk solution in the capillary becomes negligible. Here the concentration gradient in the capillary is expected to extend only a rather short distance from the liquid/vapor surface. Another observations that confirms the depletion explanation is illustrated in Fig. 1, where virtually the same value of h_{ne} is reached at 0.03 cmc as at 0.9 cmc.

The fact that the surface tension calculated from h_e coincides to a large extent with the equilibrium surface tension isotherm in Fig. 2 indicates that the receding contact angle is about 0°. A similar observation for the wetting of a capillary with a surfactant solution was made by Churaev et al. [9]. They explained the zero receding contact angle by the presence a thick wetting film. This was also reported in a previous publication, where we found that the contact angle becomes finite at concentrations below 0.1 cmc of the $C_{18}OE_{84}$ surfactant. Thus, at low $C_{18}OE_{84}$ concentrations the equilibrium height, h_e, is expected to be lower than suggested by the surface tension at $\theta = 0°$. The crossover to a nonzero contact angle is likely to be associated with surfactant adsorption at the solid/liquid interface (which indeed starts at this concentration) and the associated surface pressure from below the three-phase line.

Furthermore, Fig. 1c shows that the initial speed becomes smaller at (very) high $C_{18}OE_{84}$ concentrations. This is probably an effect of the higher viscosity of the more concentrated surfactant solutions (cf. Eq. 1). For the solution containing 500 cmc the viscosity is expected to be a factor of 1.22 times higher than for water as estimated from Eq. (3).

With the knowledge gained from the prewetted capillaries it is now straightforward to interpret the capillary rise in the dry capillaries. As explained previously, pure water has here slower wetting kinetics than in the prewetted capillaries owing to a higher friction coefficient. Consequently the $C_{18}OE_{84}$ solutions also give a lower initial rate of capillary rise in the dry capillaries; however, the initial speed is higher than for pure water in the dried capillaries, i.e., the surfactant increases the speed of capillary rise in the dried capillaries. This suggests that the surfactant molecules interact more strongly with the walls when the capillaries are dried, i.e. in the absence of a prewetting film. In fact we did not observe any overshoot on the time scale of our experiments, although the height at certain (low) $C_{18}OE_{84}$ concentration was higher than the expected equilibrium value. Here it should be noted that heating the glass surface of the capillary is expected to lead to a decrease in the number of free surface silanol groups. This makes the surface initially slightly more hydrophobic and the surfactant is expected to interact more strongly with the surface as it will also interact with its hydrophobic part. In this way one could imagine that the adsorption of the surfactants under some conditions can aid the rise of the capillary as it facilitates the hydration of an initially somewhat hydrophobic surface.

Conclusions

To interpret the capillary rise of a surfactant solution both equilibrium and nonequilibrium aspects of surfactant adsorption have to be considered. An overshoot of the liquid to a height much higher than that expected from the equilibrium surface tension occurs. The liquid can be pinned at this nonequilibrium height for an extended time (several minutes). Part of the explanation for this complex behavior most likely originates from the low surfactant concentration owing to the low cmc. Therefore, the concentration of surfactant in the liquid rising up the capillary can deviate substantially from the bulk composition owing to adsorption at the solid/liquid interface.

Acknowledgements This research is supported by the European Commission, Marie-Curie grant FMBICT972513.

References

1. Yaminsky V (1999) J Adhes Sci Techncol (2000) 14:187
2. Yaminsky V, Thuresson K, Ninham BW (1999) Langmuir 15:3683
3. Karlson L, Nilson S, Thuresson K (1999) Colloid Polym Sci 277:798
4. Hamraoui A, Thuresson K, Nylander T, Yaminsky V (2000) J Colloid Interface Sci 226:199
5. Tiberg F, Landgren M (1993) Langmuir 9:927
6. de Feijter JA, Benjamins J, Veer FA (1978) Biopolymers 17:1759
7. Washburn D (1921) Am Phys Soc 17:374
8. Lucas R (1918) Kolloidn Zh 23:15
9. Churaev NV, Ershov AP, Zorin ZM (1996) J Colloid Interface Sci 177:589

Progr Colloid Polym Sci (2000) 116:120–128
© Springer-Verlag 2000

B. Jachimska
P. Warszyński
K. Małysa

Effects of motion in *n*-hexanol solution on the lifetime of bubbles at the solution surface

B. Jachimska · P. Warszyński
K. Małysa (✉)
Institute of Catalysis and Surface
Chemistry, Polish Academy of Sciences
ul. Niezapominajek, 30-239 Kraków
Poland
e-mail: ncmalysa@cyf-kr.edu.pl
Tel.: +48-12-4252841 ext.284
Fax: +48-12-4251923

Abstract Lifetimes of single bubbles at the surface of a solution located at two different distances from the capillary orifice, at which the bubbles were formed, were determined for various *n*-hexanol concentrations. It was found that when the *n*-hexanol solution surface was located "far" ($L = 39.5$ cm) the average lifetimes of the bubbles were shorter than for those at the solution surface located "close" ($L = 4$ cm) to the point of the formation of the bubbles. The bubbles were formed in an identical manner, using the same capillary and the distance between the capillary and solution surface was the only different parameter in the measurements of the lifetimes of the bubbles at these two locations of the interface. This means that the "starting conditions" for the bubbles were identical in both series of experiments. Thus, the fact that the lifetimes of the bubbles at solution surface located far way from the capillary were shorter is an experimental confirmation of the correctness of the theoretical predictions that motion through the solution causes substantial lowering of the adsorption coverage on the upstream part of the floating bubble. When a bubble with lowered adsorption coverage at its upstream pole arrived at the solution surface located "far" then a nonsymmetrical foam film, with a lowered surfactant concentration at one of its interfaces, was formed. When the solution surface was "close", the distance traveled by the bubble was not long enough for such a nonuniform distribution to develop; hence, a symmetrical foam film, with equilibrium or close-to-equilibrium adsorption coverages on its both interfaces, was formed at the solution surface. Such symmetrical foam films are more stable and, therefore, the lifetimes of the bubbles were longer at the solution surface located close to the capillary orifice. Good agreement was found between the lifetimes of bubbles and the lifetimes of foam films as calculated from a simple theoretical model based on the velocity of thinning of the symmetrical and the fully nonsymmetrical foam film with one surface devoid of *n*-hexanol molecules.

Key words Bubble lifetime ·
Foam · Foam film · Adsorption
disequilibration · *n*-Hexanol

Introduction

The formation of any foam, being a gas/liquid dispersion, can be divided into three main stages:

1. The dispersing in solution of a gas phase into bubbles.
2. The motion of bubbles toward the solution surface, and

3. The formation of a foam column by bubbles gathered at the solution surface.

The foam column starts to form when the number of bubbles arriving at the solution surface is greater than the number of rupturing bubbles. The time of rupture of a bubble arriving at the solution surface is a function of the durability of the foam film formed by the top part of the bubble and the local area of the solution surface. The properties of the film depend on the kind of surfactant and the magnitude of its adsorption coverage at the film interfaces [1–3].

It should be remembered and taken into consideration that foam formation is a dynamic process during which the solution/gas interface is enormously enlarged. Thus, there is a lot of motion, disturbance and formation of "fresh" interfaces. Any motion and/or disturbance can mean a lack of adsorption equilibrium coverage at the interface. This lack of equilibrium coverage can affect the magnitude of the stabilizing forces (e.g. surface elasticity forces) and therefore foam formation and stability needs to be considered in terms of forces determined by actual (quite often nonequilibrium) surface coverage [4]. A lower surface coverage at interfaces of the foam film would result in a lower magnitude of the stabilizing forces and a shorter lifetime.

During the formation of a bubble the magnitude of the surfactant coverage at the bubble's surface depends on the mutual ratio of the velocities of the surfactant adsorption and the formation of a fresh interface. However, even if the bubble formation was so slow that there equilibrium coverages were attained the motion through a viscous medium (solution) could lead to disequilibration of adsorption coverages over the surface of the rising bubble [5, 6]. As the result of a viscous drag exerted by the continuous medium on the floating bubble the adsorption coverage on the upstream part of the bubble is expected to be significantly lower than the equilibrium coverage [7–12]. When the bubble arrives at the solution surface its upstream pole will constitute a "bottom" surface, while the local area of the solution surface will become the "top" surface of the foam film formed. A lower adsorption coverage than the equilibrium ones on the "bottom" film interface can, certainly, affect the magnitude of the forces stabilizing the foam film and, consequently, the lifetime of the bubble.

This article presents results of measurements of lifetimes of single bubbles at the n-hexanol solution surface located at two different distances from the point of the bubble formation. If the motion through the n-hexanol solution really caused a "stripping" of the upstream part of the bubble from the adsorbed n-hexanol molecules then there should be a detectable difference in the lifetimes of bubbles at the solution surface located "close" and "far" from the point of the bubble detachment. When the solution surface is "close" there

should not be any significant motion-induced deviations in the bubble surfactant coverage, while in the case when the free surface is "far" the disequilibration of the surface coverage should be the maximum possible for the system studied. Since the bubble lifetime at the solution surface depends on the state of the adsorption layers, the differences in the surface coverages at the top poles of the bubbles arriving at the solution surface should affect their average lifetime.

Theoretical model

The processes involved in the bubble formation, detachment and journey to the solution surface are illustrated schematically in Fig. 1. The formation of a bubble in the solution is accompanied by the convective and diffusive transport of surface-active substances to the freshly created surface. If there are no geometrical constraints, the surface of the bubble formed in the solution is uniformly accessible to adsorption. Additionally, if the time of bubble formation is long enough

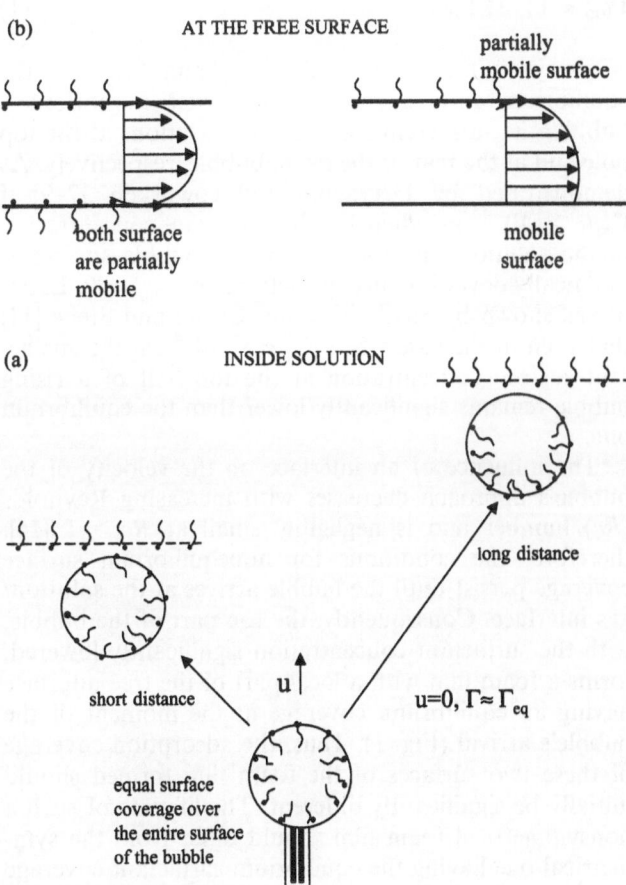

Fig. 1 Schematic illustration of the processes occurring **a** inside the solution during the bubble's motion and **b** at the solution surface when either symmetric or nonsymmetric foam films are formed

compared to the adsorption time the equilibrium coverage over the surface of the bubble is attained. Otherwise, the equilibrium may not be achieved, but surfactant molecules uniformly cover the bubble surface. When the bubble detaches and starts to move toward the solution interface the steady-state terminal velocity is attained after a certain time. That velocity is determined by the bubble's size, by the density and viscosity of the solution and by the degree of the bubble's surface retardation by surface-active agents. Numerous theoretical approaches regarding the influence of these factors on the motion of the rising bubble have been summarized in Refs. [12, 13].

In their theoretical works Frumkin and Levich [5, 6] showed that a steady motion of a bubble would induce a nonuniform adsorption coverage over its surface. The coverage by a surface-active substance is a minimum at the leading pole of the bubble and a maximum at the rear stagnation point. The difference between the minimal and maximal coverage depends on the bubble's velocity and the solute surface activity but generally the condition

$$\Gamma_{top} < \Gamma_{eq} < \Gamma_{rear} \qquad (1)$$

is always fulfilled, where Γ_{eq}, Γ_{top} and Γ_{rear} are the equilibrium surface concentration over a motionless bubble, the surfactant surface concentrations at the top pole and at the rear of the rising bubble, respectively. As demonstrated by Deryaguin and coworkers [7–9] if $\Gamma_{eq}/c > 10^{-4}$ cm, where c is the surfactant concentration in the solution, the top part of the bubble surface is practically devoid of any surfactant, i.e. $\Gamma_{top} \approx 0$. Later, it was shown by Saville [10] and Cheng and Stebe [11] that even in the case when $\Gamma_{eq}/c < 10^{-4}$ cm the surfactant surface concentration at the top half of a rising bubble remains significantly lower than the equilibrium one.

The influence of an interface on the velocity of the bubble's approach decreases with increasing Reynolds (Re) number and is negligibly small at $Re > 1$ [14]; therefore, the conditions for nonequilibrium surface coverage persist until the bubble arrives at the solution/gas interface. Consequently, the top part of the bubble, with the surfactant concentration significantly lowered, forms a foam film with a local part of the free interface having an equilibrium coverage at the moment of the bubble's arrival (Fig. 1). Thus, the adsorption coverage of these two surfaces of the foam film formed should initially be significantly different. The lifetime of such a nonsymmetrical foam film should differ from the symmetrical one having the equilibrium surfactant coverage on its two interfaces.

The lifetime of the foam film is directly dependent on the film thinning velocity. The velocity of the thinning of a film of uniform thickness is [15]

$$V = V_{Re}V_s(\Gamma_i, \Gamma_b, D, D_s, \mu, \mu_s, h)$$
$$\times V_W(\Gamma_i, \Gamma_b, D, D_s, \mu, \mu_s, \varepsilon, \lambda, R_e, h) \qquad (2)$$

where $V_{Re} = \frac{Fh^3}{2\pi\mu R_e^4}$ is the velocity of the thinning of the film d with fully immobilized surfaces (Reynolds formula), F is the total force causing the film thinning (buoyancy, negative disjoining pressure, capillary force), h is the film thickness, μ is the fluid viscosity, R_e is the effective radius of the film which can be determined from [16]

$$R_e = \sqrt{\frac{FR_b}{\pi\sigma_{eq}}}, \qquad (3)$$

where R_b is the radius of a bubble forming the film, σ_{eq} is the equilibrium surface tension, $V_s(\Gamma_i, \Gamma_b, D, D_s, \mu, \mu_s, h)$ is the correction for the Marangoni–Gibbs effect accounting for incomplete retardation of the film surfaces and is dependent on the surfactant surface concentration gradients [15, 17], Γ_i, Γ_b are the surface concentrations at the solution surface and at the top of the bubble, respectively, D is the bulk diffusion coefficient of surfactant, D_s is its surface diffusion coefficient and μ_s is the surface viscosity. $V_w(\Gamma_i, \Gamma_b, D, D_s, \mu, \mu_s, \varepsilon, \lambda, R_e, h)$ is the correction for the contribution of the surface waves to the film thinning [17] and ε, λ are the amplitude and the wavelength of the typical interfacial wave. The detailed form of these corrections was discussed previously [3].

As a first approximation we consider two limiting cases of the thinning of fully symmetrical and nonsymmetrical films (Fig. 1b). Both interfaces of the symmetrical film initially have an equilibrium surfactant surface concentration. Owing to the flow of the solution out of the thinning film the surface concentration gradients are induced and both surfaces can be partially or fully immobilized to the same extent. The nonsymmetrical film has the upper interface with a surfactant surface concentration close to equilibrium, while the lower interface formed by the top the rising bubble is assumed to be initially depleted of surfactant. This means that while the upper surface of the film can be retarded, the lower surface of the film is fully mobile. Owing to continuous adsorption, desorption, diffusion and convection of the surfactant and changes in the surface mobilities during the film thinning the real situation is between these two limiting cases. However, when the top of the surface of the rising bubble is free of surfactant its concentration in a diffuse layer adjacent to this part of the bubble surface is also significantly diminished. This makes the restoration of equilibrium coverage more difficult as surfactant molecules have to diffuse from the bulk of the solution inside the foam film. Let us estimate the time needed to reestablish adsorption equilibrium coverage at the surface of a foam film. For the effective radius of the film $R_e \approx 0.01$ cm and the surfactant diffusion coefficient $D \approx 10^{-5}$ cm^2/s this time is of the order of 5 s ($t \approx R_e^2/2D$). This time is expected to be even longer as diffusion is additionally hampered by the convection flux due to the fluid flow induced by the thinning of the foam film.

The theoretical lifetime of the foam film can be determined from

$$\tau = \int_{h_f}^{h_i} \frac{1}{V}\, \mathrm{d}h, \tag{4}$$

where h_i is an initial distance and h_f is a final distance when the film ruptures.

The lifetimes of symmetrical and nonsymmetrical films calculated using Eq. (4) with $h_i = 5000$ nm and $h_f = 100$ nm are presented in Fig. 2. The other parameters used in the calculations were as follows: $\lambda = 4 \times 10^{-5}$ m, $\varepsilon = 7$ nm, $D = 7 \times 10^{-6}$ cm^2/s. The values of the surface diffusion coefficients are not well known. In the literature the D_s values are reported to be larger than [17, 18], equal to and smaller than [19, 20] the respective bulk values. In our calculations we assumed $D_s = 5 \times 10^{-5}$ cm^2/s as the best-fitting value to the experimental data. Curve A in Fig. 2 represents the dependence of the theoretical lifetime of the symmetrical film on the concentration of n-hexanol solution taking into account all corrections considered in Eq. (2), while the dotted line (curve A') illustrates the same dependence if only the Reynolds formula is used (i.e. the contribution from the Marangoni effect and the capillary waves are neglected). It is known that the Reynolds formula alone greatly overestimates the lifetime of the foam films. Curves B (solid) and B' (dotted) show the same set of results but obtained for the fully nonsymmetrical film. In this case the full slip boundary conditions are assumed at the lower surface of the film (Fig. 1b). This leads to the

film thinning velocity being 4 times higher than given by the Reynolds formula when other corrections are neglected (cf. curves A', B').

The final distance, h_f, i.e. the critical film thickness at which the film ruptures, appears to be one of the crucial parameters determining the lifetime of a foam film. The dependence of the film lifetime on the concentration of n-hexanol solution, calculated according to Eq. (4), for the symmetrical and nonsymmetrical films is illustrated in Fig. 3. The lifetime calculations were performed for various critical film thickness. Curve A was obtained for $h_f = 80$ nm, curve B for $h_f = 100$ nm, and curve C for $h_f = 150$ nm. The critical thickness was assumed to be independent of n-hexanol concentration. All details of the calculations are given in our previous articles [3]. It was assumed here that n-hexanol adsorbs at the air/solution interface according to the Frumkin adsorption isotherm and the isotherm parameters were as follows: $a_1 = 6.4 \times 10^{-3}$ mol/dm^3, $\Gamma_\infty = 5.9 \times 10^{-10}$ mol/cm^2, $H = 2.4$ kJ/mol, where a_1 is the surface activity parameter of the isotherm, Γ_∞ is the limiting surface concentration corresponding to monolayer adsorption coverage, and H is the Frumkin isotherm parameter accounting for interaction between the adsorbed molecules. The calculated lifetime does not depend on the choice of the initial film thickness provided that $h_i > 500$ nm.

For low n-hexanol concentrations the foam film lifetime increases very abruptly with increasing solution concentration. At higher concentrations the film lifetime is almost independent of concentration for the both symmetrical and the nonsymmetrical films; however, the lifetime of the symmetrical film is approximately 4 times

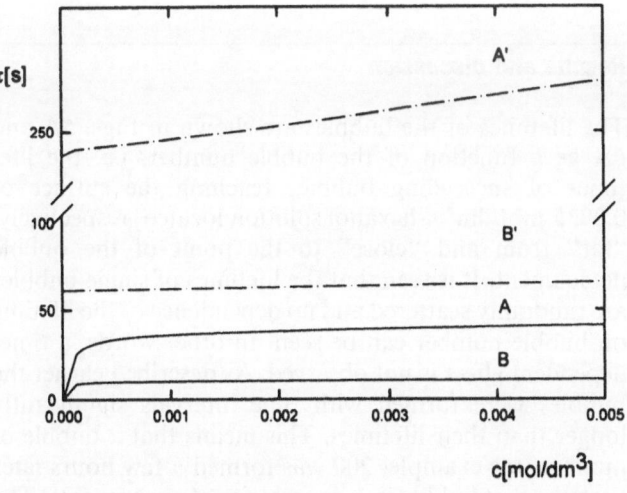

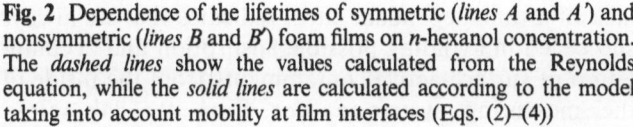

Fig. 2 Dependence of the lifetimes of symmetric (*lines A and A'*) and nonsymmetric (*lines B and B'*) foam films on n-hexanol concentration. The *dashed lines* show the values calculated from the Reynolds equation, while the *solid lines* are calculated according to the model taking into account mobility at film interfaces (Eqs. (2)–(4))

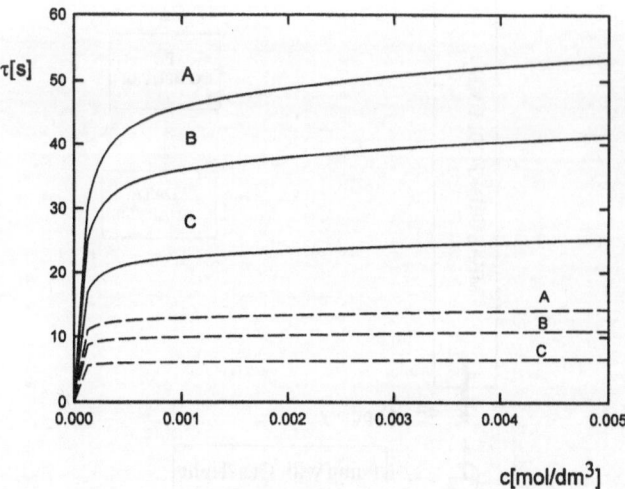

Fig. 3 Dependence of the foam film lifetime on n-hexanol solution concentration calculated according to Eqs. (2), (3), and (4) for the symmetric (*solid lines*) and nonsymmetric films (*dashed lines*). *Lines A*, $h_f = 80$ nm; *lines B*, $h_f = 100$ nm; *lines C*, $h_f = 150$ nm. Other parameters are described in the text

124

longer than the nonsymmetrical one and both decreased significantly as the critical film thickness increased.

Experimental

Apparatus

The experimental setup used for determination of the lifetimes of the bubbles at the solution surface located at different distances from the point of the bubble formation is presented schematically in Fig. 4. It consisted of the following main components: a glass column of inner diameter 52 mm having at the bottom a capillary with a 0.18-mm orifice, a gas supply system, and a data acquisition and analysis system. Details of the glass column construction, having two interchangeable upper parts of different lengths, is described elsewhere [3]. The quality of the capillary orifice was checked using a microscope. The diameters of about 100 bubbles were measured and no difference in the bubble size was detected, which showed that the reproducibility of the bubble formation was satisfactory. To have bubbles at the center of the solution surface, far away from the column walls, the solution meniscus was kept convex by placing a special Teflon ring at the top of the column [3]. For better control of the frequency of bubble formation at the capillary orifice a new gas supply system was used. Air was supplied to the capillary by an infusion/withdrawal syringe pump (Cole-Parmer Instruments) with gas-tight glass syringes of various sizes (5, 10, 50 ml). The rate of gas injection used in the experiments was within 0.1 to 0.007 ml/min. The rate of gas injection was carefully controlled and varied in a such way that in a given series of experiments the single bubbles were generated in time intervals always significantly longer than their lifetimes at the solution surface. This condition ensured that the adsorption equilibrium was always attained at the solution surface before a subsequent bubble arrived. Moreover, rupture of the bubble always causes some disturbances at the solution surface and sometime is needed for the surface to become quiescent again.

The lifetime of a single bubble was measured automatically as the time difference between the bubble appearing and disappearing at the solution surface, using an elaborate computer program. A charge-coupled-device camera coupled with a television monitor, a computer and a JVC BR-800E professional video recorder was used to monitor and record the lifetimes of the bubbles and the time intervals between the subsequent bubbles. The lifetimes of 150–400 separate bubbles were determined for every set of conditions studied and then an average lifetime was calculated. From the video recordings the sizes of about 100 bubbles was measured in every experiment. No difference in the size of the bubbles created close to and far from the solution interface was detected.

Two series of measurements of the lifetimes of single bubbles were carried out for every solution concentration. In the first series, the solution surface was located "far" from the point of the bubble detachment (capillary orifice), while in the second series it was located "close" to this point. The distance ,L, between the capillary and the solution surface was $L = 39.5$ cm and $L = 4$ cm for the locations called "far" and "close", respectively. The distance "far" was chosen to be similar to the distance between the sintered glass and the solution/foam interface in the foamability measurements [4]. The location "close" was arbitrary chosen to be close enough to the capillary and yet sufficiently far away for the solution surface not to be disturbed by the bubble departure from the capillary orifice.

The experiments were carried out at room temperature of 20 ± 1 °C.

Reagents

Four-times distilled water was used and its purity was repeatedly checked in blank experiments. The system was considered to be free of any contaminants if the lifetimes of single bubbles at the surface of the distilled water were shorter then 0.5 s. High-purity n-hexanol was used to prepare the solutions used in the experiments. Purified air was used for the formation of the bubbles.

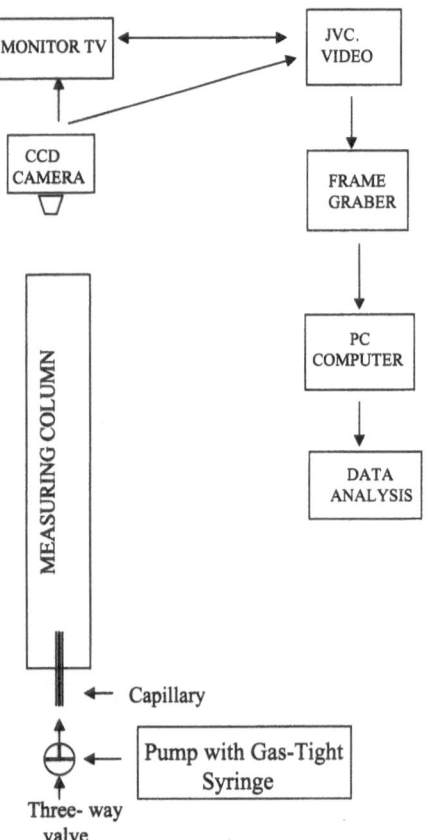

Fig. 4 Scheme of the experimental setup used for the determination of the lifetimes of the bubbles at the solution surface

Results and discussion

The lifetimes of the bubbles are shown in Figs. 5A and 6A as a function of the bubble number, i.e. the lifetimes of succeeding bubbles reaching the surface of 0.0035 mol/dm^3 n-hexanol solution located, respectively, "far" from and "close" to the point of the bubble detachment. It is seen that the lifetimes of single bubbles are randomly scattered and no dependence of the lifetime on bubble number can be seen. In other words, a time-dependent effect is not observed. As described earlier the bubbles were formed with time intervals significantly longer than their lifetimes. This means that a bubble of number, for example, 200 was formed a few hours later than the first bubble in a given series of experiments. The lack of a time-dependent effect indicates that we always succeeded in avoiding possible adsorption and accumulation of surface-active contaminants from the inside of the measuring system and/or from the laboratory

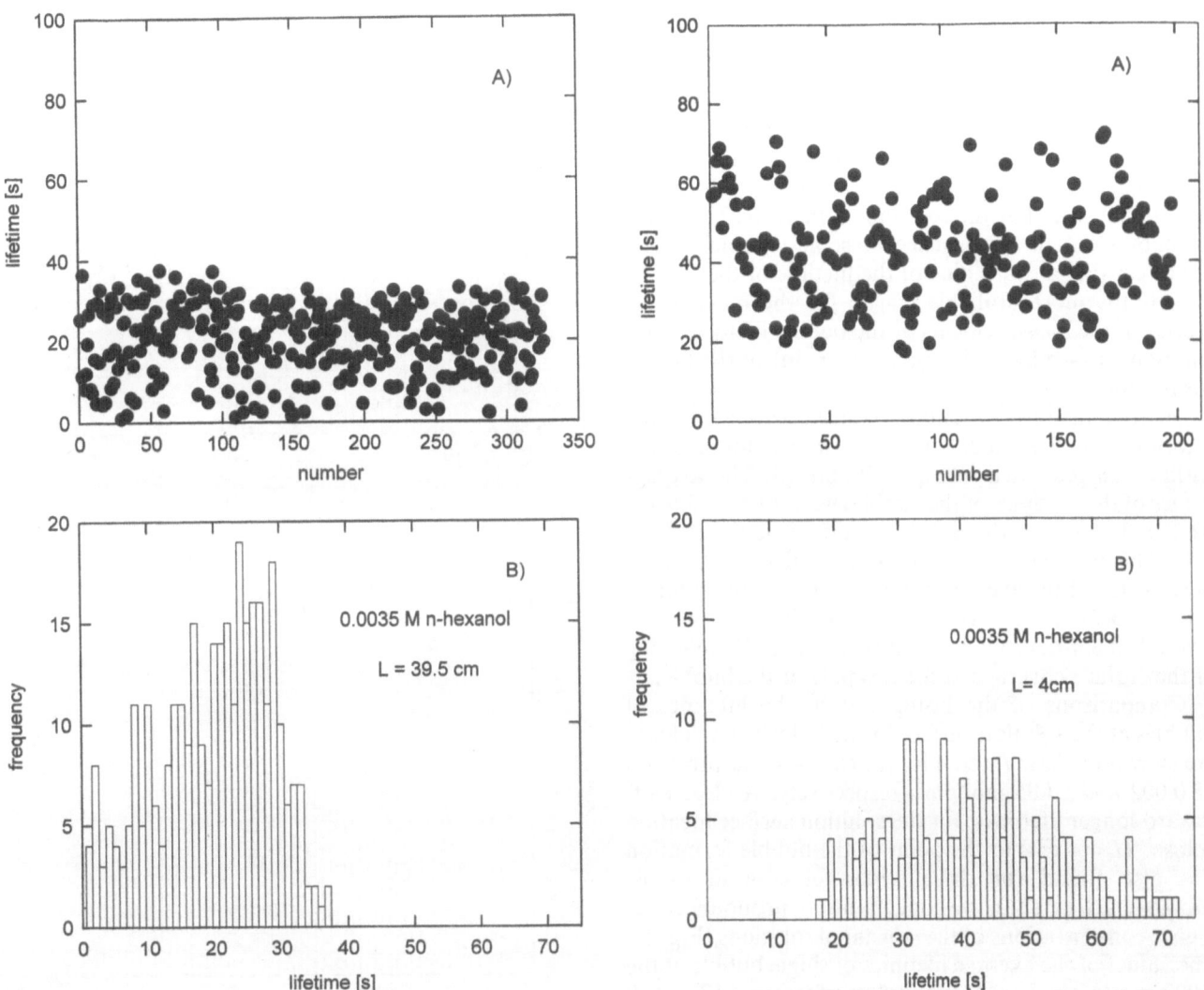

Fig. 5 A Lifetimes of the consecutive bubbles at the solution surface located 39.5 cm from the capillary orifice. **B** Histogram of the lifetimes of the bubbles for 0.0035 mol/dm³ *n*-hexanol solution

Fig. 6 The same as Fig. 5 but for the solution surface located at a distance 4 cm from the capillary orifice

atmosphere. As seen in Figs. 5 and 6 the lifetimes of single bubbles at the surface of *n*-hexanol solutions of concentration 0.0035 mol/dm³ are randomly scattered for the two locations of the interface. Moreover, a significant difference in the values of the lifetimes and the range of their scatter can be noticed for these two locations of the solution surface. Longer lifetimes and much larger scatter of the results were observed (Fig. 6A) when the solution surface was at a distance $L = 4$ cm from the capillary orifice. At this location of the *n*-hexanol solution surface, practically all the bubbles had lifetimes scattered from 20 s up to almost 80 s. When the solution surface was located far away ($L = 39.5$ cm) the bubbles showed lifetimes within the range from 0 s to almost 40 s.

A few remarks need to be added about possible reasons for the scatter of the measured lifetimes of single

bubbles. The lifetimes of the bubbles were determined under dynamic conditions. The scatter of the lifetimes was rather high but, as described earlier, fully random, which indicates that the scatter is probably related to intrinsic features of the system. Bubble formation, motion, foam film formation and thinning are dynamic processes affected by all kind of movements, mutual interactions and interdependencies. Despite all efforts to standardize the procedure of the bubble formation any, for example, fluctuations of air pressure in the gas supply system could affect the formation time and the bubble departure coverage. The state of the adsorption layer on a bubble reaching the solution surface depends on the conditions of the bubble formation, the detachment from the capillary and its motion through the solution; therefore, the scatter of experimental data was expected to be rather high and large numbers of lifetimes of single bubble were measured to get some meaningful data. Moreover, our system was not isolated from external disturbances, such as air motion in the laboratory, various kinds of ground vibrations related to outside traffic, temperature gradients, etc. Any kind of external disturbance can lower the lifetime of the bubble and the probability of this arising increases with time. The longer the bubble stayed at the solution surface, the higher the probability of the occurrence of some external disturbance causing a shortening of the lifetime of the bubble. This is the most probable reason for the much wider scatter in the measurements of the bubble lifetime at the solution surface located close to the point of the bubble formation.

Histograms of the lifetimes of single bubbles at both locations of the 0.0035 mol/dm^3 n-hexanol solution surface are presented in Figs. 5B and 6B. The average values of the lifetimes of the bubbles were 20.4 ± 8.6 and 42.2 ± 12.8 s for the solution surface located "far" (Fig. 5B) and "close" (Fig. 6B), respectively. Thus, the lower value of the average bubble lifetime at the solution surface located "far" is a clear indication that motion through the n-hexanol solution really caused a lowering of the surface coverage at the top pole of the bubble.

Comparisons of the histograms of the lifetimes of bubbles at the solution surface located "far" and "close" are shown in Figs. 7 and 8 for n-hexanol concentrations of 0.002 and 0.005 mol/dm^3, respectively. A clear shift toward longer lifetimes, for the solution surface location "close" ($L = 4$ cm) to the point of the bubble formation was also found for these n-hexanol concentrations. Moreover, this shift was much more pronounced at higher concentrations of the n-hexanol solutions (Fig. 8). The values of the average lifetimes of single bubble at the 0.002 mol/dm^3 n-hexanol solution surface were 17.3 ± 9 and 32.2 ± 12 s for surface located "far" and "close", respectively. In the case of 0.005 mol/dm^3 n-hexanol solution, the appropriate values of the average lifetimes were 18.5 ± 10 ($L = 39.5$ cm) and 48.2 ± 14 s ($L = 4$ cm). Thus, in all experiments there was a statistically meaningful difference in the lifetimes of the bubbles at the solution surface located either "far" from or "close" to the point of the formation of the bubble.

In both locations of the solution surface the bubbles were formed in the same way and using the same capillary; therefore, in both locations of the solution surface the "starting conditions" of the bubbles were identical. Moreover, we can assume that at the moment of detachment from the capillary orifice, the bubbles had equilibrium or very close to equilibrium adsorption surface coverages over their surface because they were always formed very slowly when using the syringe pump with a well-controlled gas infusion rate. Thus, the shorter lifetimes of the bubbles at the solution surface located far away from the capillary indicate that motion through n-hexanol solution really leads to disequilibration of

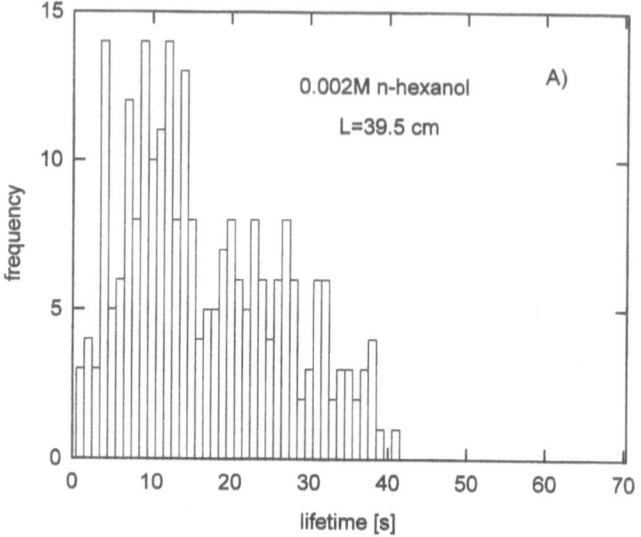

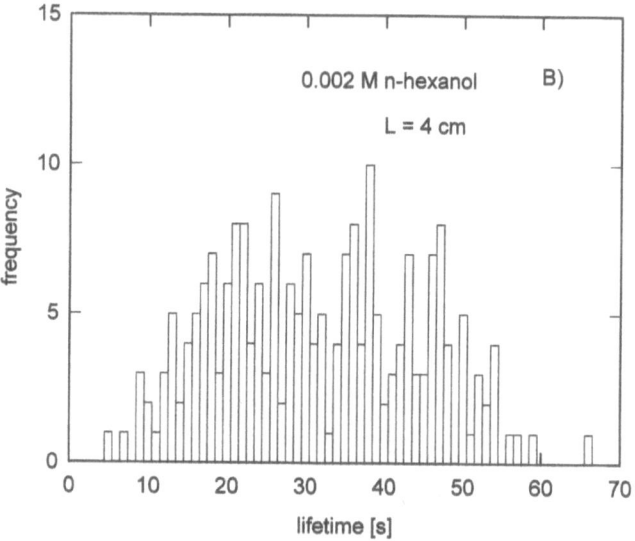

Fig. 7 Comparison of histograms of the lifetimes of the bubbles created: **A** 39.5 cm and **B** 4 cm from the surface of 0.002 mol/dm^3 n-hexanol solution

n-hexanol adsorption coverage over the surface of the moving bubble. When a bubble with lowered adsorption coverage at its upstream pole arrived at the solution surface located "far", a nonsymmetrical foam film was formed. Such a nonsymmetrical foam film, with a lowered surfactant concentration at one of its interfaces, was less stable and, therefore, the lifetimes of the bubbles were shorter.

The average lifetimes of the bubbles at solution surfaces located at distances $L = 39.5$ cm (curve 2) and $L = 4$ cm (curve 1) are shown in Fig. 9 as a function of n-hexanol concentration. The filled circles show the experimentally determined values of the average lifetime of the bubbles at the solution surface located "close" to the capillary orifice and the open circles present the data

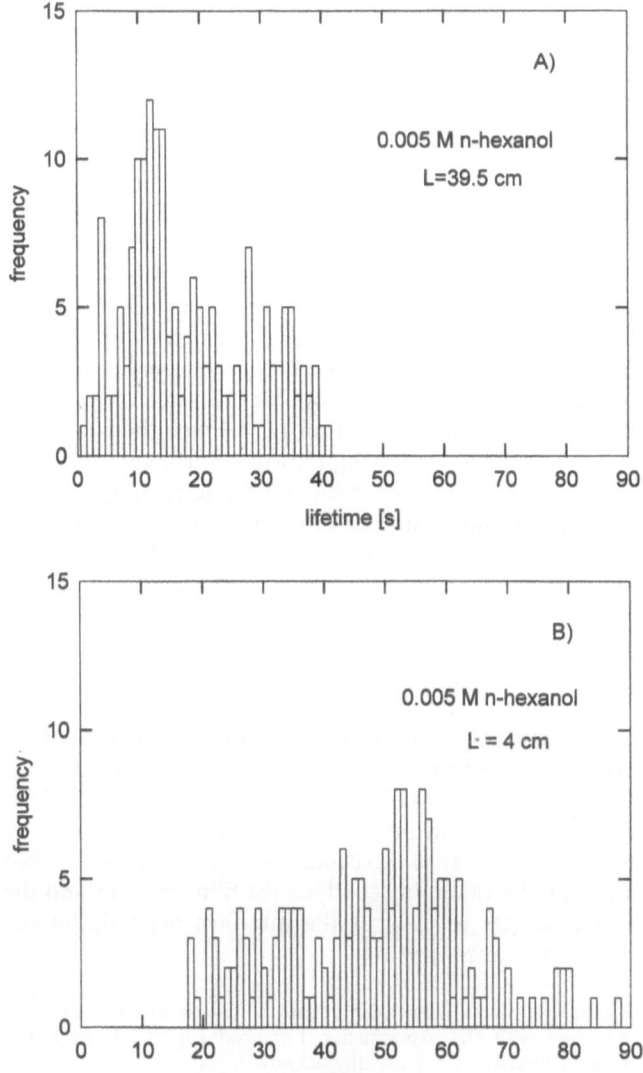

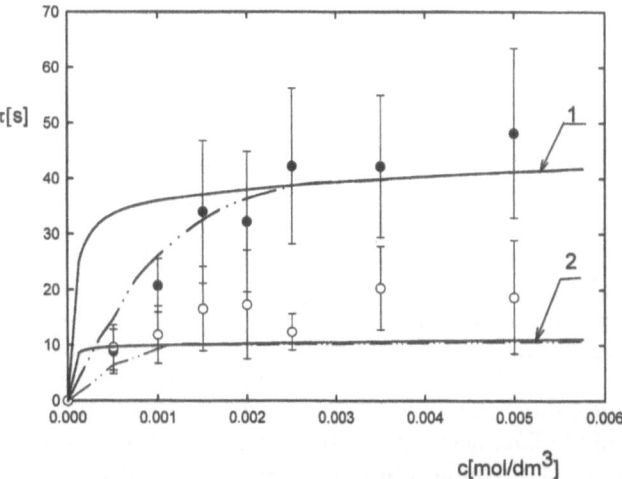

Fig. 9 Dependence of the average bubble lifetime on *n*-hexanol solution concentration for bubbles created 4 cm (●) and 39.5 cm (○) from the solution surface. The *solid lines* present results of calculations of the lifetimes according to the theoretical model (Eqs. (2)–(4)) for symmetrical (*1*) and nonsymmetrical (*2*) films for $h_f = 100$ nm. The *broken lines* are for h_f varying from 250 to 100 nm (see text)

Fig. 8 The same as Fig. 7 but for 0.005 mol/dm³ *n*-hexanol solution

for the solution surface located "far". The bars show the standard deviations of the determined values of the average lifetimes of the bubbles. It is seen that the average values of the bubble lifetime increased with increasing concentration of the *n*-hexanol solutions, and this increase was more pronounced when the solution surface was located "close" to the capillary orifice. At this location of the solution surface ($L = 4$ cm) the average bubble lifetime increased from 9 s up to almost 50 s when the *n*-hexanol concentration was changed from 0.0005 to 0.005 mol/dm³. Within the same concentration range the average bubble lifetime varied from 9 s to about 20 s only when the solution surface was at a distance $L = 39.5$ cm away from the point of the bubble detachment. Shorter lifetimes at the solution surface located "far" from the point of the bubble formation are, in our

opinion, a clear indication that motion really caused the depletion of the upstream pole of the bubble of the adsorbed *n*-hexanol molecules. As a result of the motion-induced disequilibration of *n*-hexanol coverage over the bubble surface a nonsymmetrical film was formed when the bubble arrived at the surface. When the solution surface was close to the capillary orifice a disequilibration of the surfactant coverage over a barely detached bubble hardly started to develop and a symmetrical or close-to-symmetrical film was formed. As discussed earlier, the stability of such a symmetrical foam film is higher and, therefore, the lifetimes of the bubbles at the solution surface located "close" were longer, especially at higher *n*-hexanol concentrations.

The solid lines in Fig. 9 show the dependence of the lifetime on the *n*-hexanol concentration as calculated from the theoretical model based on velocity of thinning of the symmetrical and nonsymmetrical foam films. These lifetimes were computed assuming the surface diffusion coefficient $D_s = 5 \times 10^{-5}$ cm²/s and critical film thickness $h_f = 100$ nm were independent of the surfactant concentration (see Ref. [3] for the details of the computations). The dotted lines illustrate the results of the same calculations but under the assumption that the critical film thickness changes with *n*-hexanol concentration in the range from $h_f = 250$ nm for $c = 0.0005$ mol/dm³ to $h_f = 100$ nm for $c > 0.002$ mol/dm³. It is seen that the dependence of the calculated foam film lifetime on the concentration of the *n*-hexanol solution is in good agreement with the experimentally found values of the average lifetimes of the bubbles, especially if we assume that the critical film thickness decreases with the surfactant concentration. This assumption means that

128

some minimum adsorption coverage is required to obtain a certain degree of foam film stability. The increase in the surfactant coverage at the film surfaces should stabilize it by making the film less susceptible to the fluctuations causing rupture.

The difference in the lifetimes confirms the fact that either a symmetrical or a nonsymmetrical foam film was formed depending on the location of the n-hexanol solution surface. The symmetrical foam films are more stable and, therefore, the lifetimes of the bubbles arriving at the surface from a distance $L = 4$ cm were longer. When the bubbles traveled a long distance to the interface ($L = 39.5$ cm) nonsymmetrical films having lower stability were formed and, therefore, the bubble lifetimes were shorter. In this case, the motion-induced disequilibration of the surfactant coverage over the bubble surface was fully developed. Thus, it should always be taken into account in the discussion of the mechanism of foam formation and stability that as a result of the motion the bubble approaches the solution/gas interface with a surface coverage at its upstream pole much smaller than the equilibrium one. Therefore, not an equilibrium but this actual nonequilibrium surfactant coverage can be decisive for the possibility of foam formation and can affect its stability [4, 21].

Conclusions

It was found that the average lifetimes of bubbles arriving at the n-hexanol solution surface from the distances $L = 4$ cm or $L = 39.5$ cm from the point of their formation were different. Under identical conditions of the bubble formation the lifetimes were shorter when the solution surface was located far from the point of the formation of the bubbles. Shorter lifetimes of the bubbles at the solution surface located "far" are the experimental confirmation of the theoretical predictions that the bubble motion leads to disequilibration of the adsorption coverage. As a result of the viscous drag exerted on the bubble by the continuous medium the upstream poles of the bubbles are depleted of surfactant. When the bubbles traveled a long distance ($L = 39.5$ cm) in the n-hexanol solution the disequilibration of the adsorption coverage on the rising bubble could fully develop and the nonsymmetrical foam film was formed with an adsorption coverage on the lower surface much smaller than the equilibrium one. When the bubbles traveled only a short distance through the solution ($L = 4$ cm) there was not enough time for the nonequilibrium surfactant distribution to develop and more stable symmetrical foam films, with an n-hexanol surface coverage close to the equilibrium one on both surfaces, were formed. Therefore, the lifetimes of the bubbles were longer when the solution surface was close to the point of their formation.

The reasonably good agreement between the lifetimes of the bubbles at the n-hexanol solution surface located "close" and "far" and the lifetimes of the symmetrical and fully nonsymmetrical foam films is further confirmation that when the solution surface was located far away from the point of the bubble formation the bubbles arrived at the solution surface with their upstream poles devoid of n-hexanol molecules. The results indicate that the final film thickness at which the film ruptures and the bubble ceases to exist at the interface depends on the surfactant concentration.

Acknowledgements The authors thank M. Barańska for her skillful assistance with the experiments. Financial support from KBN, grant 3T09A14715, is gratefully acknowledged.

References

1. Kruglyakov PM, Exerowa D (1990) Pena i pennye plenki. Khimya, Moscow
2. Pugh R J (1996) Adv Colloid Interface Sci 64:67
3. Jachimska B, Warszyński P, Małysa K (1998) Colloids Surf A 143:429
4. Małysa K (1992) Adv Colloid Interface Sci 40:37
5. Frumkin AN, Levich VG (1947) Zh Fiz Khim 21:1183
6. Levich VG (1962) Physicochemical hydrodynamics. Prentice-Hall, Englewood Cliffs, NY
7. Deryaguin BV, Dukhin SS, Lisichenko VA (1959) Zh Fiz Khim 33:2280
8. (a) Dukhin SS, Deryaguin BV (1961) Zh Fiz Khim 35:1246; (b) Dukhin SS, Deryaguin BV (1961) Zh Fiz Khim 35:1453
9. Deryaguin BV, Dukhin SS, Lisichenko VA (1960) Zh Fiz Khim 34:524
10. Saville DA (1973) Chem Eng J 5:251
11. Cheng J, Stebe KJ (1996) J Colloid Interface Sci 178:144
12. Dukhin SS, Kretzschmar G, Miller R (1995) In: Möbius D, Miller R (eds) Studies in interface science. Elsevier, Amsterdam, and references therein
13. Dukhin SS (1981) In: Goddard FG, Rusanov AI (eds), Modern theory of capillarity. Akademie, Berlin
14. Małysa K, Warszyński P (1995) Adv Colloid Interface Sci 56:105
15. Sharma A, Ruckenstein E (1988) Colloid Polym Sci 266:60
16. Princen HM (1969) In: Matijevic E, Eirich FE (eds) Surface and colloid science, vol 2. Interscience, New York, p 45
17. Ivanov IB, Dimitrov DS, Somasundaran P, Jain RK (1985) Chem Eng Sci 40:137
18. Manev ED, Vassilieff CS, Ivanov IB, (1976) Colloid Polym Sci 254:99
19. Clarc DC, Dann R, Mackie AR, Mingins J, Pinder AC, Purdy PW, Russell EJ, Smith LJ, Wilson DR (1990) J Colloid Interface Sci 138:195
20. Langevin A, Sonin AA (1994) Adv Colloid Interface Sci 51:1
21. Małysa K, Miller R, Lunkenheimer K (1991) Colloids Surf 53:47

Progr Colloid Polym Sci (2000) 116:129–133
© Springer-Verlag 2000

I. Ancutiene
V. Janickis

Chemical deposition of copper sulfide films in the surface of polyethylene by the use of higher polythionic acids

I. Ancutiene (✉) · V. Janickis
Department of Chemical Technology
Kaunas University of Technology
Radvilenu str. 19, 3028 Kaunas
Lithuania
e-mail: vitalijus.janickis@ctf.ktu.lt
Tel.: +370-7-456310
Fax: +370-7-451582

Abstract Films of copper sulfides of varying composition are formed in a surface matrix of polyethylene by a sorption–diffusion method using solutions of higher polythionic acids, $H_2S_nO_6$ ($n > 6$), as sulfuring agents. In the first stage of the process, elemental sulfur diffuses into the polymer when keeping it in a solution of higher $H_2S_nO_6$. In the second stage, a film of nonstoichiometric Cu_xS ($x = 1.12–1.95$) is obtained when the sulfured PE is treated with a solution of copper (I–II) salt. The physical properties of the copper sulfide film depend on the thickness and the composition of the film. The films obtained are formed from three-phases: jarowwite ($Cu_{1.12–1.18}S$), anilite ($Cu_{1.7–1.8}S$), and djurleite ($Cu_{1.91–1.95}S$). The sulfide with composition close to CuS has the highest electrical conductivity.

Key words Polyethylene · Polythionic acids · Sulfuration · Diffusion · Copper sulfide films

Introduction

New composite materials are needed for modern and advanced technology. Composites are often formed on the surface of polymeric materials, thus modifying them and imparting desired physical–mechanical and chemical properties. Sorption–diffusion is a simple and promising method for obtaining composites, when semiconductive or electroconductive films are obtained on polymers. In most cases, these films are made of metal sulfides, which are formed on the surface of a polymer as a result of the chemical reaction between diffused sulfur-containing particles and metal ions.

Sulfide films are mainly formed in two ways: by treating a polymer containing absorbed sulfuring agent with a solution of a metal salt or by sulfuration of metal compounds absorbed in a polymer. Most polymers, including polyethylene (PE), absorb various compounds of sulfur, which can act as sulfuration agents; therefore, the first method is simple and practical.

In this study binary compounds of copper with sulfur in the surface layer of PE were obtained by the sorption–diffusion method developed by us [1]. In the first stage, elemental sulfur is included into PE, and in the second stage, PE is treated with a solution of Cu (I–II) salt. For the inclusion of sulfur we used solutions of higher polythionic acids, $H_2S_nO_6$ ($n > 6$), the anions of which contain divalent sulfur atoms. These solutions are suitable for sulfuring the majority of polymers. Highly sulfured polythionic acids were synthesized for the first time in the Kaunas University of Technology Department of Inorganic Chemistry [2, 3].

Experimental

The films of copper sulfide were deposited on PE pellicles of low density (thickness, $140 \pm 5 \ \mu m$).

For the inclusion of sulfur into PE a 0.002 mol/l solution of polythionic acid was used. It was prepared according to the equation [2]

$$m(2H_2S + H_2SO_3) + 2H_2S_2O_3 \rightarrow H_2S_nO_6 + H_2S + 3mH_2O \ , \quad (1)$$

where $n = 3m + 3$.

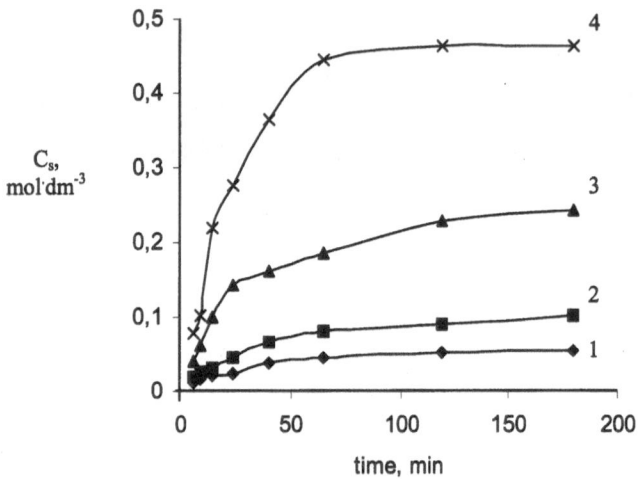

Fig. 1 Dependence of sulfur concentration in polyethylene (*PE*) on time when treating it with a solution of $H_2S_{45}O_6$. The temperature (K) of the solution is *1*, 298; *2*, 313; *3*, 333; *4*, 353

Then, on the basis of data of sulfite [3, 4] and cyanic [5] decomposition of polythionates, the average number of sulfur atoms (*n*) in a molecule of polythionic acid was established.

The samples of sulfured PE were treated with a copper salt solution consisting of 0.4 mol/l CuSO_4 and 0.1 mol/l hydroquinone (reducing agent). This mixture was established to be that of univalent and divalent copper salts, containing 0.34 mol/l Cu (II) salt and 0.06 mol/l Cu (I) salt, independent of temperature.

The amount of sulfur which diffused into PE was determined by a spectrometric cyanide method [6] using a Specord spectrophotometer ($\lambda = 450$ nm).

The amount of copper in the sulfide film after fusing in concentrated nitric acid was determined using a Perkin-Elmer atomic absorption meter ($\lambda = 325$ nm).

The phase composition of the copper sulfide films and of the sulfured PE pellicles were investigated using a DRON-3 diffractometer (Cu Kα radiation).

The transverse sections of PE with the films of copper sulfide were investigated using a JXA-50A electron microscope (Jeol).

The conductivity at constant current of the copper sulfide films was measured using an E7-8 numerical measuring instrument with special electrodes.

Results

It is known [3, 7 ,8] that polythionic acids decompose with the liberation of elemental sulfur:

$$H_2S_nO_6 \rightarrow H_2S_{n-x}O_6 + xS \ . \tag{2}$$

Our investigations showed that the decomposition of higher polythionic acids occurs during the sulfuration of PE samples; however, under certain conditions sulfur is not liberated, but it diffuses into PE.

By carrying out qualitative and X-ray analysis of sulfured PE we established that the rhombic cyclic S_8 diffuses into PE. After a quantitative kinetic study of the process of sulfur inclusion into PE from the solutions of $H_2S_nO_6$ ($n = 9$–45), we determined that the concentration of sulfur in the PE, c_s, depends on the degree of the acid

sulfurity, *n*, the temperature of the solution, and the duration of the polymer treatment. With the increase in these parameters, the concentration of sulfur in the PE increases as well. The saturation by sulfur comes to an end in approximately 2 h (Fig. 1).

The results showed that the influence of the temperature of the $H_2S_nO_6$ solution on the concentration of the sulfur absorbed (sulfur mole = 32.06 g) in the PE is much more effective than the degree of the acid sulfurity. For instance, the equilibrium sulfur concentration in PE increases only 1.4 times if PE is sulfured in a solution of $H_2S_nO_6$ at 298 K and the average number of sulfur atoms in a molecule of $H_2S_nO_6$ (*n*) is increased from 9 to 45; however, the equilibrium sulfur concentration increases 10 times when the PE is sulfured in a solution of $H_2S_{12}O_6$ and the temperature of the $H_2S_{12}O_6$ solution is increased from 298 to 353 K.

The kinetic parameters of the sulfur diffusion were determined: the apparent diffusion coefficients, the apparent thermal effects, and the apparent activation energies. The parameters of the diffusion process were defined according to Fick's law and its consequences [9–11]. The apparent diffusion coefficients (*D*) were calculated from the relation $c_{s\tau}/c_{s\infty} = f(\tau^{0.5})$, where τ is time, $c_{s\tau}$ the sulfur concentration in the sample at time τ, and $c_{s\infty}$ the equilibrium sulfur concentration in the sample.

The data obtained showed that the diffusion coefficient depends on the degree of polythionic acid sulfurity (*n*) and on the temperature. The apparent diffusion coefficient increases with an increase in these parameters. The degree of acid sulfurity has a greater influence. For instance, when *n* increases from 9 to 45 (at 313 K), *D* increases 2.5 times [$D = (4.1$–$10.5) \times 10^{-9}$ cm²/s]; when *n* increases from 12 to 45 (at 333 K), *D* increases from 5.9×10^{-9} to 15.4×10^{-9} cm²/s. However, when the PE is sulfured in a solution of $H_2S_{33}O_6$ and the temperature of the solution is increased from 298 to 353 K, the value of the diffusion coefficient increases approximately twofold, $D = (6.0$–$12.7) \times 10^{-9}$ cm²/s.

On the basis of the calculated diffusion coefficients at different temperatures for the same polythionic acids it was found out that the increase in *D* with the increase in temperature is determined by the faster decomposition of $H_2S_nO_6$. In addition, the increase in *D* with the increase in *n* is determined by the increased amount of elemental sulfur liberated during the decomposition of polythionic acid richer in sulfur.

The dependence $\ln D = f(1/T)$ in the interval 298–333 K was found to be linear and the apparent activation energies (*E*) of the diffusion process were calculated according to the Arrhenius equation. It was found that in the temperature interval 298–333 K *E* increases less than twofold ($E = 14.5$–24 kJ/mol) when the average number of sulfur atoms in a molecule of $H_2S_nO_6$ is increased from 21 to 45. Thus, the apparent activation energies of the sulfur diffusion into PE are 5–6 times lower than the activation energies of $H_2S_nO_6$ decomposition [8, 12]. Consequently, the concentration of sulfur which diffuses into PE is limited by the stage of polythionic acid decomposition.

The dependence $\ln c_{s\infty} = f(1/T)$ in the temperature interval 298–353 K is linear; therefore, the apparent thermal effects (ΔH) were calculated according to the Arrhenius equation. The values of ΔH are almost independent of the value of n of polythionic acid, $\Delta H = 34$–37 kJ/mol.

Further, the interaction of PE which had been sulfured in solutions of higher $H_2S_nO_6$ with a solution of a copper (I–II) salt was studied. On treating PE saturated by sulfur with a solution of Cu (I–II) salt, films of copper sulfide formed on the surface of the polymer. Depending on the initial concentration of sulfur in the PE and the period of treatment with the solution of the copper salt brown or even black electrically conductive copper sulfide films were obtained.

It was established that the sulfur which diffused in the PE turns into sulfide only if ions of univalent copper are present in the solution of copper salt, and only then does the heterogeneous redox reaction in the surface layer of PE take place:

$$2xCu^+ + 1/8S_8 \rightarrow Cu_xS + xCu^{2+} . \qquad (3)$$

We have established the dependence of the amount of copper in a sulfide film on the initial sulfur concentration in PE, the temperature of the Cu (I–II) salt solution, and the period of sulfured PE treatment with the solution of the copper salt. It was determined that the amount of copper in the film is strongly dependent on the concentration of sulfur which diffused in the PE, i.e. the amount of copper varies in proportion to the sulfur concentration in PE (Fig. 2).

The results obtained also indicated that the amount of copper in a sulfide film is less dependent on the period of treatment with the solution of Cu (I–II) salt: a greater influence was detected when a solution of copper salt at a lower temperature was used.

Experiments were performed by changing the temperature of the Cu (I–II) salt solution. It was determined that when the temperature of the solution increases (303, 323, 353 K), the amount of copper in the sulfide film also increases. Obviously, the diffusion of Cu^+ ions from the solution into the sulfured PE becomes faster with the increase in temperature.

The microscopic investigation of transverse sections of PE samples with copper sulfide films showed that the major part of copper sulfide (till 42 μm deep) is in the surface matrix of the polymer. A Cu_xS film of 1–2-μm thickness is deposited on the surface of PE.

Cu^+ ions and octasulfur molecules react owing to their diffusion through the film of the reaction product –

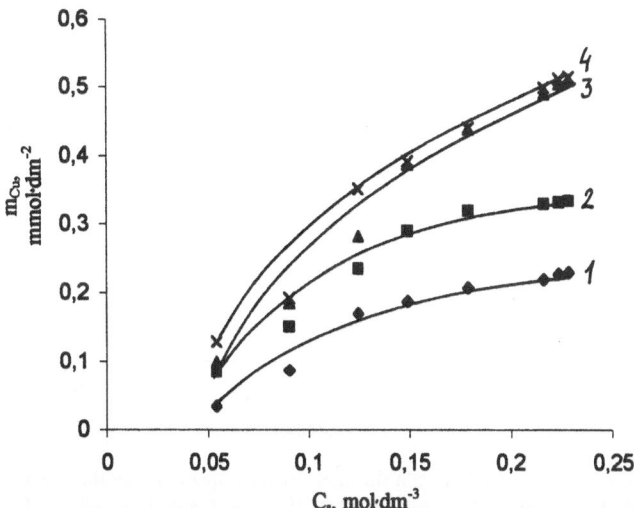

Fig. 2 Dependence of the amount of copper on the surface of PE on the sulfur concentration in PE. The temperature of the copper (I–II) salt solution is 353 K. The period of the treatment (min) with the copper (I–II) salt solution is *1*, 0.5; *2*, 1; *3*, 5; *4*, 10

the film of Cu_xS. The Cu^+ ions diffuse from the solution side through the copper sulfide film formed in the surface matrix of the PE, and the sulfur molecules diffuse from the deeper layers of PE through the sulfide film towards the surface of the polymer.

It was determined that the thickness of the Cu_xS film is in proportion with the concentration of sulfur which diffused into PE. The period of treatment with copper (I–II) salt solution was more significant only at the beginning of the deposition of the sulfide film (Fig. 3).

On treating PE samples with the diffused sulfur with the Cu (I–II) salt solution, nonstoichiometric Cu_xS films are formed on the surface of the PE and in its bulk. After having established the phase composition of Cu_xS films deposited chemically in the surface matrix of the PE by the method of diffusion, we carried out X-ray diffraction studies. The phase composition of the deposited film was established by comparing its X-ray images with those of known minerals [13–15]. The analysis of the X-ray images showed that at the beginning of the deposition Cu_xS has the structure of jarowwite ($x = 1.12$–1.18). The maxima of anilite ($x = 1.7$–1.8) appear later and their intensity increases in the X-ray images. Afterwards, new maxima corresponding to djurleite ($x = 1.91$–1.95) appear in the X-ray images. Thus, at the beginning of the deposition a film consisting of one phase – jarowwite – forms. When the period of treatment in the copper (I–II) salt solution is prolonged a film consisting of two phases (jarowwite and anilite) forms. Additionally, at least a film consisting of three phases (jarowwite, anilite, and djurleite) is obtained. Consequently, in the course of sulfured PE treatment, the phase composition of Cu_xS changes in the direction of increasing values of x.

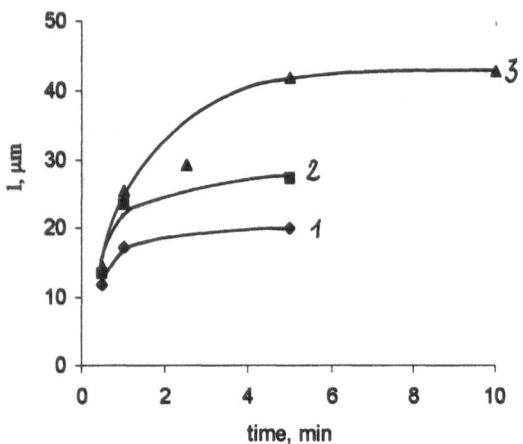

Fig. 3 Dependence of the thickness of the copper sulfide film on the period of treatment with copper (I–II) salt solution. The sulfur concentration in the PE (mol/dm^3) is *1*, 0.12; *2*, 0.15; *3*, 0.225

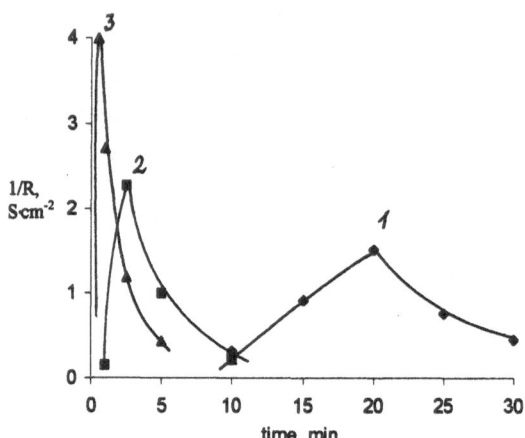

Fig. 4 Dependence of the conductivity ($1/R$) of the copper sulfide film on PE on the period of treatment with copper (I–II) salt solution. The sulfur concentration in the PE is 0.225 mol/dm^3. The temperature (K) of the solution is *1*, 303; *2*, 323; *3*, 35

The study of the electrical conductance of Cu$_x$S films has shown that the conductance greatly depends on the chemical and phase composition of the film and its thickness. It was also determined that the resistance describes the thickness and the phase composition of the copper sulfide film formed on the surface of the PE. The conductivity of the PE samples depends greatly on the conditions of PE interaction with H$_2$S$_n$O$_6$ and on the period of further interaction with Cu (I–II) salt solution. The data provided in our work show that at low concentration of sulfur which diffused in PE, thin films of copper sulfide are obtained (Fig. 3) and that the conductivity of such samples is low. When the sulfur concentration is sufficient and the time period of interaction with Cu (I–II) salt solution is rather short, more conductive films are obtained (Fig. 4).

As seen from Fig. 4, the electrical conductivity of the films depends on the temperature of the copper salt solution and different conductivity maxima are obtained: 12.5–25 min at 303 K, in about 2.5 min at 323 K, and in about 0.5 min at 353 K.

At the beginning of the deposition a very thin and uneven Cu$_x$S film appears. The conductivity of such a film is low. During the development of the process, when the thickness of the film increases, the conductivity increases as well and a compact electrically conductive copper sulfide film of the jarowwite (Cu$_{1.12–1.18}$S) phase is formed. During the consumption of sulfur which diffused into PE, the copper ions diffusing into the sulfide layer change not only the thickness of the Cu$_x$S film but its stoichiometric composition as well. Simultaneously, the x value increases. The conductivity decreases when the anilite (Cu$_{1.7–1.8}$S) phase is formed. The minimum conductivity arises when the outer layer becomes composed of the jurlite (Cu$_{1.91–1.95}$S) phase. Hence, the electrical conductivity of copper sulfide films deposited by the sorption–diffusion method decreases with increasing x values and semiconductive films are obtained.

The established regularities enable copper sulfide films of the desired conductivity to be obtained in the surface of PE.

Conclusions

1. Sulfur diffuses from solutions of higher polythionic acids, H$_2$S$_n$O$_6$ ($n > 6$), into PE in the form of cyclic S$_8$ molecules. The apparent kinetic parameters (diffusion coefficients, activation energies, thermal effects) of sulfur diffusion into PE greatly depend on the degree of sulfurity (n) of H$_2$S$_n$O$_6$.
2. On treating PE containing diffused sulfur with a copper (I-II) salt solution, a copper sulfide film is formed. The amount of copper in the sulfide film varies in proportion to the sulfur concentration in the PE.
3. The copper sulfide film formed by the sorption–diffusion method is nonstoichiometric. At the beginning of the formation the jarowwite (Cu$_{1.12–1.18}$S) phase is observed, then the anilite (Cu$_{1.7–1.8}$S) phase, and later the djurleite (Cu$_{1.91–1.95}$S) phase is formed. Thus, finally the film becomes composed of three phases.
4. The electrical conductivity of the film at constant current depends on the thickness and the phase composition of the film formed in the outer layer of the PE.
5. The value of x in Cu$_x$S increases with the prolongation of the period of sulfured PE interaction with the copper salt solution.
6. The established regularities of Cu$_x$S deposition in the surface of the PE and the composition of the films obtained enable electrically conductive or semiconductive copper sulfide films to be obtained.

References

1. Ancutienė I, Janickis V, Grevys S (1997) Lithuanian Patent LT 4111B
2. Janickis J, Valančiūnas J, Zelionkaitė V, Janickis V, Grevys S (1975) Liet TSR Mokslu Akad Darb Ser B 3:83
3. Janickis J, Valančiūnas J, Tucaitė O (1958) Zh Neorg Khim 3:2087
4. Kurtenacker A, Goldback E (1927) Z Anorg Allg Chem 166:177
5. Foss O (1950) Acta Chem Scand 4:1241
6. Babko A, Pilipenko A (1974) Photometric analysis. Method for determination of non-metals. Khimiya, Moscow
7. Grevys S, Janickis V (1975) Chemija ir cheminė technologija: material of the national scientific–technical conference, Kaunas. p 25
8. Grevys S (1976) Dissertation. Kaunas
9. Crank J, Park GS (1968) Diffusion in polymers. Academic, London
10. Rogers K (1968) Problems on the physics and chemistry of the solid state of organic substances. Mir, Moscow
11. Nikolajev N (1980) Diffusion in membranes. Khimiya, Moscow
12. Janickis V (1983) Doctoral dissertation. Kaunas
13. Žebrauskas A, Mikalauskienė A, Latvys V (1992) Chemija (Vilnius) 3:131
14. Yamammoto T, Kamigaki T, Kubota E (1987) Kobunshi Ronbunshu 44:327
15. Nomura R, Konyao K, Matsuda H (1988) Ind Eng Chem Res 28:887

Progr Colloid Polym Sci (2000) 116:134–136
© Springer-Verlag 2000

INTERFACIAL PROCESSES

R. Ivanauskas
V. Janickis

Chemical deposition of copper selenide films in the surface layer of polyamide by the use of selenopolythionates

R. Ivanauskas (✉) · V. Janickis
Department of Chemical Technology
Kaunas University of Technology
Radvilenu str. 19, 3028 Kaunas, Lithuania
e-mail: vitalijus.janickis@ctf.ktu.lt
Tel.: +370-7-456310
Fax: +370-7-766063

Abstract Inorganic binary compounds of copper selenide, Cu_xSe_y, films are formed on polyamide (PA) by a sorption–diffusion method using potassium selenotrithionate, $K_2SeS_2O_6$, as a seleniumization agent. The selenides $Cu_{0.5}Se$ orthorhombic, $Cu_{1.5}Se$ tetragonal and $Cu_{2.0}Se$ orthorhombic were matched on the polymer surface after treating the seleniumized PA films with a solution of Cu(I-II) salts. Depending on the seleniumization conditions (concentration 0.03–0.3 mol dm^{-3}, pH 2.15, 3.0 of aqueous $K_2SeS_2O_6$ solutions, 60 °C) Cu_xSe_y layers on PA of different electrical resistance (0.065–416 $k\Omega/\square$) were obtained.

Key words Copper selenide · Selenotrithionate · Polyamide · Diffusion · Seleniumization

Introduction

The chemical modification of polymers may result in the formation on their surface of semiconductive or even conductive films of binary inorganic compounds, particularly of metal chalcogenides. Polymers with a so-modified surface layer may successfully replace metallic conductors, in addition to being corrosion resistant, light, elastic and cheaper. For example, copper selenide layers have been used in microelectronics and for many other purposes [1–3].

Hydrophilic and semihydrophilic polymers are capable of absorbing ions of various electrolytes from aqueous solutions [4]. This fact enabled us to form copper sulfide layers, Cu_xSe_y, on a polyamide (PA) surface by a sorption–diffusion method [5] employing polythionates, $S_n(SO_3)_2^{2-}$, i.e. compounds containing divalent sulphur atom chains [6], in aqueous solutions.

In previous work [7] we investigated the kinetics of the diffusion of the simplest selenopolythionate, i.e. seleno-trithionate, $Se(SO_3)_2^{2-}$, from potassium selenotrithionate, $K_2SeS_2O_6$, aqueous solutions into PA. It was shown that in this case not only the $SeS_2O_6^{2-}$ anions but also the disintegration products, such as selenosulphate, $SeSO_3^{2-}$, and diselenotetrathionate, $Se_2S_2O_6^{2-}$, anions, diffuse into PA as well. The diffusion process is more intense when acidified $K_2SeS_2O_6$ solutions are used as opposed to ones containing no hydrochloric acid additive. The optimal conditions (solution pH 2.15, concentration 0.05 mol dm^{-3}, temperature 60 °C) were determined [7] under which selenium-containing-anion diffusion of sufficient intensity into PA best matches the undesirable gradual disintegration of $K_2SeS_2O_6$ solution, resulting in the liberation of elementary selenium. We have also stated [8] the optimal temperature of a solution of Cu(I-II) salts (78 °C) used in obtaining sulphide coatings on PA [5] as well as the duration of PA treatment (10 min) with this solution after the PA had interacted with $K_2SeS_2O_6$ solution to obtain a Cu_xSe_y layer of electrical conductance on the PA surface.

Our main task was to investigate the dependence of the composition and electrical properties of the selenide layers obtained by acidified and nonacidified $K_2SeS_2O_6$ solutions on some conditions of PA film treatment with these solutions.

Experimental

APA film, PK-4, of 70-mm thickness was used. Prior to the experiments, pieces of the film of size 15 × 70 mm were boiled in distilled water for 2 h to remove the remainder of the monomer. Then they were dried using filter paper and then over $CaCl_2$ for

24 h. $K_2SeS_2O_6$ was produced by the method of Ratkhe [9]. PA was treated in a thermostatic vessel using a continually stirred selenopolythionate solution. At certain time intervals, samples were removed, rinsed with distilled water, if required the selenium precipitated on the surface of the film was wiped off, dried with filter paper, left over $CaCl_2$ for 24 h and then used in further experiments and analysis.

The amount of selenium and copper in a PA sample was determined using an atomic absorption spectrometer [10].

A Cu(I-II) salt solution was made from crystalline $CuSO_4 \cdot 5H_2O$ and a reductant hydroquinone as described in Refs. [11, 12]. It is a mixture of univalent and divalent copper salts, in which independent of temperature there is 0.34 mol dm³ Cu(II) salt and 0.06 mol dm³ Cu(I) salt. After having been kept in $K_2SeS_2O_6$ solution, the sample was treated with a Cu(I-II) solution, then rinsed with distilled water, dried over $CaCl_2$ and used in subsequent experiments.

The phase composition of the copper selenide layer was analysed by means of X-ray diffraction using a DRON-3 difractometer.

Special electrodes were used to measure the conductivity to constant current of the deposited layer. An E7-8 numerical measuring instrument was used for the measurements.

Results

Aqueous 0.03–0.3 mol dm⁻³ $K_2SeS_2O_6$ solutions containing no hydrochloric acid additives (pH 3.0) and ones containing HCl (pH 2.15) were used in our investigation. The kinetic curves showing the change in Cu concentration in PA depending on solutions of different concentrations used are presented in Fig. 1. Under the experimental conditions (pH 2.15, 60 °C) with 0.05–0.30 mol dm⁻³ $K_2SeS_2O_6$ solutions the maximum Cu concentration in PA is obtained in 2.0–2.5 h at the stage of "seleniumization". The Cu concentration in PA begins to decrease 10–20 min after the beginning of the visually observed disintegration of the solution, with the liberation of elementary selenium. We have already stated [7] that disintegration of the $SeS_2O_6^{2-}$ anions diffused in the PA begins later than of those in $K_2SeS_2O_6$, i.e. in the seleniumization solution. Thus, selenium-containing anions present in the PA and capable of interacting with Cu(I-II) solution (copperizing stage) last longer; however, the elementary selenium formed in their subsequent disintegration does not react with Cu ions.

With 0.03 mol dm⁻³ $K_2SeS_2O_6$ solution at the seleniumization stage, the maximum copper concentration in the copperizing stage cannot be obtained even in a 5-h process. This might result from insufficient concentration of $SeS_2O_6^{2-}$ ions diffused into PA during the seleniumization stage. The molar ratio of Cu/Se in films obtained under these conditions ranges over a wide interval (1.78–2.74); however, the electrical resistance of these layers (97–599 $\Omega/\square$) is usually higher than that of the layers obtained with a higher concentration $K_2SeS_2O_6$ initial solution. At the beginning of the seleniumization process (up to 15 min) the ratio of Cu/Se in PA films treated with higher concentration $K_2SeS_2O_6$ solutions is 0.953–1.17

and the electrical resistance of the Cu_xSe_y layers is 65–135 $\Omega/\square$, while at the end of the process (4.5 h after the start) these values are 0.491–0.683 and 101–119 $\Omega/\square$, respectively. The change in the Cu_xSe_y electrical resistance in the layers obtained over time is related to the Cu concentration. At the beginning of the process, with the growth of Cu concentration in PA (Fig. 1) the resistance of the Cu_xSe_y layers obtained falls, then it becomes a minimum, and with the Cu concentration decreasing in PA the resistance begins to grow. This fact was checked by X-ray examination of Cu_xSe_y layers on the surface of PA. We found that after the seleniumization film treatment with Cu(I-II) salt solutions the Cu_xSe_y layer contained the following selenides: $Cu_{0.5}Se$ orthorhombic (characteristic peaks $\theta = 16.59°$, 16.98° or 22.96°), $Cu_{1.5}Se$ tetragonal (peak $\theta = 19.94°$) and Cu_2Se orthorhombic (peak $\theta = 22.09°$). The other peaks characteristic of these copper selenides are covered by the background formed by the film. The amount of copper selenide in the film increases with the continuation of the seleniumization process up to 2.5 h; however, after 4 h the $Cu_{0.5}Se$ phase cannot be detected, while the intensity of the peaks of the remaining phases decreases, i.e. the concentration of the given phases decreases. After a further 30 min the copper selenide concentration obtained in the PA films continues to decrease, as indicated by the fall in the peak intensity; however, the selenium concentration in the coating continues to grow. The decrease in the copper selenide concentration in the coating can be explained by the decomposition of the $SeS_2O_6^{2-}$ anions absorbed in the film, with nonreactive elemental selenium being isolated. Thus, increasing the concentration of elemental selenium in the copper selenide coating increases the resistance of the layers, while the amount of $SeS_2O_6^{2-}$ in the PA treated with $K_2SeS_2O_6$ solution for longer than 2.5 h is insufficient to form low-resistance coatings on the surface of the film.

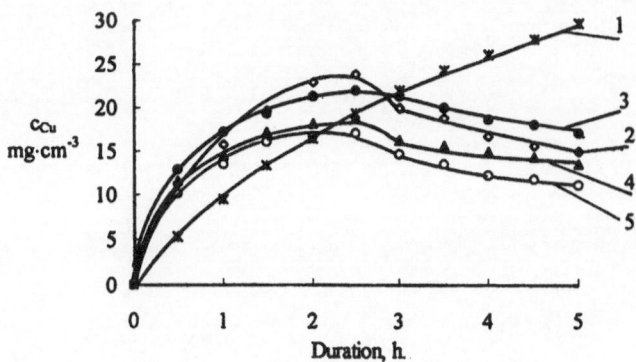

Fig. 1 The dependency of the Cu concentration in polyamide on the seleniumization solution (pH 2.15, 60 °C) concentration. The concentration of $K_2SeS_2O_6$ solution (mol dm⁻³) is *1* 0.03, *2* 0.05, *3* 0.1, *4* 0.2, *5* 0.3

The tests performed on PA films seleniumized in slightly more stable nonacidified (pH 3.0) $K_2SeS_2O_6$ solutions of the same concentration show that maximum Cu concentrations of 5.8–9.8 mg cm^{-3} are reached in films seleniumized for about 6 h. Comparison with the data Fig. 1 shows that even after seleniumization of the polymer for twice as long under the given conditions, the Cu concentration obtained is 2–4 times lower than for the polymer treated with acidified (pH 2.15) $K_2SeS_2O_6$ solution. The Cu/Se molar ratio in such films is considerably lower as well (0.344–0.475) and the resistance of the Cu_xSe_y layers is considerably higher (0.385–416 k$\Omega/\square$).

The data show that by changing the conditions of PA film treatment with $K_2SeS_2O_6$ solutions (concentration, pH) copper selenide layers of different composition and electrical resistance can be obtained on the surface of the polymer.

Conclusions

1. The maximum Cu concentrations in PA (17–23 mg cm^{-3}) seleniumized in 0.05–0.3 mol dm^{-3} acidified $K_2SeS_2O_6$ aqueous solutions (pH 2.15) for 2.0–2.5 h at 60 °C can be obtained by treating the given PA films with Cu(I-II) aqueous solution.

2. X-ray investigation of Cu_xSe_y layers obtained on PA surfaces have shown that their chemical and phase composition (orthorhombic $Cu_{0.5}Se$, tetragonal $Cu_{1.5}Se$ and orthorhombic Cu_2Se) as well as electrical resistance are highly dependent on the duration of seleniumization in acidified (pH 2.15) $K_2SeS_2O_6$ solution. The optimal concentration of $K_2SeS_2O_6$ solution is 0.05 mol dm^{-3}, enabling Cu_xSe_y layers of low electrical resistance (65 $\Omega/\square$) to be obtained on the surface of the polymer after treating 2.5-h seleniumizated films with Cu(I-II) salt solutions at 78 °C for 10 min.

3. By using $K_2SeS_2O_6$ solutions of the same concentration and temperature, but lower acidity (nonacidified), the Cu concentrations obtained on PA are considerably lower, while the resistance of the Cu_xSe_y layers formed is markedly higher (0.385–416 k$\Omega/\square$).

Acknowledgements This work was supported by the Lithuanian State Science and Studies Foundation.

References

1. Glazov VM, Burhanov AS, Krestikov AN (1982) Review of electronic techniques vol 2. pp 3–48 (in Russian)
2. Glazov VM, Burhanov AS (1980) Inorganic materials, vol 16. pp 565–585 (in Russian)
3. Žebrauskas A (1996) Chemical technology. pp 39–44 (in Lithuanian)
4. Zaikov GE, Yordanskiy AL, Markin VS (1984) Diffusion electrolytes into polymers (in Russian). Moskow
5. Lithuanian (1998) Patent LT 4402 B
6. Foss O (1960) In: Emeleus HJ, Sharpe AG (eds) Advances in inorganic chemistry and radiochemistry, vol 2
7. Ivanauskas R, Janickis V (1998) Chemistry. pp 3–12 (in Lithuanian)
8. Ivanauskas R, Janickis V (1999) Uk Chem J 65:49–56
9. Rathke BI (1895) J Pr Chem 95:1–48
10. Perkin-Elmer (1973) Analytical methods for atomic absorbtion spectrometry Perkin-Elmer 503, Perkin-Elmer
11. Ancutienė I (1995) Doctoral thesis. Kaunas
12. Zelionkaitė V, Janickis V, Maciulevičius R (1992) In: The problems of inorganic chemistry technical electrochemistry. Kaunas, pp 83–87 (in Lithuanian)

Progr Colloid Polym Sci (2000) 116:137–142
© Springer-Verlag 2000

J. Kulys
R. Vidžiūnaitė

The role of micelles in mediator-assisted peroxidase catalysis

J. Kulys (✉) · R. Vidžiūnaitė
Institute of Biochemistry
Mokslininkų 12
2600 Vilnius, Lithuania
e-mail: jkulys@bchi.lt

Discussed at the First Nordic–Baltic Meeting on Surface and Colloid Science, 21–25 August, Vilnius, Lithuania

Abstract To elucidate the role of micelles in mediator-assisted peroxidase catalysis the oxidation of *N*-benzoyl leucomethylene blue (BMB) was performed in the presence of Triton X-100, sodium dodecyl sulfate (SDS) or cetyltrimethylammonium bromide (CTAB) at pH 8.5. Recombinant *Coprinus cinereus* peroxidase was used as an enzyme. It was shown that oxidation of BMB proceeded at high peroxidase concentration. The addition of 10-phenothiazine propionic acid (PPA) increased the oxidation rate of BMB tremendously and the rate was almost linearly dependent on PPA concentration. The action of PPA was explained by mediation of BMB oxidation with the cation radical of PPA. An increase in the surfactant concentration decreased the rate of both the direct and the mediator-assisted BMB oxidation. The results were analyzed by assuming an enzymatic reaction in the aqueous pseudo-phase. The calculated distribution coefficient between the micellar and the aqueous phases varied from 28 to 1770 and from 11 to 2200 for BMB and PPA, respectively. The constant of PPA oxidation in the aqueous phase of Triton X-100 micelles fitted an independently determined value in buffer solution, whereas in SDS micelles it was 4 times less. The decrease by 2 orders of magnitude of the constant in CTAB micelles was explained by micelle/substrate interaction and the PPA concentration in the micellar phase.

Key words *N*-Benzoyl leucomethylene blue · 10-Phenothiazine propionic acid · Triton X-100 · Sodium dodecyl sulfate · Cetyltrimethylammonium bromide

Introduction

Micelles are inherent components of many natural, technical and biotechnological processes. In many biological systems biopolymers are incorporated into water-soluble micelles, membranes or lipoprotein particles. In techniques and biotechnology micellar technology has been used for biopolymer solubilization, extraction and purification [1]. Biotransformation of compounds of low solubility, such as polycyclic aromatic hydrocarbons (PAHs), is also associated with micellar systems [2, 3]. Surfactants are known to increase the solubility of PAHs, but surfactants can also influence biocatalytic activity and enzyme stability.

The kinetics studies of the function of enzymes in micellar systems are very limited. Micellar effects and the mechanism of peroxidase-catalyzed oxidation of *n*-alkyl-ferrocenes have been explored very recently [4]. The kinetics of peroxidase-catalyzed and mediator-assisted substrate oxidation has, to the best of our knowledge, not been evaluated before though the importance of such reactions in lignin degradation, dye bleaching and other bioprocesses [5, 6].

The task of this investigation was the evaluation of recombinant *Coprinus cinereus* peroxidase (rCiP)-catalyzed and mediator-assisted *N*-benzoyl leucomethylene blue (BMB) oxidation in the presence of micelles with special emphasis on the influence of the structure of the micelles on the enzymatic rate. As a mediator 10-phenothiazine propionic acid (PPA) was used. Three different detergents, i.e. Triton X-100 (uncharged), sodium dodecyl sulfate (SDS) (negatively charged) and cetyltrimethylammonium bromide (CTAB) (positively charged), were used in this work.

Materials and methods

Commercial recombinant fungal peroxidase from *C. cinereus* was heterogously expressed in *Aspergillus oryzae* [7] and was additionally purified by anion-exchange chromatography (Novo Nordisk, Denmark). The enzyme was homogenous as assessed by SDS polyacrylamide gel electrophoresis. The concentration of the enzyme was determined spectrophotometrically at 405 nm by using an extinction coefficient of 108 mM^{-1} cm^{-1} [8].

BMB was from Tokyo Kasei Kogyo Co and PPA was purchased from Novo Nordisk. The structures of these compounds are depicted in Fig. 1. Tris(hydroxymethyl)aminomethane (Tris) was obtained from Fluka, Triton X-100 was purchased from Aldrich and SDS and CTAB were products of Sigma.

Hydrogen peroxide solution was prepared in water from 30% Perhydrol (Reachim, Russia). The concentration of hydrogen peroxide in water was determined spectrophotometrically by using an extinction coefficient of 39.4 M^{-1} cm^{-1} at 240 nm [9]. A concentrated solution of BMB was prepared in acetonitrile. The final amount of acetonitrile in the system was 2% (v/v).

Kinetics and electrochemical measurements

The kinetics measurements of BMB oxidation were performed at 25 °C in 0.05 M Tris–HCl buffer solution pH 8.5. The pH was chosen owing to the great practical importance of peroxidase-catalyzed reactions at alkaline pH [5, 6]. Moreover, rCiP shows almost maximal activity at this pH if PPA is used as a substrate [10]. The change in absorbance was measured at 661 nm by using a computer-assisted spectrophotometer (LKB Biochrom, Ultrospec II) [11]. The reaction mixture contained 0.1 mM BMB, 0.1 mM H$_2$O$_2$ and an appropriate concentration of detergent. The reaction was started by the addition of 0.1 ml rCiP holding different concentrations of the enzyme. The kinetics of direct BMB oxidation or of BMB oxidation in the presence of PPA was carried out at 164–175 nM or 11.5 nM of rCiP, respectively.

The redox activity of BMB and PPA was established by cyclic voltammetry. The cyclic voltammetry was performed by using an electroanalytical system (Cypress Systems, USA) and a glass carbon electrode (model CS-1087, Cypress Systems, USA). A saturated calomel electrode (SCE, saturated with KCl, model K-401, Radiometer, Denmark) was used as a reference electrode and a Pt wire (diameter 0.2 mm, length 4 cm) mounted on the end of the reference electrode was used as an auxiliary electrode. The measurements were performed in acetonitrile/50 mM Tris–HCl buffer solution pH 8.5 (3/1 v/v) containing 0.1 M tetraethylammonium tetrafluoroborate (Aldrich) at room temperature. Before the measurements the glass carbon electrode was polished with aluminum oxide slurry and treated in an ultrasonic bath for 5 min. The concentration of BMB was 1.3 mM and that of PPA was 2.1 mM. The potential scanning rate was 12, 25, 50 and 100 mV/s. The formal redox potential was calculated as $E = (E_{p,a} + E_{p,c})/2$, where $E_{p,a}$ and $E_{p,c}$ are the peak potentials of anodic and cathodic conversion.

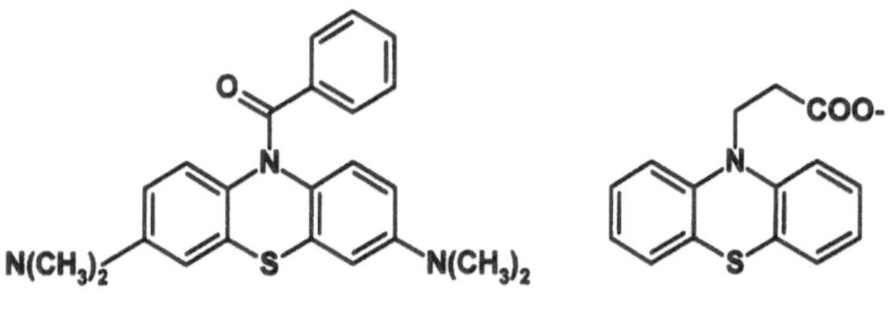

Calculations

The initial rate of the product formation was calculated by linear or cubic polynomial approximation of the kinetics curves. The dependence of the reaction rate (V) on the surfactant concentration was plotted in the coordinates of Eq. (1):

$$V = V_w/[1 + Pv_m(c - \text{cmc})] \ , \tag{1}$$

where V_w is reaction rate in water solution, P is the partition coefficient of the substrate, v_m is the molar volume of the micelles (0.3 M^{-1} [4]), c is a total concentration of surfactant and cmc is the critical micelle concentration. The cmc was assumed to be 0.3, 8.1 and 0.92 mM for Triton X-100, SDS and CTAB, respectively [1]. k_w was calculated as $k_w = V_w/[E][S]$.

Results

In the presence of 5–35 mM Triton X-100, rCiP catalyzed the oxidation of BMB at rather high peroxidase concentration. The reaction product was MB$^+$, which absorbed at 661 nm. The MB$^+$ production increased linearly over a 0–6-min period (data not shown). The increasing Triton X-100 concentration decreased the oxidation rate (Fig. 2). The addition to the system of 7.2 μM PPA increased the oxidation rate of BMB tremendously; therefore, the MB$^+$ production was measured at a much lower peroxidase concentration (Fig. 3). As in the case of the PPA-free system, MB$^+$ was produced at a constant rate at the beginning of the process and the concentration stopped changing at prolonged incubation times. The saturation level of the MB$^+$ concentration was proportional to the initial substrate concentration and was caused by complete BMB conversion; therefore, the saturation level of the absorbance was used for the calculation of the MB$^+$ extinction coefficient. The calculated value of the coefficient was 28 mM^{-1} cm^{-1} at 661 nm.

Fig. 1 Structures of *N*-benzoyl leucomethylene blue (*BMB*) and 10-phenothiazine propionic acid (*PPA*)

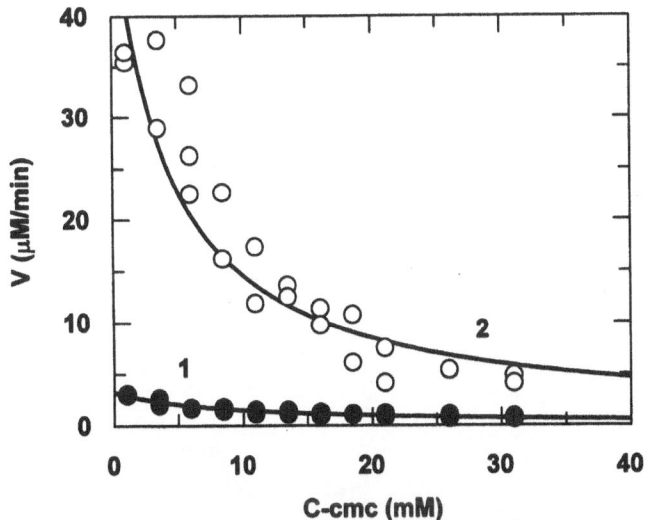

Fig. 2 The dependence of the rate of direct (*1*) and PPA-assisted BMB (*2*) oxidation on Triton X-100 concentration. Concentrations: BMB 0.1 mM, PPA 7.2 μM, recombinant *Coprinus cinereus* peroxidase (*rCiP*) 164 nM (*1*) and 11.5 nM (*2*), H_2O_2 0.1 mM

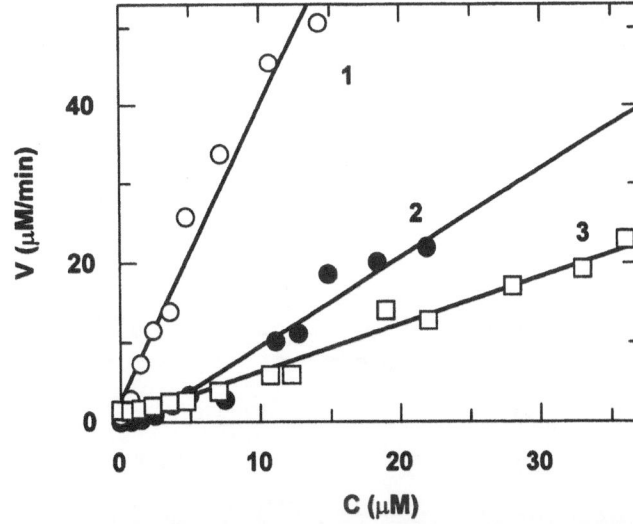

Fig. 4 The dependence of BMB oxidation rate on PPA concentration in the presence of sodium dodecyl sulfate (*SDS*) (*1*), Triton X-100 (*2*) and cetyltrimethylammonium bromide (*CTAB*) (*3*). Concentrations: BMB 0.1 mM, rCiP 11.5 nM, H_2O_2 0.1 mM, SDS 20 mM, Triton X-100 30 mM, CTAB 10 mM

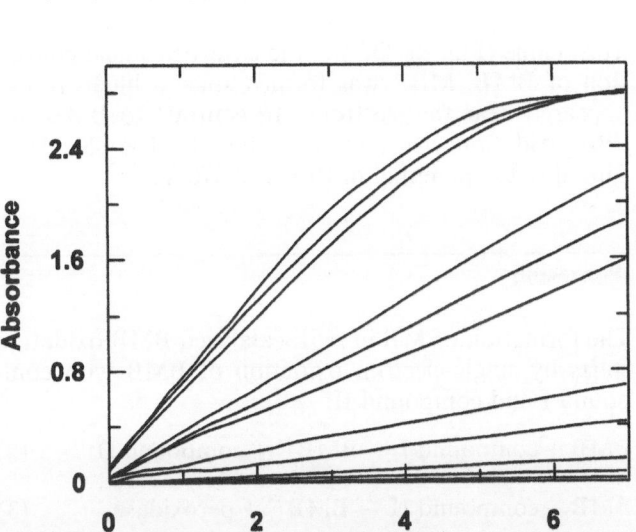

Fig. 3 Kinetics of PPA-assisted BMB oxidation at different PPA concentrations. Concentrations: BMB 0.1 mM, rCiP 11.5 nM, H_2O_2 0.1 mM, Triton X-100 30 mM. The reaction rate increased when the concentration of PPA increased in the range 0.75, 1.5, 2.5, 3.7, 4.9, 7.5, 11.1, 12.7, 14.8 and 18.4 μM

The increase in Triton X-100 concentration from 5 to 35 mM decreased the PPA-assisted MB^+ production rate 8.9 times (Fig. 2). The MB^+ production rate was almost linear at 5–25 μM PPA (Fig. 4).

In the presence of SDS the direct BMB oxidation rate was also much less compared to the PPA-assisted substrate oxidation rate. The rate was dependent on PPA concentration (Fig. 4); however, the MB^+ production rate varied little during the SDS concentration

change for direct as well as for mediator-assisted BMB oxidation (Fig. 5).

In the solution containing CTAB and during direct BMB oxidation MB^+ production showed saturated kinetics curves (Fig. 6). In contrast, the initial rate of MB^+ production was linear in the presence of PPA (data not shown). The rate of PPA-assisted BMB oxidation was directly proportional to 7–37 μM PPA, but was very low at low amount of the mediator

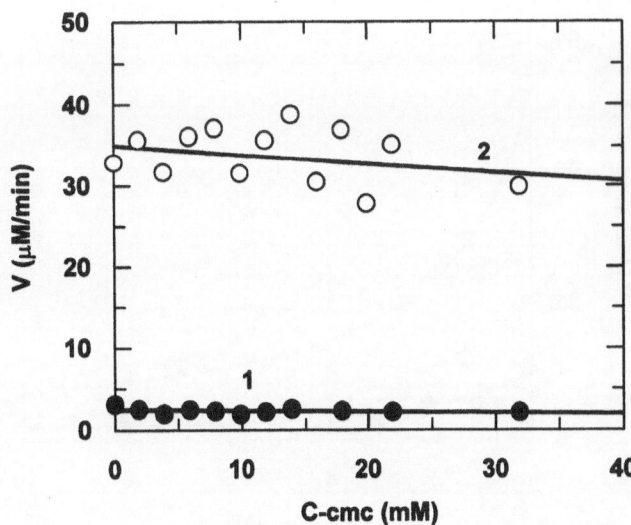

Fig. 5 The dependence of the rate of direct (*1*) and PPA-assisted BMB (*2*) oxidation on SDS concentration. Concentrations: BMB 0.1 mM, PPA 7.2 μM, rCiP 175 nM (*1*) and 11.5 nM (*2*), H_2O_2 0.1 mM

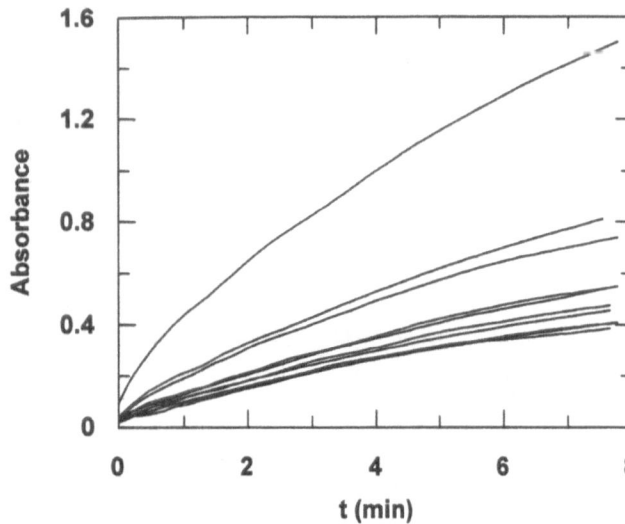

Fig. 6 Kinetics of BMB oxidation at different CTAB concentrations. Concentrations; BMB 0.1 mM, rCiP 164 nM, H_2O_2 0.1 mM. The initial rate decreased when the concentration of CTAB increased in the range 1, 2, 3, 4, 5, 6, 7, 8, 9 and 10 mM

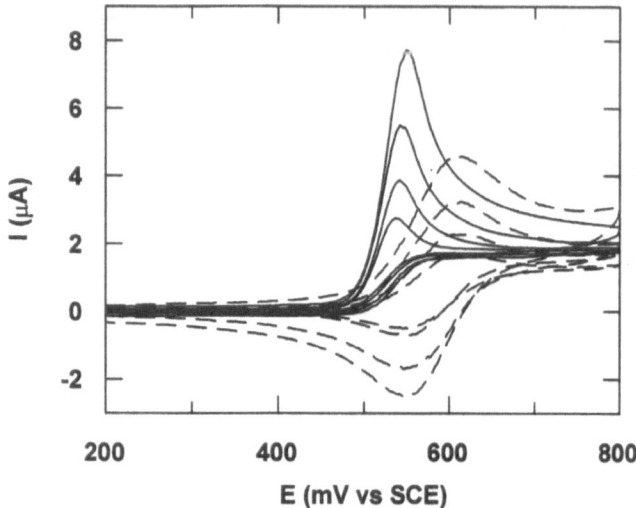

Fig. 8 Electrochemical conversion of BMB and PPA. The *solid curves* correspond to BMB and the *dotted curves* correspond to PPA. The potential scan rate was 12, 25, 50 and 100 mV/s; other experimental conditions are given in the text

(Fig. 4). The reaction rate decreased on increasing the CTAB concentration from 1 to 10 mM for direct as well as for mediator-assisted BMB oxidation (Fig. 7). The most important difference for the CTAB system compared to the Triton X-100 and SDS systems was the much smaller variance between direct and mediator-assisted BMB oxidation.

BMB and PPA are electrochemically active compounds. BMB was oxidized at an electrode potential higher than 0.5 V versus the SCE, but the reaction was

irreversible (Fig. 8). During the electrochemical conversion of BMB, MB^+ was formed since a blue product appeared near the electrode. In contrast to BMB, the PPA oxidation was reversible (Fig. 8). The calculated formal redox potential of PPA is 0.576 V.

Discussion

The formation of MB in rCiP-catalyzed BMB oxidation starts by single-electron oxidation of BMB with compound I and compound II:

$$BMB + compound\ I \rightarrow BMB^{+\bullet} + compound\ II \qquad (2)$$

$$BMB + compound\ II \rightarrow BMB^{+\bullet} + peroxidase \qquad (3)$$

where compound I and compound II represent doubly and singly oxidized peroxidase.

The intermediate ($BMB^{+\bullet}$) reacts with water, possibly, by forming benzoic acid (BA) and a MB radical, which, in turn, can be easy oxidized by compound I and compound II by forming methylene blue (MB^+):

$$BMB^{+\bullet} + H_2O \rightarrow BA + MB^{\bullet} + H^+ \qquad (4)$$

$$MB^{\bullet} + compound\ I \rightarrow MB^+ + compound\ II \qquad (5)$$

$$MB^{\bullet} + compound\ II \rightarrow MB^+ + peroxidase \qquad (6)$$

In the presence of PPA the reaction starts with PPA oxidation (Eqs. (7), (8)) since the oxidation rate of this substrate is much higher compared to BMB [10]. The limiting step of PPA oxidation at turnover is a reaction of compound II [10].

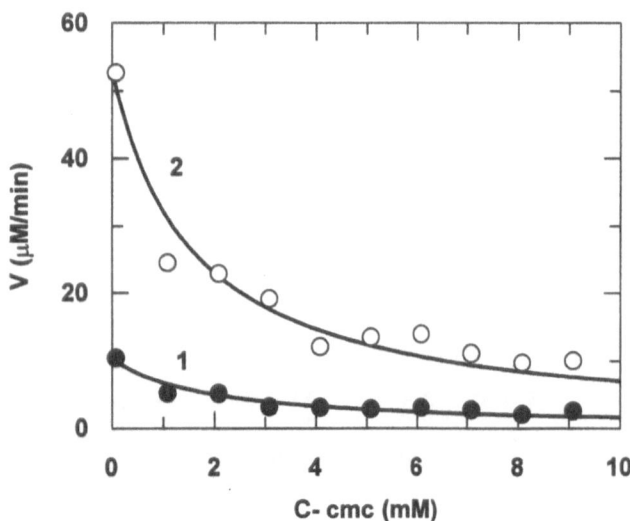

Fig. 7 The dependence of the rate of direct (*1*) and PPA-assisted BMB oxidation (*2*) on CTAB concentration. Concentrations: BMB 0.1 mM, PPA 20 μM, rCiP 164 nM, H_2O_2 0.1 mM

PPA + compound I → PPA$^{+\bullet}$ + compound II (7)

PPA + compound II → PPA$^{+\bullet}$ + peroxidase (8)

The cation radical of PPA (PPA$^{+\bullet}$) formed during peroxidase-catalyzed reaction oxidizes BMB:

BMB + PPA$^{+\bullet}$ → BMB$^{+\bullet}$ + PPA (9)

The rate of PPA$^{+\bullet}$ interaction with BMB should be high since the redox potential of PPA is higher than the potential of BMB oxidation (Fig. 8). The singly oxidized intermediate reacts with water as in the case of direct BMB oxidation (Eq. (4)) by forming BA and a MB radical, which is easily oxidized by other molecule of PPA$^{+\bullet}$ (Eq. (10)) or interacts with compound I or compound II as described by Eqs. (5) and (6).

MB$^{\bullet}$ + PPA$^{+\bullet}$ → MB^{+} + PPA (10)

Following this scheme PPA acts as a mediator. The mediator is regenerated during the reaction of PPA$^{+\bullet}$ with BMB and MB$^{\bullet}$. The absence of an initiation period of MB^{+} production (Fig. 3) indicates that the intermediate in the BMB oxidation is unstable and does not accumulate in a significant concentration during the reaction. The linear dependence of the BMB oxidation rate on PPA concentration (Fig. 4) implies PPA oxidation is the limiting step in the mediator-assisted reaction; therefore, the mediator-free as well as the mediator-assisted BMB oxidation is associated with BMB or PPA oxidation and the role of the micelles, which contain different charges and structure, can be analyzed separately for these two substrates.

RCiP is a hydrophilic enzyme. The isoelectric point of rCiP is approximately 3.5 and rCiP contains, on average, two glycosamines and 10–12 mannoses per molecule [12]. Therefore, to examine the kinetics data it was assumed that the enzyme is dissolved in water phase, that the substrate distributes between water and the micellar phase and that rCiP catalyzes the oxidation of the substrate that is dissolved in water solution. Following these assumptions the dependence of the experimentally determined rate on the surfactant concentration should be parabolic as expressed by Eq. (1) [4]. The data presented in Figs. 2, 5 and 7 show rather good agreement between the experiment and the model.

The value of k_w for PPA is similar to the independently measured substrate oxidation rate, $(5.3 \pm 1.2) \times 10^7$ M^{-1} s^{-1}, in pH 8.5 borate buffer solution with the assumption of two-electron transfer [10]. This confirms the suggested model of the enzyme action in the micellar system. The calculated apparent bimolecular constant of BMB oxidation in water (k_w) was 4 orders of magnitude less compared to that for PPA oxidation (Table 1). The large difference of the constants for BMB and PPA is associated with the specificity of the rCiP-

Table 1 The rate constants (k_w) and the partition coefficients (P) of *N*-benzoyl leucomethylene blue (*BMB*) and 10-phenothiazine propionic acid (*PPA*) at pH 8.5 and 25 °C in Triton X-100, sodium dodecyl sulfate (*SDS*) and cetyltrimethylammonium bromide (*CTAB*)

Substrate	Micellar phase	k_w (M^{-1} s$^{-1)}$)	P
BMB	Triton X-100	$(6.0 \pm 1.2) \times 10^3$	770 ± 200
PPA	Triton X-100	$(3.1 \pm 1.4) \times 10^7$	2100 ± 1100
BMB	SDS	$(2.4 \pm 0.2) \times 10^3$	28 ± 19
PPA	SDS	$(7.1 \pm 0.3) \times 10^6$	11 ± 11
BMB	CTAB	$(1.05 \pm 0.07) \times 10^4$	1770 ± 110
PPA	CTAB	$(2.7 \pm 0.2) \times 10^5$	2200 ± 330

catalyzed reaction. Similar effects were indicated in other dye/PPA systems [10, 13].

The results presented in Table 1 show that in Triton X-100 micelles the distribution coefficient of PPA is larger than that of BMB. This result can be explained by the "amphiphilic" (having a dual attraction, i.e. containing both a lipid-soluble phenothiazine core and a water-soluble propionate residue) property of PPA. The decrease in P for PPA in the case of the SDS micellar system is a result of micelle and substrate charge repulsion; however, the reason for the decrease in the distribution coefficient of BMB in these micelles is not clear. The unexpectedly low value of k_w in the case of SDS micelles indicates a complex surfactant and oxidized mediator interaction. It is possible that the structure of the SDS micelles is disturbed in solutions containing 2% acetonitrile and that the solubility of BMB in the SDS/acetonitrile pseudophase is much less than, for example, in the Triton X-100 one. The decrease in k_w may also be associated with inhibition of enzyme activity with surfactant molecules, as was indicated in Ref. [2].

An increase in the partition coefficient in the case of PPA in CTAB can be explained by charge attraction. An increase in P of 2.3 times for BMB compared to Triton X-100 is possible owing to the greater hydrophobicity of the CTAB pseudophase.

The low oxidation rate of PPA-assisted BMB oxidation in the CTAB system strongly supports the model under development: in the CTAB system almost all PPA is associated with the micellar pseudophase and the enzyme-catalyzed rate in water is low. The low rate at a small mediator concentration is also in agreement with micellar model: at a low concentration of the mediator most of the substrate has been absorbed by the micellar pseudophase.

Acknowledgements The authors express sincere thanks to Palle Schneider and Anders Hjelholt Pedersen (Novo Nordisk, Denmark) for valuable discussions on the problem, the donation of the enzymes and other chemicals. We also acknowledge John A. Berges for providing the computer program for the control of the spectrophotometer.

References

1. Von Jagow G, Schägger H (1994) A practical guide to membrane protein purification. Academic, San Diego
2. Kotterman MJJ, Rietberg H-J, Hage A, Field JA (1998) Biotechnol Bioeng 57:220–227
3. Rouse JD, Sabatini DA, Suflita JM, Harwel JH (1994) Crit Rev Environ Sci Technol 24:325–370
4. Ryabov AD, Goral VN (1997) JBIC 2:182–190
5. Schneider P, Pedersen AH (1995) Int Pat Appl WO 95/01426
6. Schneider P, Ebdrup S (1994) Int Pat Appl WO 94/12621
7. Cherry JR, Lamsa MH, Schneider P, Vind J, Svendsen A, Jones A, Pedersen AH (1999) Nature Biotech 17:379–384
8. Farhangrazi ShZ, Copeland BR, Nakayama T, Amschi T, Yamazaki I, Powers LS (1994) Biochemistry 33:5647–5652
9. Nelson DP, Kiesow LA (1972) Anal Biochem 49:474–478
10. Kulys J, Krikstopaitis K, Ziemys A (2000) J Biol Inorg Chem 5:333–340
11. Berges JA, Virtanen C (1993) Comput Biol Med 23:131–141
12. Tams JW, Vind J, Welinder KG (1999) Biochem Biophys Acta 1432:214–221
13. Danhus T, Kulys J, Schneider P (1994) Abstracts Eurobic II. Metal ions in biological systems, Florence, Italy, 30 August–3 September. p 101

Progr Colloid Polym Sci (2000) 116:143–148
© Springer-Verlag 2000

A. Ramanavicius
K. Habermüller
J. Razumiene
R. Meškys
L. Marcinkeviciene
I. Bachmatova
E. Csöregi
V. Laurinavicius
W. Schuhmann

An oxygen-independent ethanol sensor based on quinohemoprotein alcohol dehydrogenase covalently bound to a functionalized polypyrrole film

A. Ramanavicius (✉)
Department of Analytical and
Environmental Chemistry,
Vilnius University, Naugarduko 22
2006 Vilnius, Lithuania
e-mail: arman@bchi.lt
Tel.: +370-2-729068
Fax: +370-2-29196

K. Habermüller · W. Schuhmann
Analytische Chemie – Elektroanalytik &
Sensorik, Ruhr-Universität Bochum
Universitätsstrasse 150
44780 Bochum, Germany

A. Ramanavicius · J. Razumiene
R. Meškys · L. Marcinkeviciene
I. Bachmatova · V. Laurinavicius
Laboratory of Bioanalysis,
Institute of Biochemistry, Mokslininku 12
2600 Vilnius, Lithuania

E. Csöregi
Department of Biotechnology
University of Lund
P.O. Box 124, 221 00 Lund, Sweden

Abstract In the present work the characteristics of a phenazine methosulphate mediated alcohol biosensor based on a newly isolated quinohemoprotein alcohol dehydrogenase are described. The enzyme was covalently linked at a functionalized polypyrrole film which had been electrochemically deposited on the surface of a platinum-black electrode. The biosensor architecture developed was characterized with regard to sensitivity, selectivity, and long-term operational stability. Owing to the inherent properties of the new enzyme the related biosensors are oxygen-independent and exhibit improved selectivity to ethanol in contrast to alcohol biosensors based on alcohol oxidase or on cationic nicotinamide adenine dinucleotide dependent alcohol dehydrogenase.

Key words Quinohemoprotein · Alcohol dehydrogenase · Polypyrrole · Alcohol · biosensor · Redox mediator

Introduction

Determination of alcohol concentration is very important in bioprocess monitoring, the pulp industry, and food/beverage control. The presently used methods for alcohol monitoring involve distillation–oxidation [1], osmometry [2], gas chromatography [3], and enzymatic methods [4, 5]. To increase the selectivity and to reduce the analysis costs, biosensors with immobilized ethanol-oxidizing enzymes have been developed previously [6–11] based on alcohol oxidases (AOx) [6] or cationic nicotinamide adenine dinucleotide (NAD$^+$) dependent alcohol dehydrogenases (ADH) from various sources; however, these biosensors have some drawbacks, which are mainly related to the specific properties of the biological recognition elements used. In the case of oxidases, the sensor response is dependent on the availability of molecular oxygen and the oxidation of enzymatically produced H_2O_2 is only possible at high electrode potentials. In addition, no suitable redox mediators to facilitate the electron transfer between the active site of the enzyme and the electrode are known. Thus, the AOx-based biosensors so far described are based on coupled enzyme systems, requiring the need of complex electrode designs [12]. In the case of the NAD$^+$-dependent ADH one may successfully circumvent the problems imposed by molecular oxygen [7]; however, the coenzyme has to be added to the analyte solution [8], which is only possible in specifically designed flow-injection systems [13], or has to be incorporated into a graphite paste electrode matrix [14].

Therefore, the use of pyrroloquinoline quinone (PQQ)-dependent enzymes is promising, since these enzymes are oxygen-independent and their PQQ-cofactor is usually tightly bound within the enzyme's active site [9].

Previously described ethanol biosensors based on PQQ-dependent ADH (PQQ-ADH) required the use of free-diffusing or covalently bound redox mediators to allow a fast electron transfer between the enzyme and the electrode surface [10, 11]. For this purpose the ferrocyanide/ferricyanide couple has often been used [15].

A recently isolated and purified quinohemoprotein ADH (QH-ADH) [11] contains several cofactors: one PQQ and three heme c groups, which allow an internal bridging ("wiring") between the cofactors and hence an improved interaction with artificial redox mediators [16]. In addition, it is known that this type of enzyme can directly transfer electrons to gold and carbon surfaces [17] or to platinum electrodes (via conducting polymer chains of polypyrrole, Ppy) [11] involving the heme c moieties [15, 16]. Thus, at least one of the heme c groups has to be localized close to the surface of the protein, consequently increasing the probability for an efficient interaction between the enzyme's active site and the electrode or with an artificial redox mediator.

Covalent binding of enzymes is often preferred owing to an expected improvement of the sensor lifetime [18, 19]. Owing to the hydrophobic nature of electrochemically generated functionalized Ppy films it can be supposed that covalent binding of the membrane protein QH-ADH may be advantageous for stable immobilization. Therefore, in this work covalent binding of QH-ADH at modified Ppy films is considered as one possible way to develop an oxygen-independent alcohol biosensor.

Materials and methods

Enzyme and chemicals

QH-ADH (specific activity 32.2 Umg^{-1} protein) was purified from *Gluconobacter* sp. 33 as described previously [20]. KCl and Cu(NO$_3$)$_2 \cdot$3H$_2$O (analytical grade), acetonitrile, NaOH, HCl, H$_2$SO$_4$, and acetic anhydride (chemical grade) were obtained from Reachim (Kiev, Ukraine). Acetonitrile and acetic anhydride were used after distillation. Cu(NO$_3$)$_2 \cdot$3H$_2$O was dried under vacuum for 24 h before use. Methanol, ethanol, 1-propanol, 1-butanol, 2-methyl-1-propanol, H$_2$PtCl$_6$, and sodium acetate were purchased from Merck (Darmstadt, Germany). Phenazine methosulphate (PMS), dichlorophenol indophenol (DCPIP), K$_3$[Fe(CN)$_6$], 1-ethyl-3-(3-dimethylaminopropyl) carbodiimide (EDAC) and tetrabutylammonium perchlorate (TBAP) were obtained from Sigma (St. Louis, USA). Pyrrole (97%) was purchased from Fluka (Neu-Ulm, Germany) and purified prior to use by passing it through a neutral Al$_2$O$_3$ column (Sigma, St. Louis, USA). All solutions were prepared using high-performance liquid chromatography grade water purified in a Purator-B (Glas Keramic, Berlin, Germany) if not otherwise specified.

Instruments

All electrochemical experiments were performed using a conventional three-electrode system containing of a platinum disk (1-mm diameter) as the working electrode, a platinum wire as the counter electrode, and a Ag/AgCl/3 M KCl reference electrode. Amperometric measurements and cyclic voltammetry were carried out in a 3-ml volume electrochemical cell using a model PA-2 polarografic analyzer from Laboratorny Pristroje (Prague, Czech Republic) connected to an ENDIM 622.01 XY-recorder from Elektrogeräte Apolda (Schlotheim, Germany), while sensor preparation was performed in a specially designed and previously described microcell, which allows work in a very small volume (50 μl) under strict exclusion of oxygen, using a homebuilt potentiostat [21]. Photometrical enzyme-activity measurements were made with a Perkin-Elmer 550 spectrophotometer (Perkin-Elmer, Friedrichshafen, Germany).

Sensor preparation

The sensors were prepared following previously described optimized procedures [22–25]. The platinum disk electrode was polished with alumina paste from Reachim (Kiev, Ukraine) and cleaned electrochemically in an oxygen-free solution of 0.5 M H$_2$SO$_4$ following a procedure described elsewhere [26]. The pretreated platinum electrode was platinized by applying three voltammetric cycles between +500 and –400 mV versus Ag/AgCl with a scan rate of 10 mVs^{-1} in an oxygen-free solution of H$_2$PtCl$_6$ (4 gl^{-1}) to obtain a reproducible and active electrode surface. Finally, the electrode was carefully rinsed with oxygen-free water and acetonitrile to remove any traces of H$_2$PtCl$_6$.

Electrochemical deposition of the Ppy film was performed in a degassed solution containing 50 mM pyrrole and 100 mM TBAP in acetonitrile. Polymerization was achieved by application of a potentiostatic pulse profile consisting of 20 consecutive pulses of +900 mV for 1 s and +300 mV for 5 s [25]. The thickness of the electrochemically synthesized Ppy films was determined by means of a PMT-3 optical microscope from LOMO (St. Petersburg, Russia). The Ppy film was carefully punctured with a diamond prism until the surface of platinum was reached and the width of the hole created was measured as the basis for the estimation of the polymer-film thickness.

For functionalization (nitration) of the Ppy film, the electrode was immersed into an oxygen-free solution of 100 mM Cu(NO$_3$)$_2 \cdot$3H$_2$O in acetic anhydride. Acetyl nitrate generated in situ nitrated the Ppy chains in the 3-position of the pyrrole ring. The resulting nitro groups were reduced electrochemically to amino groups by applying three voltammetric cycles between +900 and –2100 mV versus Ag/AgCl at a scan rate of 10 mVs^{-1} in 100 mM TBAP/CH$_3$CN [22, 23].

For enzyme immobilization the carboxylic side chains of QH-ADH (3 gl^{-1} in 0.1 M acetate buffer (pH 5.0) containing 5 mM ethanol) were activated in the presence of 100 mM EDAC. Then, the modified Ppy electrode was immersed into this solution for 2 h to obtain covalent binding via amide groups formed between the enzyme's activated carboxylic residues and the amino functions introduced at the conducting polymer film. Finally, the electrode was kept in 100 mM KCl solution under stirring for 30 min to remove any adsorbed enzyme molecules. Control electrodes were prepared in the same way but without nitration of the polymer film.

Electrochemical characterization of the biosensor

The response of the prepared biosensors and the related control electrodes to increasing concentrations of ethanol was investigated under potentiostatic conditions at +200 mV versus Ag/AgCl in a stirred solution of 0.1 M KCl containing 10 mM PMS in 0.1 M acetate buffer, pH 6.0.

Spectrophotometrical measurements

The activity of immobilized QH-ADH was determined independently by a photometric test. The electrodes were immersed in a

2-ml cuvette (l = 1 cm) containing a solution of acetate buffer, pH 7.3, 10 mM ethanol, 0.033 mM PMS, and 0.066 mM DCPIP for 5 h. The absorption of DCPIP was measured at 600 nm ($\varepsilon = 16778$ M^{-1} cm^{-1} at pH 7.3) at 30 °C under thermostatic conditions.

Results and discussion

The schematic view of the oxygen-independent alcohol sensor is depicted in Fig. 1. A conducting polymer film is electrochemically formed on a platinized electrode surface and subsequently functionalized to obtain amino functions at the polymer film. QH-ADH is covalently linked after carbodiimide activation to the modified electrode. The response in the biosensor configuration obtained was investigated for increasing ethanol concentrations in the presence of 10 mM PMS, which was used as a free-diffusing redox mediator. The calibration of the biosensor in the presence and in the absence of the oxygen was done, and any influence of oxygen was observed (data not shown). The calibration graph for ethanol was linear up to a concentration of 2.0 mM (Fig. 2). The highest steady-state current (8 nA) was observed in the presence of 5 mM ethanol. When the concentration of PMS was increased to 20 mM the observed steady-state currents versus concentration of ethanol increased by only about 5%. This

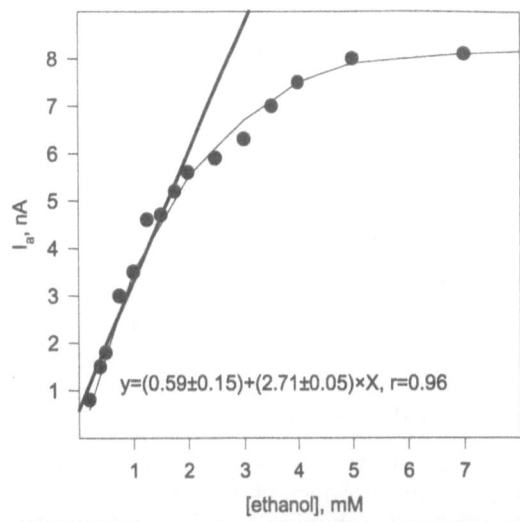

$$y = (0.59 \pm 0.15) + (2.71 \pm 0.05) \times X, \ r = 0.96$$

Fig. 2 Steady-state current of the QH-ADH-based alcohol biosensor versus ethanol concentration (200 mV versus Ag/AgCl; 10 mM phenazine methosulphate, *PMS*, stepwise addition of ethanol aliquots)

suggests that the mediator concentration, and hence its diffusional mass transport, does not limited the overall reactions under these conditions. Simultaneously, control

Fig. 1 Schematic representation of the reaction sequence for the modification of an electrode with an amino-functionalized polypyrrole (*Ppy*) film and subsequent covalent binding of quinohemoprotein alcohol dehydrogenase (*QH-ADH*)

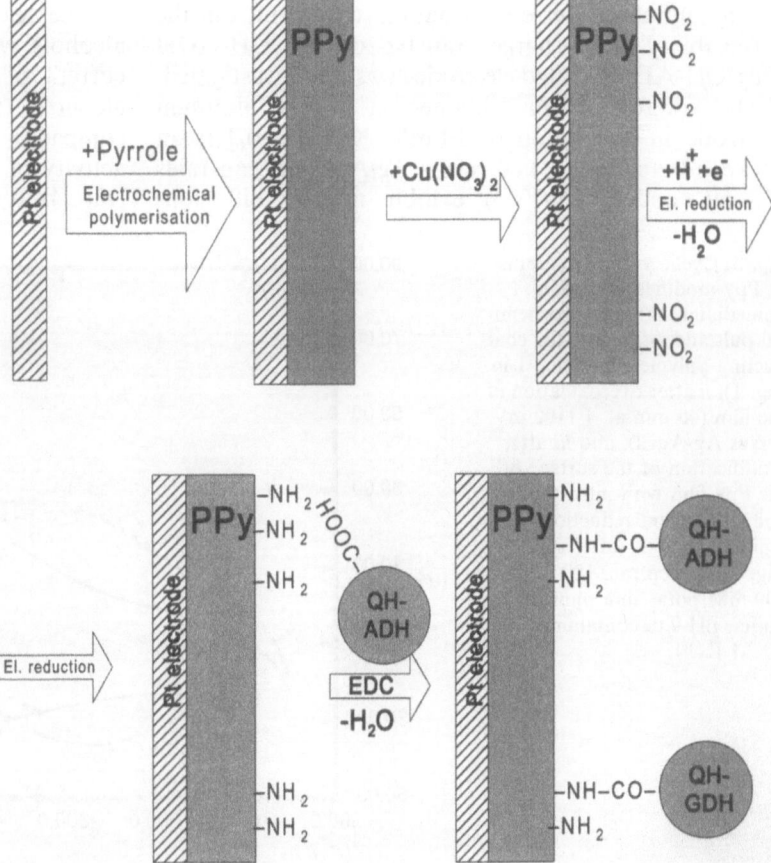

electrodes (QH-ADH adsorbed on untreated Ppy film) displayed 8 times lower steady-state currents. The spectrophotometrical measurements of immobilized QH-ADH activity showed enzyme activity 11 times smaller for these electrodes compared with those prepared with the covalently bound QH-ADH. This confirms that QH-ADH was covalently bound at the Ppy surface.

However, as could be expected from the sensor architecture, direct electron transfer between QH-ADH via the Ppy chain was negligible in contrast to previously described electrodes in which QH-ADH was entrapped within the polymer network during its electrochemical formation. One can assume that at least two factors are responsible for these findings:

1. The enzyme is predominantly bound at the outer surface of the polymer film, which has a relatively high thickness (about 5.5 μm). Consequently, the electron-transfer distance would be too long to ensure an efficient direct electron-transfer pathway via the polymer chains.
2. The reduction–oxidation procedure (for the generation of the amino functions at the polymer chains) might modify the structure of the Ppy film. This results in cyclic voltammograms for the modified Ppy film that are similar to those obtained using overoxidized Ppy (Fig. 3) [27].

The influence of overoxidation treatment on the permeability and charge transfer of Ppy/QH-ADH (Ppy/QH-ADH) coated electrodes was also investigated. Cyclic voltammograms obtained at a bare platinum electrode in a solution of 10 mM $K_3[Fe(CN)_6]$ (scan between 0 and +500 mV versus Ag/AgCl at scan rates 20, 50, and 100 mVs^{-1}) exhibit normal diffusion-controlled redox peaks corresponding to reversible redox processes of $K_3[Fe(CN)_6]$. The cyclic voltammograms of this electrode modified with Ppy and QH-ADH in the same solution shows very small peaks; less than 5% of the original current at the bare platinum electrode can be detected. In contrast the voltammetric and amperometric responses for 1 mM cationic PMS at the same modified electrode was 2.3 times higher than at the bare platinum electrode. These results are in good agreement with results obtained by Gao et al. [28], and they demonstrate clearly that the overoxidized Ppy film effectively excludes anions from the electrode but that it increases the sensitivity for cations. This was because electron-rich groups are introduced onto pyrrole units during overoxidation of the upper Ppy layer, and the conducting upper layer of Ppy was converted into an ion-exchange polymer [28]. That means that a partially oxidized Ppy film with a well-defined interface between the conducting and insulating zones was obtained [29], and the redox processes of the Ppy/QH-ADH electrode are easy owing to the high conductivity of the inner Ppy layer [30].

The response to addition of 10 mM PMS for Ppy/QH-ADH was 2.7 times longer than for the bare platinum electrode. That illustrates that the overoxidized Ppy layer forms some diffusion barriers for penetration of PMS and that the biosensor response time was extended.

The selectivity of the biosensor obtained to various alcohols was analyzed by comparing the steady-state current of the biosensor in the presence of different alcohols (5 mM each). The selectivity pattern was compared with the spectrophotometrically detected activity of free QH-ADH towards the same substrates (Fig. 4). It was found that the sensitivity of the biosensor

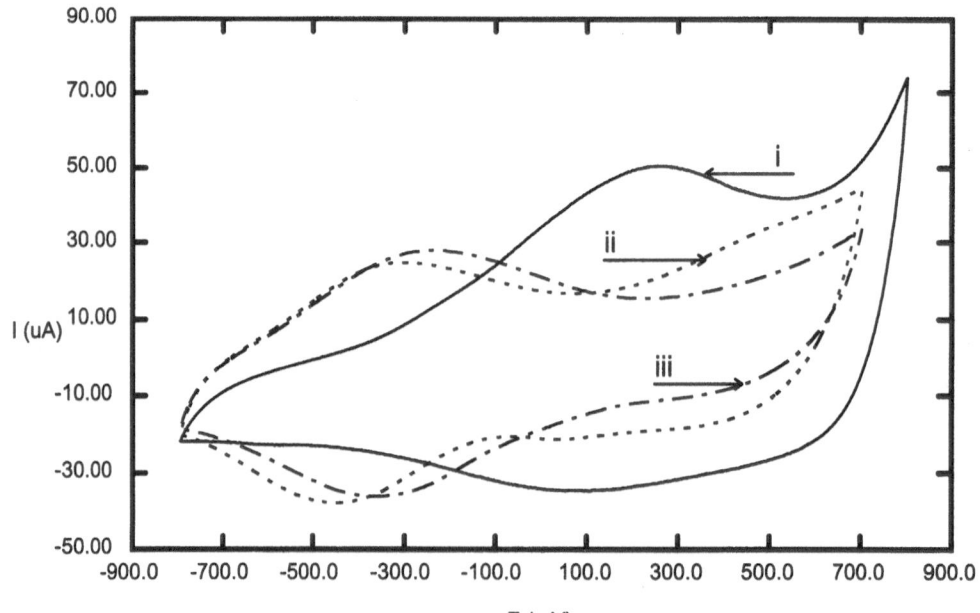

Fig. 3 Cyclic voltammograms of Ppy-modified electrodes: *i* immediately after electrochemical pulse-deposition of the conducting polymer film (step 1 in Fig. 1), *ii* after overoxidation of the film (40 min at +1100 mV versus Ag/AgCl), and *iii* after modification of the surface of the Ppy film with nitro groups and subsequent reduction to amino functions (after step 3, Fig. 1). (sweep rate 100 mVs^{-1}; 100 mM potassium phosphate buffer, pH 7.0, containing 0.1 M KCl)

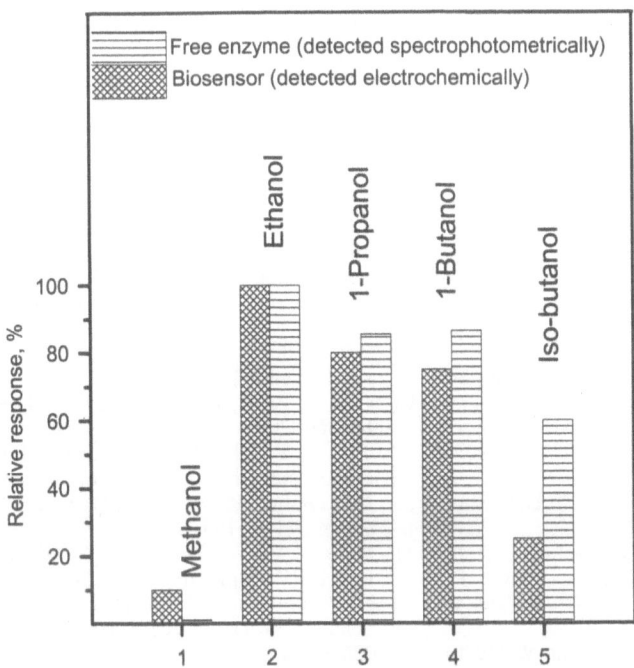

Fig. 4 Relative response of QH-ADH biosensors for 5 mM of different alcohols normalized to the ethanol response of the same sensor (sodium acetate buffer, pH 6.0, 10 mM PMS). The response to ethanol was considered to be 100%

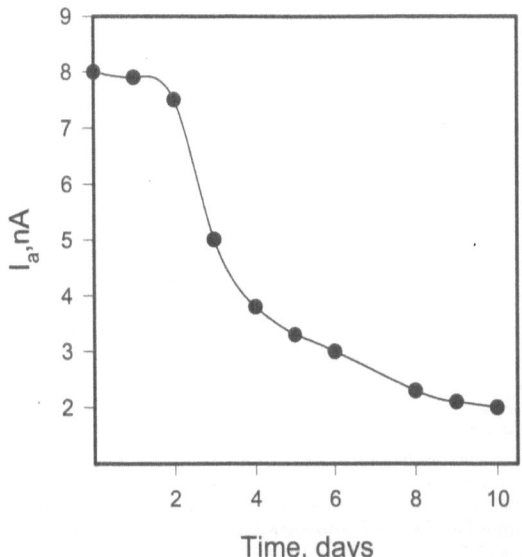

Fig. 5 Storage stability (day-to-day response) of the alcohol biosensors developed (5 mM ethanol; sodium acetate buffer, pH 6.0, 10 mM PMS)

was higher for methanol and lower for *n*-propanol or *n*-butanol compared with the native QH-ADH in solution. These data together with a prolonged linear region of the steady-state currents versus concentrations for ethanol (about 10 times longer than expected from the K_m value of native enzyme for ethanol (0.2 mM) indicated that the biosensor partially works in a diffusion-limited mode.

The sensitivity of the biosensor decreased with decreasing chain length for aliphatic alcohols (except methanol, for which the selectivity of the enzyme is the most determining factor) and is distinctly lower for secondary alcohols (e.g. 2-methyl-1-propanol). This fact illustrates that the developed biosensor exhibits improved selectivity for the determination of ethanol in the presence of other alcohols. The sensor configuration obtained seems to be superior in this particular aspect to biosensors based on QH-ADH from *Comamonas testosteroni* coimmobilized within a poly(vinylpyridine) redox polymer [31].

The storage stability of the QH-ADH biosensor at 20 °C is illustrated in Fig. 5. Twenty measurements in the concentration range between 0.5 and 5 mM ethanol were carried out per day; the electrode was kept in sodium acetate buffer, pH 6.0, at +20 °C between the measurements. The half-life of the biosensor was about 4 days. The shape of the deactivation curve obtained suggests that during the first 2 days of operation the response of the biosensor was limited by substrate diffusion, while the decrease in sensitivity observed later was invoked by either the inactivation of the enzyme or the destruction of the polymer matrix. The stability of the biosensor was the same in the presence and in the absence of PQQ. That means that inactivation of native QH-ADH and the biosensor is not limited by the loss of PQQ, and is in agreement with Refs. [32–34], where it was reported that PQQ in PQQ-dependent enzymes is bound owing to ion-pair interactions of three carboxylic groups of PQQ with Ca^{2+}/Mg^{2+} and other ions found in the active site of the enzyme.

Conclusions

QH-ADH can be covalently immobilized on amino-functionalized Ppy films. Compared with control electrodes obtained by adsorption of the enzyme the sensitivity of the biosensor increases by 8 times. In the presence of PMS as a free-diffusing redox mediator, efficient electron transfer between the enzyme and the working electrode can be obtained. It allows the biosensor to be used for the determination of alcohols with improved selectivity for ethanol. As a matter of fact, the need of a free-diffusing redox mediator limits the application of the sensor to stationary measurement or flow-injection systems.

Acknowledgements This work was financially supported by the European Commission (ERB IC15C15CT96-1008 and ERB IC15CT98-0907), NUTEK, the Lithuanian State Science and Studies Foundation, and SJSF (E.C.). The authors thank B. Kurtinaitienė for technical help.

148

References

1. Widmark EMP (1915) Scand Arch Physiol 32:85–96
2. Cravely RH, Jain NC (1974) J Chromatogr Sci 12:209–211
3. Machata G (1962) Mikrochim Acta 6:91–95
4. Roos KJ (1971) Clin Chim Acta 31:285–288
5. Matzinger D, Evin KR, Philips R (1984) Clin Chem 30:1029–1031
6. Gorton L, Jönsson-Pettersson G, Csööregi E, Johanson K, Dominguez E, Marko-Varga G (1992) Analyst 117:1235–1239
7. Gorton L, Bremle G, Csööregi E, Jonsson-Pettersson G, Person B (1991) Anal Chim Acta 249:43–48
8. Gorton L (1995) Electroanal 7:23–45
9. Duine JA, Frank J, Jongejan JA (1986) FEMS Microbiol Rev 32:165–178
10. Graham D, Hill HAO, William J, Aston I, Higgins J, Turner APF (1983) Enzyme Microb Technol 5:383–388
11. Ramanavicius A, Habermüller K, Csöregi E, Laurinavicius V, Schuhmann W (1999) Anal Chem 71:3581–3586
12. Butter T, Johnson K, Gorton L (1993) Anal Chem 65:2628–2632
13. Prinzing U, Ogbomo I, Lehn C, Schmidt H-L (1990) Sens Actuators B 1:542–545
14. Ikeda T, Hamada H (1986) Anal Sci 2:501
15. Karube I, Yokoyama K, Kitagawa Y (1993) In: Davidson VL (ed) Principles and applications of quinoproteins. Dekker, New York, pp 439–444
16. Frebortova J, Matsushita K, Arata H, Adachi O (1997) Biochim Biophys Acta 1363:24–34
17. Ikeda T, Kobayashi D, Matsushita K, Sagara T, Niki KJ (1993) J Electroanal Chem 361:221–228
18. Kulys JJ, Samalius AS (1984) Bioelectrochem Bioenerg 13:163–169
19. Razumas V (1984) Bioelectrochem Bioenerg 12:297–322
20. Marcinkeviciene L, Bachmatova I, Sėmenaitė R, Rudomanskis R, Braženas G, Meškienė R, Meškys R (1999) Biologija 2:24–29
21. Habermüller K, Schuhmann W (1998) Electroanal 10:1281–1284
22. Schuhmann W (1991) Synth Met 41:429–432
23. Schuhmann W (1991) Sens Actuators B 4:41–49
24. Schuhmann W (1998). In: Cass T, Ligler F (eds) Immobilised biomolecules in analysis. A practical approach. Oxford University Press, Oxford, pp 187–210
25. Kranz C, Wohlschläger H, Schmidt HL, Schuhmann W (1998) Electroanal 10:546–552
26. Schuhmann W, Lammert R, Uhe B, Schmidt H-L (1990) Sens Actuators B 1:537–541
27. Witkowski A, Brajter-Toth A (1992) Anal Chem 64:635–641
28. Gao Z, Zi M, Chen B (1994) J Electroanal Chem 373:141–148
29. Tezuka Y, Ishii T, Aoki K (1996) J Electroanal Chem 402:161–165
30. Maksymiuk K (1994) J Electroanal Chem 373:97–106
31. Somers VAC, Stinger ECA, Hartinsveldt W, Lugt JP (1999) Appl Biochem Biotechnol 75:151–162
32. Schmidt B (1997) Clin Chim Acta 226:33–37
33. Oubrie A, Rozenboom HJ, Kalk KH, Olsthoorn JJ, Duine JA, Dijkstra BW (1999) EMBO J 18:5187–5194
34. Jongejan A, Machado SS, Jongejan JA (2000) J Mol Catal B 8:121–163

Progr Colloid Polym Sci (2000) 116:149–153
© Springer-Verlag 2000

R. Šimkus
R. Meškys
E. Csöregi
P. Corbisier
B. Mattiasson

Millihertz waveband oscillations in an unstirred bacterial culture

R. Šimkus (✉) · R. Meškys
Institute of Biochemistry
Mokslininkų 12, Vilnius 2600, Lithuania
e-mail: simkus@bchi.lt
Fax: +370 2 729196

E. Csöregi · B. Mattiasson
Department of Biotechnology
Lund University
P.O. Box 124, 221 00, Sweden

P. Corbisier
Environmental Division
VITO, Boeretang 200
2400 Mol, Belgium

Abstract Recently the oscillatory bioluminescence of an unstirred and exposed-to air-culture of lux-gene modified, copper-sensitive *Alcaligenes eutrophus* (a bacterium preferring the liquid–gas interface as a habitat) was discovered (Šimkus et al. (1999) *Biotechnol Tech* 13: 529). Here we report the spectral analysis of the bioluminescence oscillations recorded after illumination of the unstirred bacterial culture. The millihetz waveband spectra obtained were close to the spectrum of a frequency continuum or contained well-resolved sharp peaks, which were attributed to a certain resonance phenomenon in the optimized cultures.

Key words Lux genes · Bacterial bioluminescence · Liquid–gas interface · Oscillations

Introduction

Although the accumulation of certain bacteria in the vicinity of an air–liquid interface is evident [1], dynamical features of this process, which is tightly related to the taxis of bacteria and oxidation–reduction reactions in the water phase, have not been investigated intensively.

Genetic engineering methods have allowed the development of noninvasive tools for investigation of bacteria. Recently, marked interest has been shown for the use of microorganisms with fused lux genes [2–4]. These genetically modified microorganisms produce light in the presence of appropriate compounds named inducers. If a genetically constructed bacterium is used for analytical purposes then the term "a whole cell biosensor" is convenient [2–4]. These "artificial" cells contain an internal source of light – the luciferase-catalyzed bioluminescence reaction inducible by an appropriate analyte. Bacterial luciferase constitutes a shunt of the electron transport pathway, which theoretically could involve some kind of feedback regulation with cyclic bursts [5]. Recently, the oscillatory luminescence of lux-gene-engineered, copper-sensitive *Alcaligenes eutrophus*, a bacterium preferring the liquid–gas interface as a habitat, was demonstrated for the first time [6]. Sustained oscillations were observed in the case of optimized-concentration and light-treated cultures which were exposed to air and allowed to stay without mixing for several hours. In the present work we report the spectral analysis of these recently discovered biochemical oscillations.

Materials and methods

Initiation of oscillations

The initiation and detection scheme of the bioluminescence oscillations in the culture is shown in Fig. 1. In this work the strain of *A. eutrophus* AE1239 containing the luxCDABE genes of *Vibrio fisheri* placed under the control of a copper-inducible promoter [6, 7] was used. *A. eutrophus* was grown in a Luria broth (LB) medium containing 20 mg/l tetracycline hydrochloride (tet) (Duchefa, Harlem, the Netherlands) to prevent the cells from loosing their plasmid. The LB medium was prepared by mixing the following chemicals per liter of Millipore-quality water: 10 g tryptone (Duchefa), 5 g yeast extract (Duchefa) and 10 g sodium chloride (Merck, Darmstadt, Germany). Glycerol stock (20% v/v) of AE1239 was made by adding glycerol (Merck) to LB + tet containing a culture with an optical density of about 1. Optical density measurements (UV-120–02, Shimadzu, Kyoto, Japan) at 600 nm were used to monitor cell concentrations. To start the

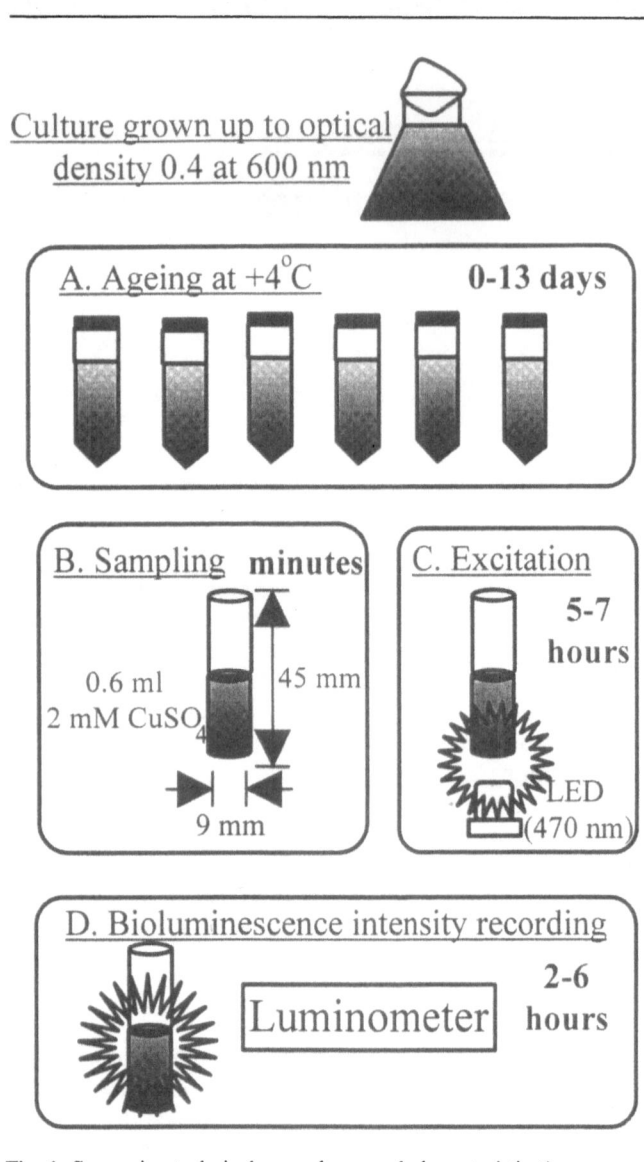

Fig. 1 Successive technical procedures and characteristic times were used for initiation and recording of oscillatory bioluminescence (*A–D*). The bacteria were allowed to "self-organize" without any mechanical treatment, except for the sampling procedure (*B*), when copper sulfate stock solution was added to the aged culture

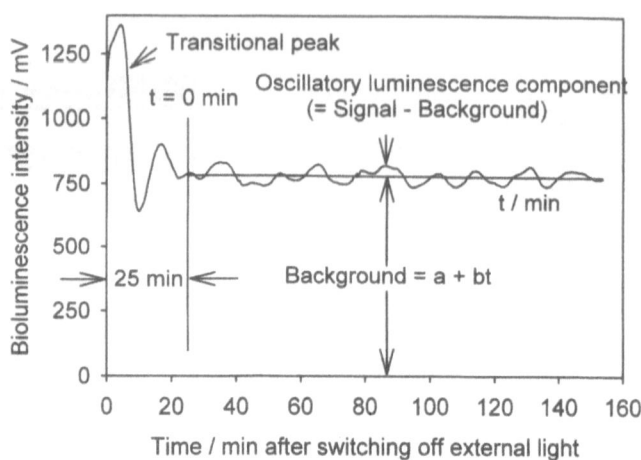

Fig. 2 Signal processing scheme (exemplified by signal processing of a 4-day-old culture)

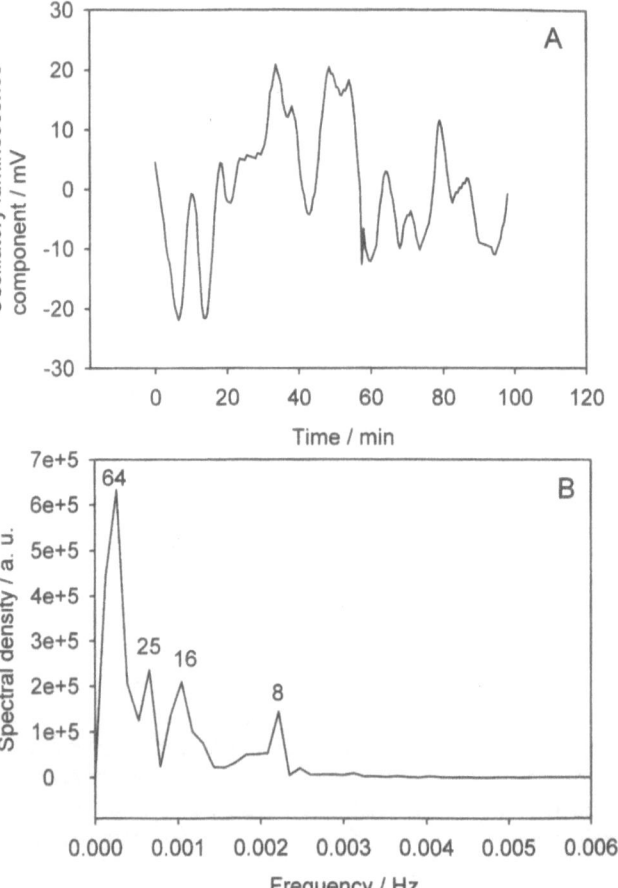

Fig. 3 **A** Oscillatory luminescence component and **B** corresponding spectrum of a harvesting day (0-day-old) culture. Optical density (*OD*) 0.428. Duration of bacteria exposure under light, T_L, 5.78 h. The background signal was $-(258 + 0.227t)$ mV

growth 1 ml glycerol stock was added to 100 ml LB + tet medium. The cells were cultured aerobically in baffled Erlenmeyer flasks closed with cotton stoppers on a rotary shaker at 30 °C. The bacteria were grown until the optical density reached 0.4. The culture was kept in closed vials at 4 °C. Vials of bacteria suspensions were taken from the refrigerator 30 min before preparing samples for the bioluminescence measurements. The optical densities were measured before the addition of copper. In each case 480 µl suspension was mixed with 120 µl 10 mM $CuSO_4$ (Merck) stock solution in LB + tet. A glass test tube (inner diameter 8.9 mm, height 45.2 mm) was used as a reaction vial. The bioluminescence was measured after incubation of the sample for 5–7 h under the weak light (about 10 lx) of a light-emitting diode (470 nm). The duration of bacteria exposure under light is denoted T_L. The bioluminescence intensity expressed in arbitrary units (millivolts) was monitored using an LKB type 1250 luminometer (LKB Wallac, Bromma, Sweden). The measurements were carried out at 20 °C.

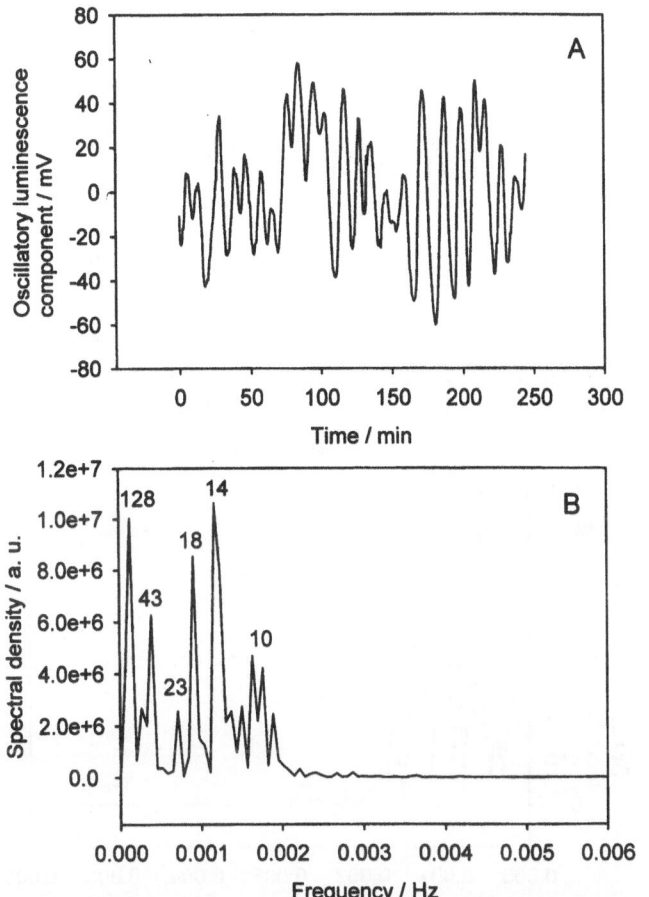

Fig. 4 A Oscillatory luminescence component and **B** corresponding spectrum of a 1-day-old culture. OD = 0.494. T_L = 5.07 h. The background signal was $-(611 + 0.521t)$ mV

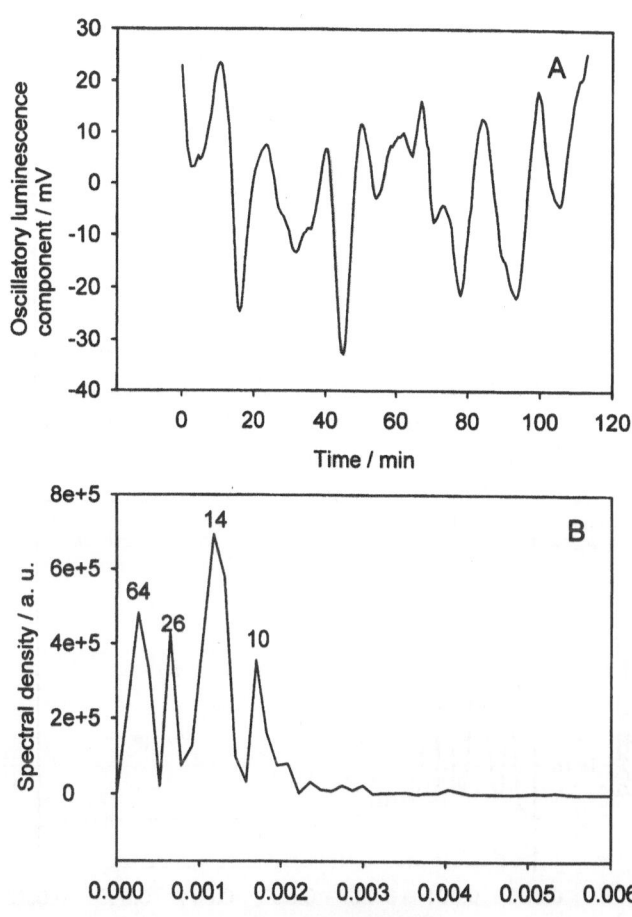

Fig. 5 A Oscillatory luminescence component and **B** corresponding spectrum of a 2-day-old culture. OD = 0.540. T_L = 4.58 h The background signal was $-(485-0.094t)$ mV

Signal processing

The signal processing scheme is shown in Fig. 2. The oscillatory luminescence component was defined as the difference between the measured signal and a linear plot, obtained after fitting the data, which excludes 25-min-duration transitional responses. The best-fit plot was considered to be the background signal. The best-fit linear plot (Fig. 2) represents a linear approximation of the monotonous (nonoscillating) component of the output signal. Other approximation functions (exponential, higher polynomial, etc.) could be used instead of linear approximation; however, in practice, the monotonous components were close to constant or possessed a low slope (Fig. 2). Thus, for the sake of simplicity, the linear approximation of the background signal was preferred. The analysis of the oscillatory components of the bioluminescence defined as shown in Fig. 2 was carried out using the SigmaPlot 4.0 program.

Results and discussion

The time dependencies of the oscillatory component of the luminescence and the corresponding spectra are shown in Figs. 3–7. The corresponding periods in minutes are shown above the spectral peaks. There are

many possible processes which could perturb the output flux of photons from the dense and exposed-to-air culture, for example, the oxidation–reduction of certain dyes, chemiluminescence reactions, division of cells, changes of multiple light scattering characteristics due to motion of bacteria (photo- and chemotaxis). Considering the scarce information and the complex nature of the system, it is reasonable to approach the dynamical features of the bacterial culture from the analysis of the observed spectra.

The spectra recorded on the day of harvesting (Fig. 3) and those observed after prolonged storage (Fig. 7) possess multiple but not highly expressed peaks covering the wide range of periods. The processes resulting in complicated spectra are not known; however, some preliminary speculative assumptions can be made. The spectra cover a range with periods exceeding 8–10 min. It was concluded, therefore, that the period of the processes responsible for the complicated spectra is in the range 10–100 min. Thus, it is a slow physiological process, which could include a periodic transcription of

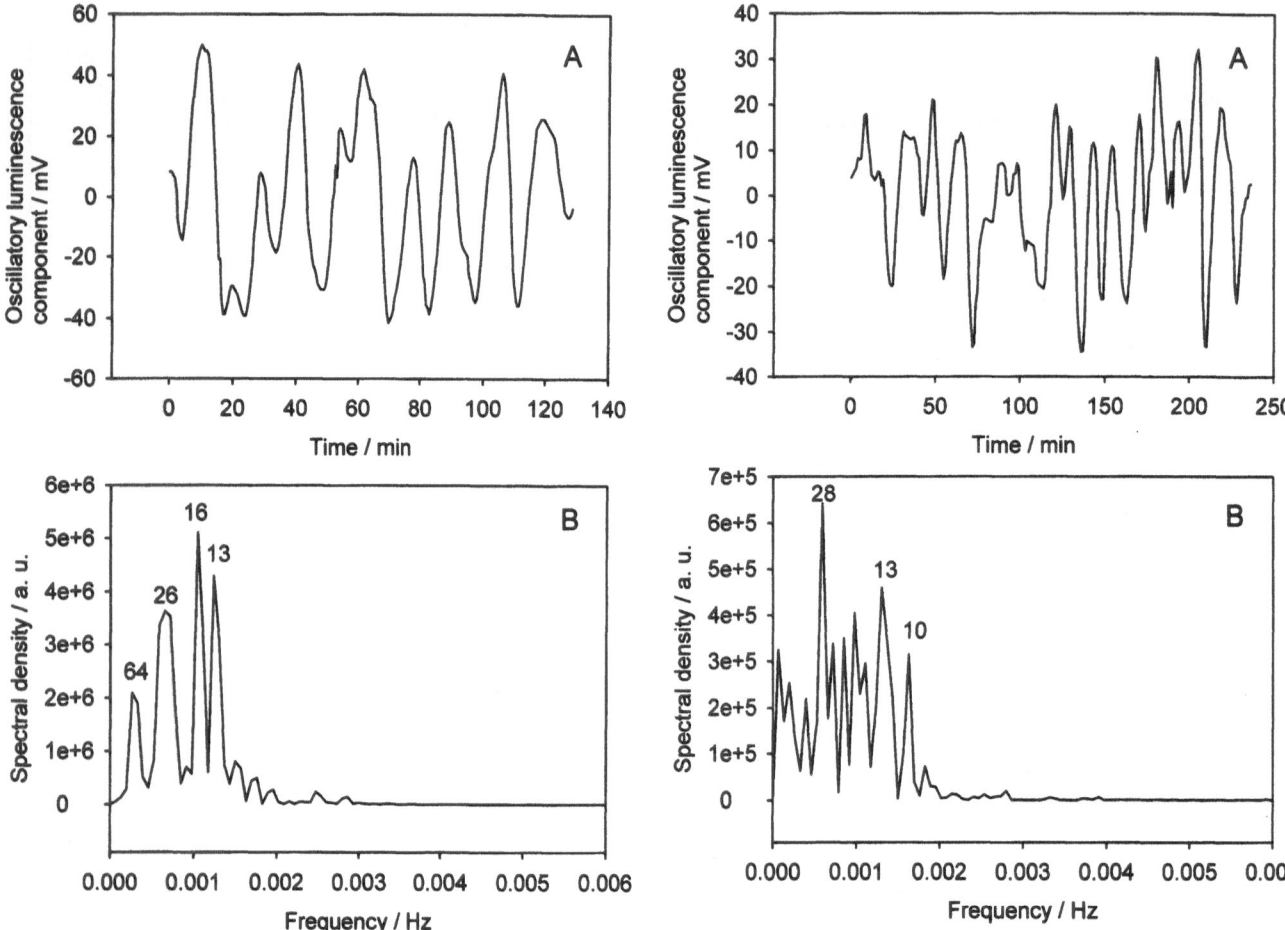

Fig. 6 **A** Oscillatory luminescence component and **B** corresponding spectrum of a 4-day-old culture. OD = 0.439. T_L = 6.33 h The background signal was −(782−0.062t) mV

Fig. 7 **A** Oscillatory luminescence component and **B** corresponding spectrum of a 13-day-old culture. OD = 0.165. T_L = 6.0 h The background signal was −(330 + 0.666t) mV

DNA, translation, and/or a slow mechanical motion (translocation and/or rotation) and/or synchronous respiration of bacteria. All these processes are energy-consuming. Thus, these wide frequency range oscillations could reflect periodic changes in the intracellular concentration of substrates for bioluminescence reactions (O_2, reduced flavine mononucleotide, aldehyde [2–7]) and adenosine 5′-triphosphate required for aldehyde formation [4].

The spectra recorded after 1–4 days of culture storage possess resolved peaks (Figs. 4, 5, 6). The resolution of the spectral peaks was dependent on the level of the background signal. The best-resolved spectra were observed at the highest levels of background luminescence (Figs. 4, 6). It is worth noting the resonant character of the spectra (Figs. 4, 6) displaying unexpectedly high stability of millihertz waveband frequencies in the optimized cultures.

Conclusions

It was observed that lux-gene-harboring bacteria excited and synchronized by light generate oscillatory bioluminescence, which was analyzed in terms of the spectral density. Despite the complexity of the system, the spectral analysis of the output bioluminescence indicates a temporal order in cultures of genetically constructed bacteria under certain conditions. Construction and evaluation of bioluminescence oscillators is a promising, new area for investigation of the morphology of unstirred bacterial cultures as a dynamical and self-organized structure.

Acknowledgements The authors thank the European Commission (ENV4-CT95-0141), the Swedish Institute (R.Š. and E.C.) and the Lithuanian State Science and Studies Foundation (grant No. 366)(R.Š. and R.M.) for financial support.

References

1. Schlegel HG, Jannasch (1992) In: Balows A, Trüper HG, Dworkin M, Harder W, Schleifer K-H (eds) The prokaryotes, 2nd edn, vol 1. Springer, Berlin Heidelberg, New York, pp 75–126
2. King JMH, DiGrazia PM, Applegate B, Burlage B, Sanseverino J, Dunbar P, Larimer F, Sayler GS (1990) Science 249:778
3. Ramanathan S, Ensor M, Dounert S (1997) Trends Biotechnol 15:500
4. Ulitzur S (1997) J Biolumin Chemilumin 12:179
5. Wilson T, Hastings JW (1998) Annu Rev Cell Dev Biol 14:197
6. Šimkus R, Csöregi E, Leth S, Corbisier P, Diels L, Mattiasson B (1999) Biotechnol Tech 13:529
7. Corbisier P, Thiry E, Diels L (1996) Environ Toxicol Water Qual 11:171

Progr Colloid Polym Sci (2000) 116:155–156
© Springer-Verlag 2000

Progr Colloid Polym Sci (2000) 116:157
© Springer-Verlag 2000